Fundamental and Applied Catalysis

Series editors

Martyn Twigg, Royston, UK
Michael Spencer, Cardiff, UK

More information about this series at http://www.springer.com/series/5964

Rinaldo Poli
Editor

Effects of Nanoconfinement on Catalysis

Springer

Editor
Rinaldo Poli
Laboratoire de Chimie de Coordination
(LCC)
Toulouse Cedex 4
France

and

Ecole Nationale Supérieure des Ingénieurs
en Arts Chimiques et Technologiques
(ENSIACET)
Toulouse Cedex 4
France

and

Institut Universitaire de France
Paris Cedex 05
France

ISSN 1574-0447
Fundamental and Applied Catalysis
ISBN 978-3-319-50205-2 ISBN 978-3-319-50207-6 (eBook)
DOI 10.1007/978-3-319-50207-6

Library of Congress Control Number: 2016958495

Printed on acid-free paper

This Springer imprint is published by Springer Nature
The registered company is Springer International Publishing AG
The registered company address is: Gewerbestrasse 11, 6330 Cham, Switzerland

Preface

Enormous progress has been made in the recent years in the synthesis of nanostructured functional materials with controlled size, shape, and functionality. These are either molecular and macromolecular materials of nanoscopic dimensions or extended tridimensional materials with nanoscopic cavities where specific functionalities are installed with controllable density and at desired locations. These materials include, but are not limited to, well-defined molecular cages, self-assembled supramolecules, dendrimers, molecularly imprinted polymers, polymers with a variety of elaborated and sophisticated structures made by living or controlled polymerization, artificial enzymes, metallorganic frameworks, porous organic frameworks, carbon nanotubes, mesoporous inorganic solids, nanocapsules, and hybrids thereof. The multiple applications that these materials offer are revolutionizing industrial sectors such as energy, electronics, sensors, biomedicine, and separation technology.

This book highlights some of the recent advances and state of the art in the use of functionalized nanostructured environments for catalysis with a particular focus on organic and metallorganic cavities. The area of catalysis within structured cavities of 3-dimensional materials such as zeolites and mesoporous oxides has not been included in this book because, though quite important in heterogeneous catalysis, it is already amply covered in other recent books. Numerous research groups show growing interest in benefitting from the nanostructure control to gain advantages or bring to light new phenomena, facing the omnipresent challenges of catalysis: higher activities, improved selectivities, catalyst stabilization, cooperativity effects, simplified protocols for cascade syntheses, better catalyst recovery, and recyclability. These advantages can be equally applied to all types of catalytic transformations, be they based on molecular complexes, stabilized nanoparticles, organocatalysts, or enzymes. The book provides contributions covering all these aspects.

Chapter "Endohedral Functionalization of Molecular Cavities for Catalysis in Confined Spaces" by Martinez et al. and Chapter "Self-Assembled Coordination Cages and Organic Capsules as Catalytic Supramolecular Reaction Vessels" by Bolliger highlight catalyst confinement in the cavity of small molecular objects, of a covalent nature for the former and constructed via supramolecular strategies in the latter. In Chapter "Artificial Metalloenzymes," Trindler and Ward give an overview of the area of metalloenzymes, combining metal catalysis with the enzymatic scaffold ability to provide shape control and shielding. The shape selectivity concept is further developed in Chapter "Metal Complexes and Imprinted Polymers for Shape-Selective Catalysis" by Mirata and Resmini with an overview of the molecular imprinting technology and the use of molecularly imprinted polymers as catalysts and photocatalysts. In Chapter "Catalysis Inside Folded Single Macromolecules in Water," Artar and Palmans bring supramolecular interactions back to the spotlight, but this time focusing on how to create nanoconfinement via folding single macromolecules and mimicking the enzyme functionality, different from the generation of well-defined small cavities of Chapter "Self-Assembled Coordination Cages and Organic Capsules as Catalytic Supramolecular Reaction Vessels." The book continues with two chapters dedicated to star polymers topologically related to micelles. Chapter "Microgel Star Polymer Catalysts as Active and Functional Nanoreactors for Organic Reactions and Polymerizations" by Terashima and Sawamoto is focused on microgel star polymers built by atom transfer radical polymerization in a homogeneous phase, whereas Chapter "Core-Cross-Linked Micelles and Amphiphilic Nanogels as Unimolecular Nanoreactors for Micellar-Type, Metal-Based Aqueous Biphasic Catalysis" by us with collaborators presents related catalytic nanoreactors made by reversible addition–fragmentation chain-transfer radical polymerization in emulsion. Aspects such as biphasic catalysis and recycling are highlighted in these chapters. Then, Caminade et al. present recent advances made with catalytic dendrimers in Chapter "Catalysis Within Dendrimers." Finally, we conclude the book with a survey, in Chapter "Site Isolation for Non-orthogonal Tandem Catalysis in Confined Nanospaces," of catalyzed tandem processes achieved in one pot by site-isolating incompatible catalysts in different compartments of the same nanoreactors.

I hope that the selected material provides a balanced and useful reference in this rapidly expanding area to all readers: a useful introduction for the nonexperts, useful references for the practitioners, and especially perspectives and opportunities for future development for those who wish to embrace this area in their future research work.

Toulouse, France Rinaldo Poli

Contents

Endohedral Functionalization of Molecular Cavities for Catalysis in Confined Spaces

Laure Guy, Jean-Pierre Dutasta and Alexandre Martinez

Abstract The confined space of molecular cavities provided by synthetic organic receptors presents a great opportunity to mimic enzymatic systems. They can be used as artificial receptors or molecular vessels to perform catalytic reactions, combining confined environment around the reactive site and significantly increase substrate selectivity. Understanding reactivity changes and catalytic effects observed in enzyme catalysis is thus of great significance for the design of efficient catalysts desiring to mimic naturally occurring systems. The specific size, shape, and chemical environment encountered in a confined nanospace induce a change in reactivity and selectivity on an encapsulated guest molecule by imposing specific orientations and conformations. Nanosized reaction chambers incorporating a catalytic site have been successfully synthesized and the endohedral location of the reactive functions proved to have an important impact on catalytic transformations carried out therein. This review article summarizes recent advances in the design of covalent molecular cages combining both hydrophobic cavities and endohedral functionalities. Comparison with non-endohedral catalysts or model compounds, lacking cavity, will also be presented, in order to highlight the interest of such an approach.

1 Introduction

Biological systems are fascinating because of their high degree of complexity and the remarkable tasks they are capable to perform. The specific design of supramolecular structures found in nature and the resulting non-covalent interactions

L. Guy · J.-P. Dutasta
Laboratoire de Chimie, École Normale Supérieure de Lyon,
CNRS, UCBL, 46 allée d'Italie, 69364 Lyon, France

A. Martinez (✉)
AMU iSm2 Service 462, Campus Scientifique de St Jérôme,
13397 Marseille, Cedex 20, France
e-mail: alexandre.martinez@centrale-marseille.fr

R. Poli (ed.), *Effects of Nanoconfinement on Catalysis*,
Fundamental and Applied Catalysis, DOI 10.1007/978-3-319-50207-6_1

and pre-organization are responsible for the efficiency and selectivity encountered in biological systems. In particular enzymes display outstanding catalytic activity and selectivity: the folding of a protein chain results in a pocket around the catalytic center with the surrounding amino acids dictating a specific orientation and conformation of the incoming substrate, thus pre-organizing the substrate which leads to the observed reactivity. The high efficiency of such systems has led chemists to mimic them and therefore, nanoreactors presenting a well-defined cavity capable of including both a catalytic site and the substrate have been designed [1–14]. However, the synthesis of such complicated object is far from trivial, and true endohedral functionalization of a molecular cavity has been rarely described. Moreover, supramolecular catalysts—i.e., catalysts presenting a well-defined cavity just above a catalytic center—can suffer from product inhibition: the product formed in the confined space of the molecular cavity tends to display a high affinity for the host and is difficult to remove from the cavity, thus preventing any catalytic turnover [15, 16]. Molecular containers containing endohedral functional groups can be divided into two main classes: self-assembled cage-like structures obtained from smaller subcomponents and covalent (organic) capsules and macrocycles. The self-assembled cages developed for instance by J. Nitchke, K.N. Raymond, M. Fujita, J.N. Reek, and others are reviewed by J. Bolliger in Chapter "Self-Assembled Coordination Cages and Organic Capsules as Catalytic Supramolecular Reaction Vessels" of this book, and thus will not be presented here. In the present chapter, we will focus on covalent host molecules exhibiting truly endohedral functionalization of their inner space. As it will emerge throughout this chapter, such examples are extremely rare due to the difficulties encountered in their synthesis and the challenges faces in their use as catalysts. Covalent molecular cavities including a catalytic metal center will be described in the first section, whereas organocatalysts encapsulated in the confined space of a cage compounds are the topic of the second half of this contribution. The presented systems will be analyzed with respect to the benefits of confinement by comparison with model catalysts lacking a cavity, thereby highlighting the gain in selectivity and activity arising from encapsulation.

2 Metallic Catalysis in Confined Spaces

Many supramolecular catalysts containing a well-defined cavity just above a metallic center have been designed in order to increase the concentration of a substrate close to the catalytic center, thus mimicking enzymatic catalysis. Recent reviews describe the most frequently occurring systems [17–20]. Most metal-containing supramolecular catalysts are based on calixarenes, resorcinarenes, or cylclodextrins (CD) scaffolds. This is mainly due to their excellent guest binding properties but also because they offer many possibilities to introduce metal ligands at the rim of the cavity.

Here, we have chosen to describe recently published examples based on endohedral functionalization of a covalent cavity. A metal catalyst is described as being in an "endohedral location" means that the metal is encapsulated in a three-dimensional cavity where also the catalytic transformation must take place. Figure 1 compares supramolecular catalysts where the active catalytic center is, or not, considered as endohedral. Compounds **1** [21], **2** [22], and **3** [23] are calix[4] arenes acting directly as a metal ligand in **1**, or functionalized either by two linked arms leading to a planar coordination site in **2** or by three independent arms in **3**. Compared to calix[6]arene **4** [24], where the rim is equipped with a TPA (tris (2-pyridylmethyl)amine) unit, **1–3** are not able to ensure the reaction to occur inside the cavity. Similarly, C_4 symmetrical calix[4]arenes bonded to porphyrins like **5** and **6** [25] are considered as endohedral supramolecular catalysts only if the outward face of the porphyrins is hindered by an *exo* ligand like in compound **6**.

Finally, literature reports are limited to only a few structures containing endohedrally functionalized molecular cavities and their applications as catalysts remain to a large extent unexplored. One reason is that many of these encapsulated metal catalysts are copper complexes and are naturally dedicated to oxidation reactions. Unfortunately, under the reaction conditions ligand oxidation is often favored over substrate oxidation, thus leading to catalyst degradation [18].

In 1990, Kuroda et al. [26] presented the catalytic activity of a Fe^{II} porphyrin complex **7** sandwiched by two cyclodextrins (Fig. 2) as a water-soluble cytochrome P-450 mimic. Compared to the model compound **8**, the binding of substrates by the cyclodextrin units in **7** significantly increases the yield of the epoxidation of less reactive substrates such has cyclohexene (55% for **7**, <2% for **8**).

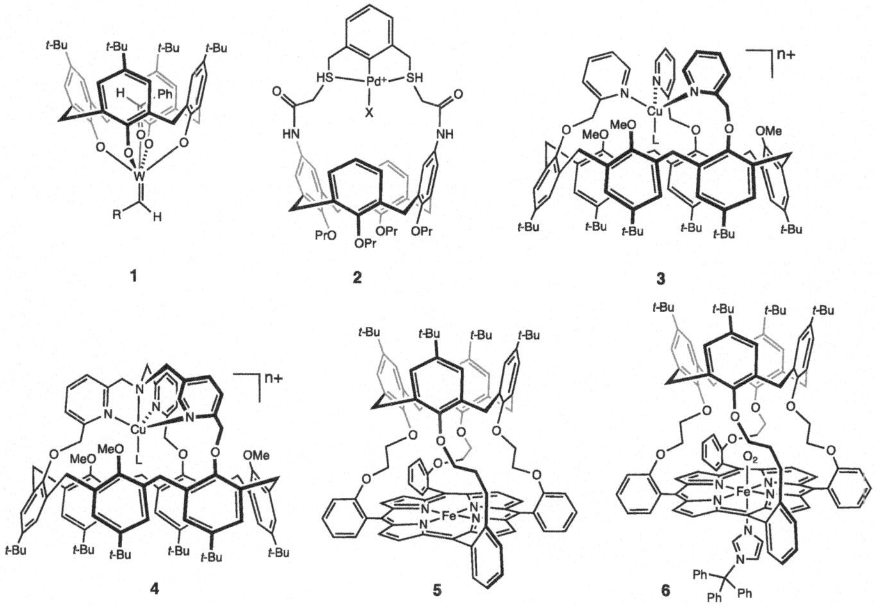

Fig. 1 Non-endohedral supramolecular catalysts versus endohedral molecular catalysts

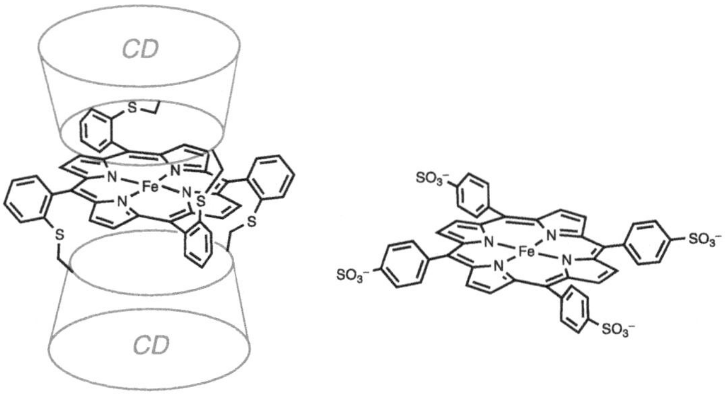

7

8

Fig. 2 Fe^{II} complex of cyclodextrin sandwiched porphyrin **7** complex and the related model compound **8** used as catalyst for the epoxidation of cyclohexene

Two years later, Nolte's group published the first version of a metal-containing cage Rh@**9** (Fig. 3) where the rhodium(I) center converts allyldihydroxybenzenes into their isomerized and hydrogenated products [27]. As a result of the cleft conformation of the organic scaffold, an endohedrally functionalized cavity containing the metal center is formed; external reactions of the rhodium center are prevented by a hindered acetylacetonate ligand. This example of supramolecular catalysis shows a clear relationship between the measured binding constants of the substrates by the cyclodextrin cavities and the observed rate of their conversion.

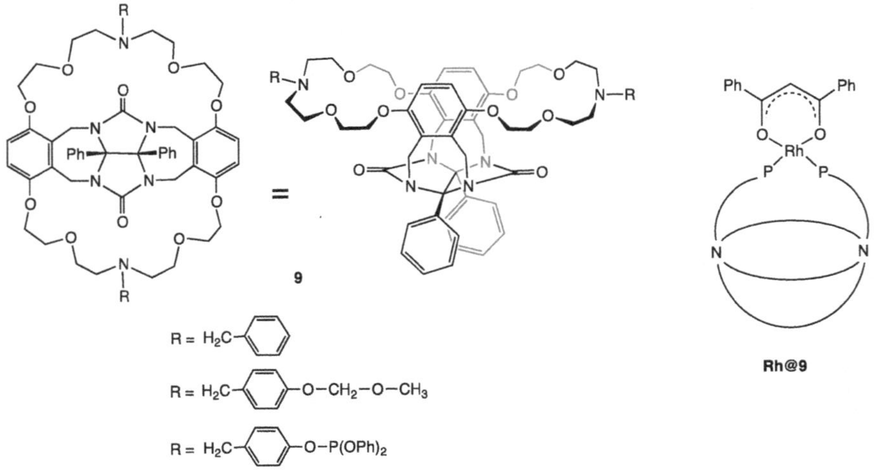

Fig. 3 Capped rhodium complex used for the hydrogenation and isomerization of allyldihydroxybenzenes compounds

The Mn^{III}-porphyrin rotaxane version of the previous system was published 10 years later [28]. The authors clearly showed that the $Mn^{V} = O$ generated during the catalytic reaction was pointing into the cavity. This rotaxane catalyst (**10**, Fig. 4) acts as an efficient progressive oxidative supramolecular catalyst on a polybutadiene polymer blocked capped by bulky terminal groups, thus mimicking natural enzymes such as T4 DNA polymerases. The mechanism and the catalytic properties of this system have been investigated in depth over the past decades [29–31].

In their recent work, Sollougoub et al. describe NHC-capped cyclodextrins (CD) as ligands for gold complexes which are fully encapsulated in the cyclodextrin cavity (gold complexes **11** and **12**; Fig. 5) [32]. The confinement of the catalytically active metal resembles the active center of metallozymes in which the metal is protected by the surrounding protein. The confinement of the above-mentioned gold complexes in **11** and **12** has been confirmed by NMR experiments which indicate the partial inclusion of the capping imidazolium ring inside the cavity. Such sterically shielded metals also exhibit resistance toward electron transfer during electrochemical experiments when compared for example to metals placed at the entrance of a CD cavity. These β-CD derived gold catalysts show unexpected selectivity toward the 6-membered ring cyclization of enyne substrates. Moreover, the same catalysts form a bicyclic product with promising enantioselectivity when a differently substituted enyne substrate is employed (59% *ee* with Au(I)@**12**).

We have reported the synthesis of hemicryptophane cages containing an endohedral location of an oxidovanadium moiety (see for instance compound **13**, Fig. 6) [33]. These complexes proved to be efficient and highly selective catalysts for the oxidation of thioanisol with yields up to 95% (Fig. 6) and were found to proceed with a turnover number up to 180, demonstrating that the hemicryptophane structure can be used to build efficient supramolecular transition metal-based catalysts. Furthermore, a direct comparison between hemicryptophane complex **13** and the model complex **14**, which lacks the cavity, was performed. The cage structure in **13** was found to strongly enhance the catalytic activity since yields are better and kinetic constants were estimated up to six times higher with **13** than those obtained with the model molecule **14**.

The Zn(II)@hemicryptophane complex **15**, which has a zinc(II) center embedded into the cavity (Fig. 7), was synthesized and characterized by Makita and

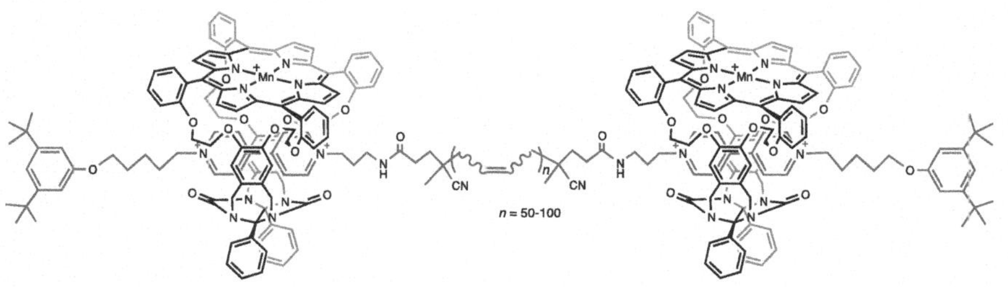

10 10

Fig. 4 The Mn^{III}-porphyrin rotaxane **10** synthesized by Nolte and colleagues

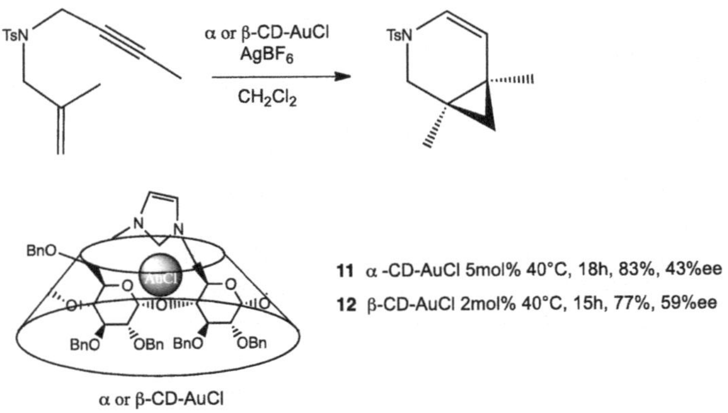

Fig. 5 Au(I)@cyclodextrin complexes synthesized by Sollogoub and colleagues

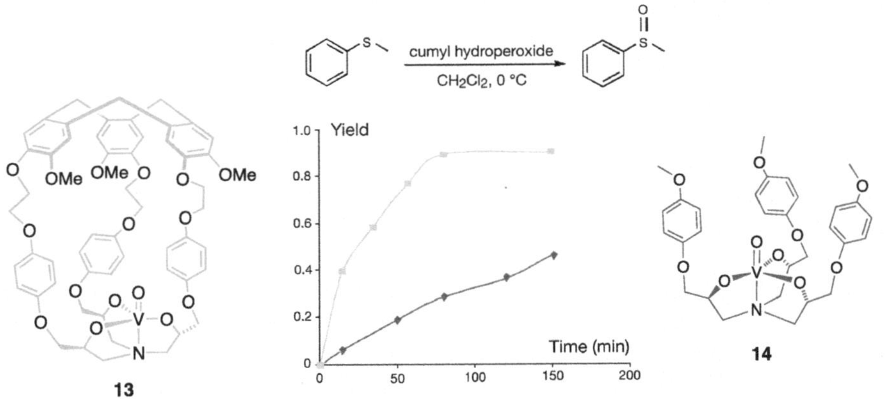

Fig. 6 Comparison of the kinetics of oxidation of thioanisol by oxidovanadium@hemicryptophane supramolecular catalyst **13** and the related model compound **14**

colleagues [34, 35]. The catalytic activity of **15** was tested in the hydrolysis of methyl *para*-nitrophenyl carbonate (MPC). A direct comparison between the supramolecular catalyst **15** and the model complex **16** demonstrated that the cage structure clearly enhanced the catalytic activity. Indeed, the kinetic studies reveal a *k*(**15**)/*k*(**16**) ratio of up to 2.2. DFT calculations suggest that the higher catalytic activity of **15** is correlated both to substrate encapsulation within the cavity and control of solvent access to the zinc complex inside the cavity, thus avoiding inhibition by DMSO due to coordination to the catalytically active metal center.

Among the reactions catalyzed by transition metal complexes, C–H oxidation has recently been gaining increasing interest as an efficient and cheap route to directly access valuable functionalized products required by the chemical industry from readily available alkanes and other constituents of natural gas and petroleum —one of the most ambitious goals yet to be achieved in transition metal catalysis.

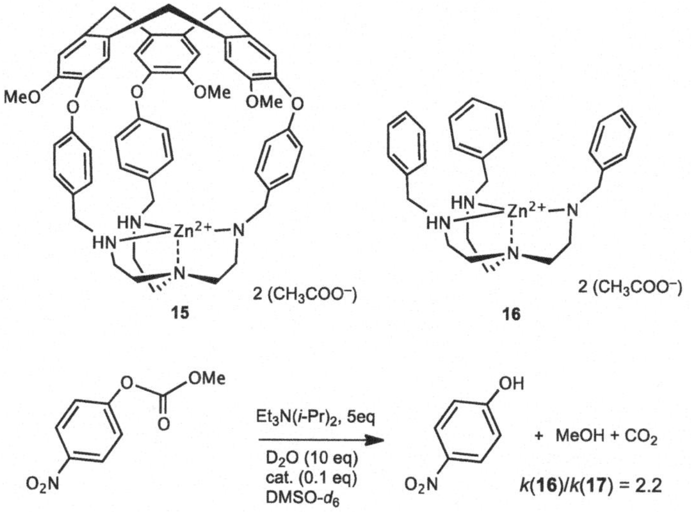

Fig. 7 Zn@hemicryptophane supramolecular catalyst **15** and the model compound **16** used for the hydrolysis of MPC

Although much attention has been paid to design supramolecular–copper complexes with a well-defined cavity above the reactive center, their use as catalysts for the efficient and selective oxidation of alkanes still remains a largely unexplored area of research. In this context, we have reported an unprecedented and efficient cyclohexane oxidation using supramolecular Cu@hemicryptophane complex **17** (Fig. 8) [36, 37]. We have shown that bio-inspired catalysts with elaborated binding pockets are able to catalyze this reaction under mild conditions using hydrogen peroxide as clean oxidant. Moreover, turnover frequencies and yields are doubled when compared to the model molecule **18**, which lacks the cavity of **17**, demonstrating the great potentiality of such an approach for C–H oxidation. Furthermore, the ability of **17** to discriminate cyclohexane from adamantane resulting in a selective oxidation of the former molecule emphasizes the influence of the cavity on the catalytic properties. This is a promising development in the field of supramolecular catalysis as it opens up new pathways for selective for C–H bond oxidation involving copper catalysts in a confined environment.

In order to improve the reaction rate and substrate selectivity, the water-soluble parent hemicryptophane **19** was designed (Fig. 9) [38]. Incorporation of three carboxylate moieties on the CTV unit renders Cu@hemicryptophane **19** soluble at pH 7. The spectroscopic and electrochemical properties of this complex in water were investigated, showing that in water the geometry around the metal center remains unchanged—Cu remains five-coordinate with the copper ion bound to the four nitrogen atoms of the *tren* unit and to one solvent molecule. On the other hand, the solvent strongly affects the redox properties of **19**, compared to the previous Cu (II)@hemicryptophane complex **17**.

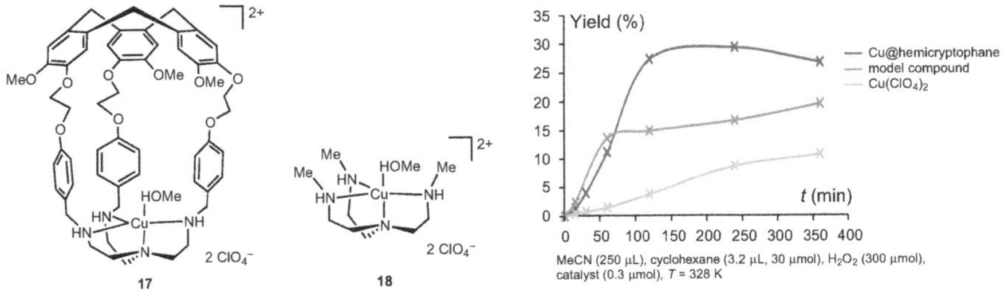

17 18

Fig. 8 Cu@hemicryptophane **17** supramolecular catalyst and its model counterpart **18** used for the oxidation of cyclohexane to cyclohexanone and cyclohexanol

Fig. 9 Structure of the water-soluble Cu(II)@hemicryptophane complex **19**

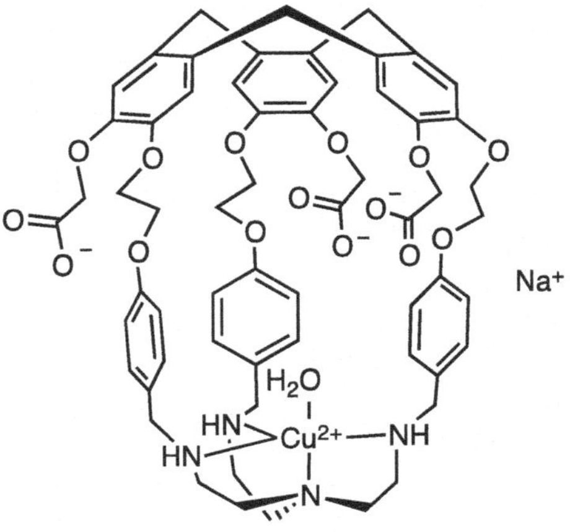

19

3 Organocatalysis in Confined Spaces

Similarly, organocatalysts have been encapsulated into covalent molecular cages, leading to supramolecular catalysts with endohedral functionalization of the inner space of their cavity. The consequences of encapsulation on the rate and the selectivity of the promoted reactions have been investigated, as well as the stability of the encapsulated catalysts. Most reported examples of supramolecular organocatalysts involve the ability of the cavity to stabilize transition states, thus leading to more efficient product formation. As in the above chapter, only examples based on covalent host molecules surrounding the catalytic center will be presented.

Rebek and colleagues have described the synthesis of cavitand **20** which has a deep, open-ended cavity functionalized with a Kemp's triacid derivative (Fig. 10) [39]. This cage compound displays a well-defined cavity with an inwardly directed carboxylic acid. Cavitand **20** was found to catalyze the cyclization reaction of alcohol **21** into epoxide **22** with remarkable regioselectivity. When compared with the related model compound **23**, the confinement provided a >50-fold rate acceleration. The specific shape of the space around the carboxylic acid induces the coiling of the substrate inside the cavitand, bringing the reactive centers of the epoxyalcohols into close proximity. Such folding of the substrates into conformations resembling the transition state structures of the cyclization reactions can account for the improvement of the reaction rate and provides a nice example of a covalent molecular cavity bearing truely endohedral functionalization resulting in a remarkable cage effect in catalysis.

A proazaphosphatrane catalyst, also known as Verkade's superbase, has been encaged in a hemicryptophane structure to study how confinement can modify its stability and reactivity (Fig. 11) [40, 41]. Indeed, Verkade's superbases display remarkable properties as nucleophilic and/or basic catalyst, leading to shorter reaction times enhanced selectivity, and milder reaction conditions, compared to most of the neutral basic catalysts. For instance they were found to efficiently catalyze the Henry reaction, transesterifications, silylations….

We have shown that the encapsulation Verkade's superbase in a hemicryptophane does not alter the strong basicity of the proazaphosphatrane moiety, but dramatically decreases the rate of proton transfer (Fig. 12) [41]. The protonation of the entrapped superbase **24** was found to be one hundred time slower than its model compound **25** despite the latter being seven times less basic. This leads to the conclusion that as a consequence of encapsulation, the thermodynamics and

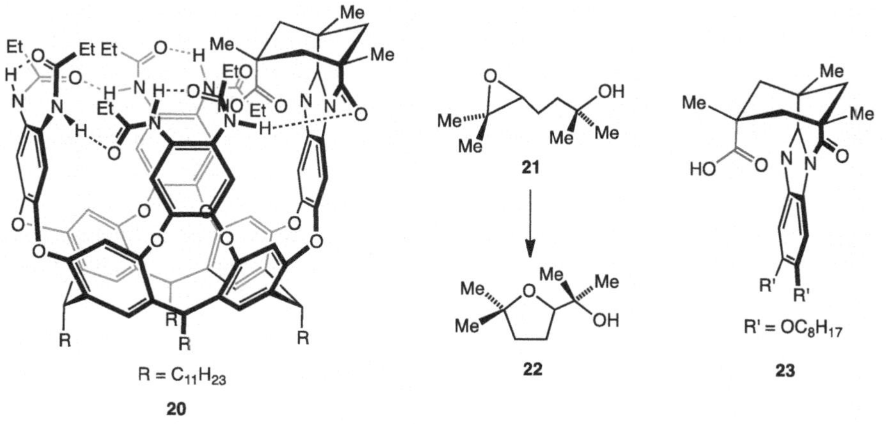

Fig. 10 Structure of cavitand **20** and the related model compound **23** used for the reaction allowing the regioselective transformation of **21** into **22**

Fig. 11 Structure of azaphosphatranes and pro-azaphosphtranes (Verkade's superbase)

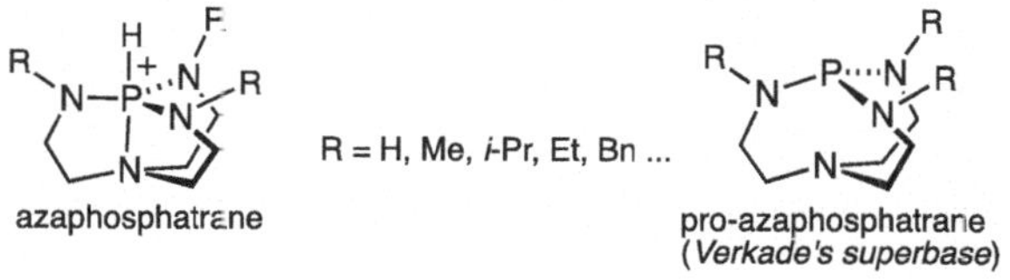

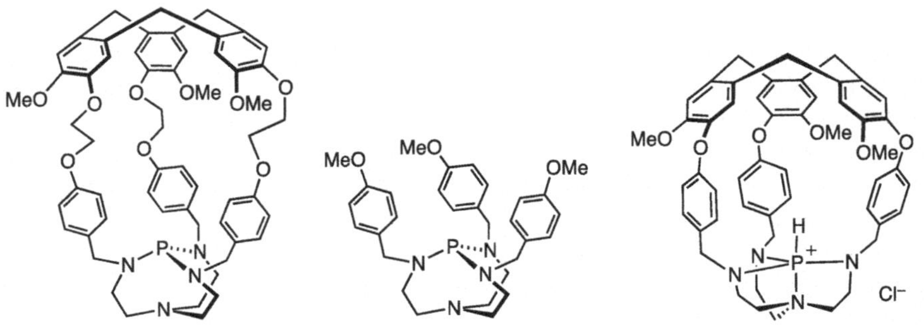

24 25 26

Fig. 12 Structures of encaged superbase **24**, model superbase **25**, and azaphosphatrane **26**

kinetics of the proton transfer are significantly altered. Interestingly, another cyclotriveratrylene (CTV)-capped azaphosphatrane **26**, which contains an endohedral proton within the cavity of the azaphosphatrane, was also synthesized by Makita et al. [42]. In this case the endohedral proton was shielded by the CTV-capped structure from the strongly basic bulk solution and, thus impossible to deprotonate as has been demonstrated by the authors numerous attempts involving various basic conditions.

The role of the cage structure on the rate and thermodynamic of the protons transfer was further investigated by synthesizing two other encapsulated superbases **27** and **28**, exhibiting different sizes and shapes of their cavity (Fig. 13) [43]. It was found that the confinement in these specific architectures affects the pK_a values: depending on the hemicryptophane structure, more or less basic systems can be obtained. For example, the encaged superbase **24** is more than seven times more basic than the model molecule **25** (K_a of 1.03 × 10^{-33} and 7.25 × 10^{-33}, respectively), whereas **27** is more than 30 times less basic than its related model compound **29** (K_a of 4.42 × 10^{-32} and 1.26 × 10^{-33}, respectively). Interestingly, superbase **28** with naphthyl linkers was found to be more than one hundred times

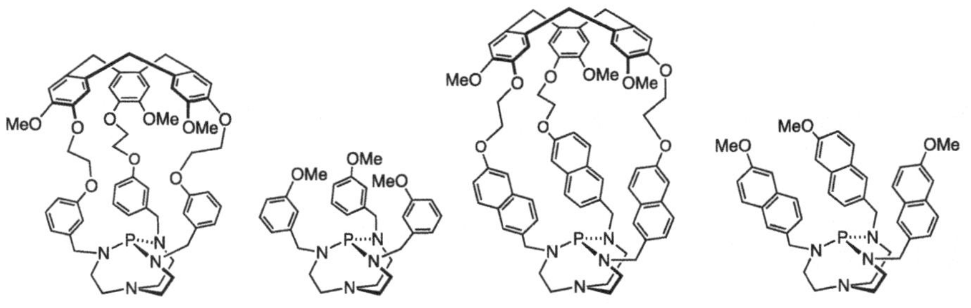

27 29 28 30

Fig. 13 Structure of model and encapsulated superbases presenting various cavity sizes and shapes

more basic than the model counterpart **30**. The rate of proton transfer decreases with increasing cavity size of the host: a strong decrease of the rate constant was observed from **27** to **24** and **28**, highlighting how the shape of the space around the endohedral functionality can affect the kinetics of this reaction. The rate of protonation of the highly basic specie **28** was found to be 500 000 times slower than that of its related model compound. Accurate inspection of the X-ray molecular structures of the host compounds reveals that the naphthalene linkers block the access to the reactive center, thus proposing a strong correlation of the kinetic and thermodynamic data with the space available around the basic center.

These confined organocatalysts have subsequently been employed in catalytic transformations: the encapsulated proazaphosphatrane superbase **24** was tested as catalyst in the base-catalyzed Diels–Alder reactions (Scheme 1) [44] where it shows good to high catalytic activity: 100% yield was obtained in the cycloaddition between anthrone and dimethylfumarate—for comparison, employing triethylamine as base only affords 38% yield under the same conditions. When compared to the model compound **25** which lacks the cavity, kinetic studies reveal that encapsulation induces a drop of the reaction rate by a factor of two which in this case is less pronounced compared to the proton transfer observed in acetonitrile. This constitutes a key point in the development of supramolecular catalysts based on engaged Verkade's superbases, since a decrease of reactivity similar to that previously observed for the proton transfer would strongly limit the interest of such systems. When tested in the Diels–Alder reaction between 3-hydroxy-2-pyrone and *N*-methylmaleimide, a much higher diastereomeric excess is observed with the supramolecular catalyst **24** than with the corresponding model **25** (77 and 43%,

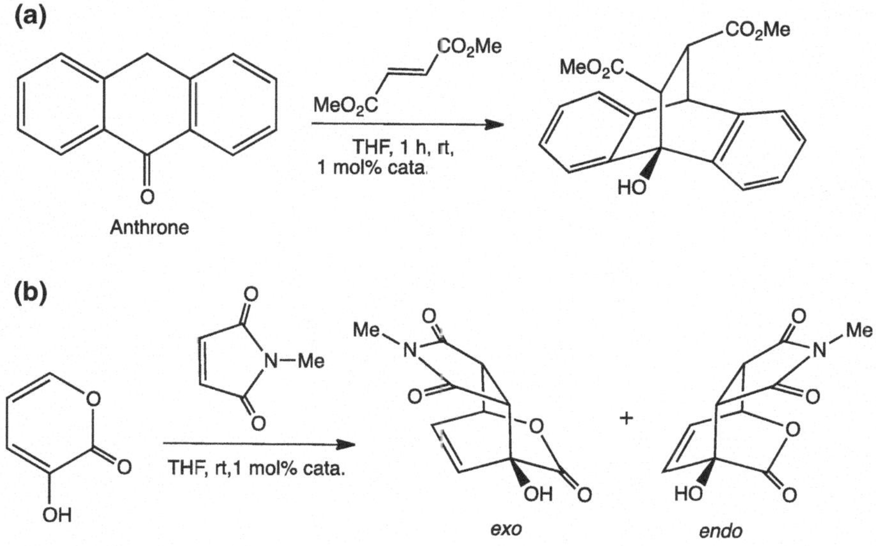

Scheme 1 Diels–Alder reactions between **a** anthrone and dimethylfumarate, and **b** 3-hydroxy-2-pyrone and *N*-methylmaleimide, catalyzed by **24** or **25**

respectively, both in favor of the *endo* product). Previous reports suggest that catalysts in possession of a deep and narrow cavity should favor the formation of the *endo* product which is in agreement with this experimental result [45, 46]. It can be concluded that the encapsulation of an organocatalysts into the confined space of a hemicryptophane can improve the stereoselectivity of the catalyzed reaction: the tight space around the reactive center can limit transition state geometries, thus favoring the more compact ones and leading to an overall improvement of the selectivity.

Azaphosphatranes, the conjugated acids of Verkade's superbases, have up to date received little attention. However, we have shown recently that these robust cations can act as efficient and original catalysts for the conversion of CO_2 into carbonates in the presence of epoxide, under mild conditions (1 bar, 100 °C) (Scheme 2) [47].

Azaphosphatrane@hemicryptophane compounds **31–34**, displaying endohedral functionalization of their molecular cavity, and the model counterparts **35–38**, have been synthesized and tested as catalysts for the above-mentioned reaction (Fig. 14) [48]. As a general trend, it was observed that the confinement of the catalytic

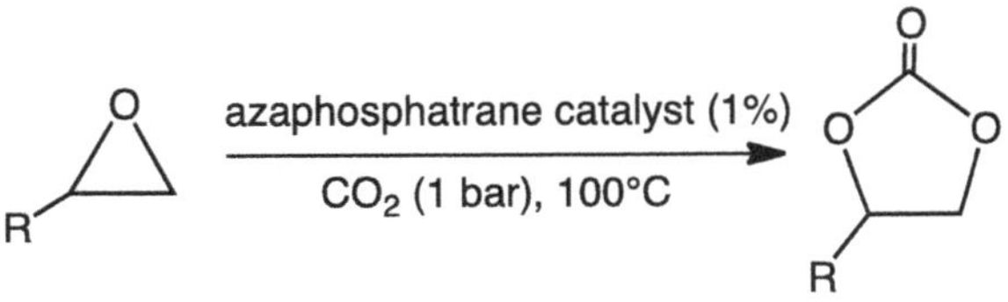

Scheme 2 Conversion of epoxyde into cyclic carbonate with azaphosphatranes catalysts

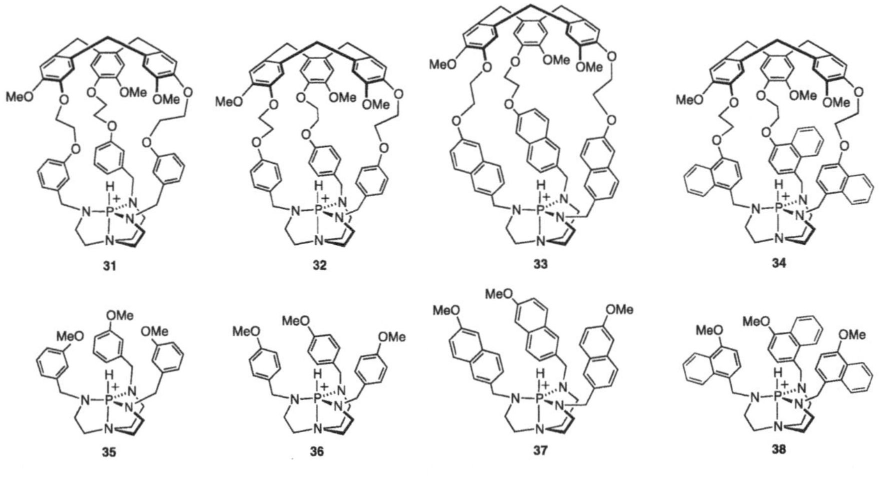

Fig. 14 Structure of encapsulated azaphosphatranes presenting various cavity sizes and shapes and their model parents, used as catalyst for CO_2 conversion. Chlorine counterion is omitted for clarity

species improves both its reactivity and stability. However, supramolecular catalyst **33**, which bears bulky naphthyl linkers, was far less active than its model counterpart **37**. Whereas model catalyst **37** was found to display behavior that was similar to that of other model catalysts, the catalytic activity of **33** was strongly affected by the confinement. This highlights how the inner-cavity size and the accessibility of the PH^+ site affect the reactivity of the encaged azaphosphatrane. Interestingly, for all other supramolecular catalysts, no drop in catalytic activity was observed over time, whereas those of the related model compounds decrease strongly with time; hence, much higher TON can be reached with the former: TON up to 700 were achieved, making **31**, **32,** and **34** remarkable supramolecular catalysts. This behavior underlines the relevance of hemicryptophane hosts for designing efficient modular supramolecular catalysts. Indeed, most of the known supramolecular catalysts suffer from product inhibition and have to be used in stoichiometric amounts. On the contrary, hemicryptophane compounds are rather flexible, as previously observed, allowing for greater lability of the host–guest association. Therefore, the easy release of the carbonated product may account for the remarkable efficiency for the catalytic CO_2 conversion observed here.

In summary, we have described covalent molecular capsules displaying endohedral functionalization of their molecular cavity. The confined space around the catalytic site can induce an improvement on the reaction rate, stability of the catalyst, or selectivity of the catalyzed reaction. Remarkable cage effects were obtained with inwardly orientated reactive groups, thus highlighting the relevance of such an approach. However, examples of truly endohedral functionalization of molecular hosts are still limited. Given the high potential of this strategy, the recent achievements presented in this review may promote a renewed interest in this area of research.

References

1. Rebek J Jr (2009) Molecular behavior in small spaces. Acc Chem Res 42:1660–1668
2. Lehn JM (2004) Supramolecular chemistry: from molecular information towards self-organization and complex matter. Rep Prog Phy 67:249–265
3. Yoshizawa M, Klosterman JK, Fujita M (2009) Functional molecular flasks: new properties and reactions within discrete, self-assembled hosts. Angew Chem Int Ed 48:3418–3438
4. Koblenz TS, Wassenaar J, Reek JNH (2008) Reactivity within a confined self-assembled nanospace. Chem Soc Rev 37:247–262
5. Feiters MC, Rowan AE, Nolte RJM (2000) From simple to supramolecular cytochrome P450 mimics. Chem Soc Rev 29:375–384
6. Kirby AJ (1996) Enzyme mechanisms, models, and mimics. Angew Chem Int Ed Engl 35:707–724
7. Fiedler D, Leung DH, Bergman RG, Raymond KN (2005) Selective molecular recognition, C-H bond activation and catalysis in nanoscale reaction vessels. Acc Chem Res 38:349–358
8. Murakami Y, Kikuchi J, Hisaeda Y, Hayashida O (1996) Artificial enzymes. Chem Rev 96:721–758

9. Wiester MJ, Ulmann PA, Mirkin CA (2010) Enzyme mimics based upon supramolecular coordination chemistry. Angew Chem Int Ed 50:114–137
10. Pluth MD, Bergman RG, Raymond KN (2007) Making amines strong bases: thermodynamic stabilization of protonated guests in a highly-charged supramolecular host. J Am Chem Soc 129:11459–11467
11. Smith PB, Dye JL, Cheney J, Lehn JM (1981) Proton cryptates. Kinetics and thermodynamics of protonation of the [1.1.1] macrobicyclic cryptand. J Am Chem Soc 103:6044–6048
12. Ciampolini M, Nardi N, Valtancoli B, Micheloni M (1992) Small aza cages as "fast proton sponges" and strong lithium binders. Coord Chem Rev 120:223–236
13. Chambron JC, Meyer M (2009) The ins and outs of proton complexation. Chem Soc Rev 38:1663–1673
14. Zarra S, Wood DM, Roberts DA, Nitschke JR (2015) Molecular containers in complex chemical systems. Chem Soc Rev 44:419–432
15. Sanders JKM (1998) Supramolecular catalysis in transition. Chem Eur J 4:1378–1383
16. Yoshizawa M, Tamura M, Fujita M (2006) Diels-Alder in aqueous molecular hosts: unusual regioselectivity and efficient catalysis. Science 312:251–254
17. Raynal M, Ballester P, Vidal-Ferran A, van Leeuwen PWNM (2014) Supramolecular catalysis. Part 2: artificial enzyme mimics. Chem Soc Rev 43:1734–1787
18. Rebilly JN, Colasson B, Bistri O, Over D, Reinaud O (2015) Biomimetic cavity-based metal complexes. Chem Soc Rev 44:467–489
19. Homden DM, Redshaw C (2008) The use of calixarenes in metal-based catalysis. Chem Rev 108:5086–5130
20. Leenders SHAM, Gramage-Doria R, de Bruin B, Reek JNH (2015) Transition metal catalysis in confined spaces. Chem Soc Rev 44:433–448
21. Giannini L, Solari E, Dovesi S, Floriani C, Re N, Chiesi-Villa A, Rizzoli C (1999) Genesis, redox, and acid-base relationships among W–C, W=C, and W≡C functionalities over an oxo surface modeled by calix[4]arene. J Am Chem Soc 121:2784–2796
22. Cameron BR, Loeb SJ, Yap GPA (1997) Calixarene metalloreceptors. Synthesis and molecular recognition properties of upper-rim functionalized calix[4]arenes containing an organopalladium binding Site. Inorg Chem 36:5498–5504
23. Blanchard S, Le Clainche L, Rager MN, Chansou B, Tuchagues JP, Duprat AF, Le Mest Y, Reinaud O (1998) Calixarene-based copper(I) complexes as models for monocopper sites in enzymes. Angew Chem Int Ed 37:2732–2735
24. Izzet G, Zeng X, Akdas H, Marrot J, Reinaud O (2007) Drastic effects of the second coordination sphere on neutral vs. anionic guest binding to a biomimetic Cu(II) center embedded in a calix[6]aza-cryptand. Chem Commun 810–812
25. Kobayashi N, Mizuno K, Osa T (1994) A 'calix[4]arened' porphyrin as a new host and an oxygen carrier model. Inorg Chim Acta 224:1–3
26. Kuroda Y, Hiroshige T, Ogoshi H (1990) Epoxidation reaction catalysed by cyclodextrin sandwiched porphyrin in aqueous buffer solution. J Chem Soc Chem Commun 1594–1595
27. Coolen HKAC, van Leeuwen PWNM, Nolte RJM (1992) A Rh^{I}-centered cage compound with selective catalytic properties. Angew Chem Int Ed Engl 31:905–907
28. Thodarson P, Bijsterveld EJA, Rowan AE, Nolte RJM (2003) Epoxidation of polybutadiene by a topologically linked catalyst. Nature 424:915–918
29. Deutman ABC, Monnereau C, Elemans JAAW, Ercolani G, Nolte RJM, Rowan AE (2008) Mechanism of threading a polymer through a macrocyclic ring. Science 322:1668–1671
30. Deutman ABC, Cantekin S, Elemans JAAW, Rowan AE, Nolte RJM (2014) Designing processive catalytic systems. Threading polymers through a flexible macrocycle ring. J Am Chem Soc 136:9165–9172
31. Thomassen PJ, Varghese S, Bijsterveld EJA, Thordarson P, Elemans JAAW, Rowan AE, Nolte RJM (2015) A double-cavity-containing porphyrin host as a highly stable epoxidation catalyst. Eur JOC 5246–5253
32. Guitet M, Zhang P, Marcelo F, Tugny C, Jiménez-Barbero J, Buriez O, Amatore C, Mouriès-Mansuy V, Goddard JP, Fensterbank L, Zhang Y, Roland S, Ménand M,

Sollogoub M (2013) NHC-capped cyclodextrins (ICyDs): insulated metal complexes, commutable multicoordination sphere, and cavity-dependent catalysis. Angew Chem Int Ed 52:7213–7218
33. Martinez A, Dutasta JP (2009) Hemicryptophane-oxidovanadium(V) complexes: lead of a new class of efficient supramolecular catalysts. J Catal 267:188–192
34. Makita Y, Sugimoto K, Furuyosho K, Ikeda K, Jujiwara SI, Shin-ike T, Ogawa A (2010) A zinc(II)-included hemicryptophane: facile synthesis, characterization, and catalytic activity. Inorg Chem 49:7220–7222
35. Makita Y, Ikeda K, Sugimoto K, Fujita T, Danno T, Bobuatong K, Ehara M, Jujiwara SI, Ogawa A (2012) Enhancement of catalytic reactivity of zinc(II) complex by a cyclotriveratrylene-capped structure. J Organomet Chem 706:26–29
36. Perraud O, Sorokin AB, Dutasta JP, Martinez A (2013) Oxidation of cycloalkanes by H_2O_2 using a copper–hemicryptophane complex as a catalyst. Chem Commun 49:1288–1290
37. Perraud O, Tommassino JB, Robert V, Albela B, Khrouz L, Bonneviot L, Dutasta JP, Martinez A (2013) Hemicryptophane-assisted electron transfer: a structural and electronic study. Dalton Trans 42:1530–1535
38. Schmitt A, Collin S, Bucher C, Maurel V, Dutasta JP, Martinez A (2015) Synthesis and physico-chemical properties of the first water soluble Cu(II)@hemicryptophane complex. Org Biomol Chem 13:2157–2161
39. Shenoy SR, Crisostomo FRP, Iwasawa T, Rebek J Jr (2008) Organocatalysis in a synthetic receptor with an inwardly directed carboxylic acid. J Am Chem Soc 130:5658–5659
40. For a review on Verkade's superbases, see: Kisanga PB, Verkade JG (2001) Synthesis of new proazaphosphatranes and their application in organic synthesis. Tetrahedron 57:467–475
41. Dimitrov-Raytchev P, Martinez A, Gornitzka H, Dutasta JP (2011) Encaging the Verkade's superbases: thermodynamic and kinetic consequence. J Am Chem Soc 133:2157–2159
42. Makita Y, Furuyoshi K, Ikeda K, Fujita T, Fujiwara S, Ehara M, Ogawa A (2011) Synthesis and characterization of a cyclotriveratrylene-capped azaphosphatrane. Tetrahedron Lett 52:4129–4131
43. Chatelet B, Gornitzka H, Dufaud V, Jeanneau E, Dutasta JP, Martinez A (2013) Superbases in confined space: control of the basicity and reactivity of the proton transfer. J Am Chem Soc 135:18659–18664
44. Chatelet B, Dufaud V, Dutasta JP, Martinez A (2014) Catalytic activity of an encaged Verkade's superbase in a base-catalyzed diels–alder reaction. J Org Chem 79:8684–8688
45. Hatano M, Ishihara K (2012) Conformationally flexible chiral supramolecular catalysts for enantioselective Diels-Alder reactions with anomalous endo/exo selectivities. Chem Commun 48:4273–4283
46. Hatano M, Mizuno T, Izumiseki Usami R, Asai T, Akarura M, Ishihara K (2011) Enantioselective Diels-Alder reactions with anomalous endo/exo selectivities using conformationally flexible chiral supramolecular catalysts. Angew Chem Int Ed 50:12189–12192
47. Chatelet B, Joucla L, Dutasta JP, Martinez A, Szeto KC, Dufaud V (2013) Azaphosphatranes as structurally tunable organocatalysts for carbonate synthesis from CO_2 and epoxides. J Am Chem Soc 135:5348–5351
48. Chatelet B, Joucla L, Dutasta JP, Martinez A, Dufaud V (2014) Azaphophatrane organocatalysts in confined space: cage effect in CO_2 conversion. Chem Eur J 20:8571–8574

Self-Assembled Coordination Cages and Organic Capsules as Catalytic Supramolecular Reaction Vessels

Jeanne L. Bolliger

Abstract Host-guest chemistry has undergone an enormous development since the discovery of cyclodextrins more than 100 years ago which has culminated in the preparation of many artificial host molecules that are not only capable of encapsulating a variety of guests but also of promoting reactions inside their cavities. As the environment dramatically influences the behavior of chemical systems, recent years have seen increased interest in the use of the shielded inner phases of synthetic hosts to stabilize reactive species, shift equilibria, or achieve otherwise unfavorable conformations of guest species. Confinement inside hosts has been used to lower the symmetry of guests, thereby creating new means to control the outcomes of asymmetric reactions in the same way that biological systems make extensive use of tailored microenvironments to promote stereospecific reactions by destabilizing the ground state and stabilizing certain transition state geometries. This chapter will focus on the use of self-assembled coordination cages and organic capsules as homogeneous catalytic supramolecular reaction vessels. Modulation of the cavity environment and binding selectivity is relatively easily achieved because small changes to the geometries of building blocks can lead to much larger changes in the structures and properties of the hollow polyhedral coordination cages formed upon self-assembly. As the reaction medium influences the binding of the reactants and products in subtle but important ways, control over host solubility through host framework charge and substituent effects provides further means to control guest binding strengths, selectivity, and dynamics, and thereby a possible way to overcome product inhibition which is often encountered in supramolecular catalysis. A review will be provided over unusual selectivity observed in reactions carried out in metal-organic capsules as a result of structural constraints. Similarly, rate enhancements in bimolecular reactions due to an increase in effective molarity and

J.L. Bolliger (✉)
Department of Chemistry, Oklahoma State University, 107 Physical Science, Stillwater, OK 74078, USA
e-mail: jeanne.bolliger@okstate.edu

J.L. Bolliger
Department of Chemistry, University of Cambridge, Lensfield Road, Cambridge CB2 1EW, UK

R. Poli (ed.), *Effects of Nanoconfinement on Catalysis*,
Fundamental and Applied Catalysis, DOI 10.1007/978-3-319-50207-6_2

stabilization of the transition state as well as transformations carried out under unusual conditions—for example, the acid catalyzed hydrolysis of orthoformates under neutral or basic conditions—will be discussed in this chapter. Furthermore, self-assembled coordination cages based on chiral ligands are of particular interest because they provide an asymmetric microenvironment for promoting stereoselective reactions by purely non-covalent interactions. Particular emphasis will be laid on the hydrolysis of organophosphorus species: As an example of the author's work, the catalytic degradation of the insecticide dichlorvos by a $[Fe_4L_6]^{8+}$ cage molecule will be presented, and this report will also include the up-to-date unpublished results obtained from experiments with other organophosphorus insecticides.

1 Introduction

Supramolecular catalysis has evolved into an extremely broad field ranging from the use of non-covalent interactions as a tool for building and modifying homogenous catalysts to artificial enzyme mimics [1, 2]. This chapter aims to provide a thorough review over a small proportion of this field, namely the use of self-assembled coordination cages and organic capsules as homogeneous catalytic supramolecular reaction vessels [3–6]. These container molecules provide defined hydrophobic cavities which have the potential to mimic binding pockets in enzymes and achieve similar selectivity and catalytic rate enhancements as encountered in biological systems. Self-assembly as a tool for the synthesis of hollow polyhedral coordination cages and organic capsules (see Sect. 2) provides the means for facile modulation of the cavity environment and binding selectivity because small changes to the geometries of building blocks can lead to much larger changes in the structures and properties of the resulting host assemblies. Additionally, small changes to the structures of the host molecules can significantly influence binding constants of both substrates and products as well as the rates achieved in host-guest catalysis. In order to gain a better understanding of the difficulties to overcome when designing host materials for catalytic applications, the thermodynamic and kinetic basics underlying successful host-guest catalysis will be reviewed (see Sect. 3). Eventually, examples of host-guest catalysis will be provided (Sect. 4). Applications of container molecules in catalysis include the promotion of hydrolysis reactions at unusual pH ranges (Sect. 4.1), pre-organization of guest molecules or reactive intermediates leading to regioselective deprotonations (Sect. 4.2), stabilization of cationic transition states in unimolecular and bimolecular reactions (Sect. 4.3 and Sect. 4.4, respectively), and rate enhancements of cycloaddition and condensation reactions (Sect. 4.5). These examples include self-assembled coordination cages based on chiral ligands which are of particular interest as they provide an asymmetric microenvironment for promoting stereoselective reactions by purely non-covalent interactions. Furthermore, encapsulated transition metal complexes will be covered in this review but are limited to non-covalently bound host-guest systems (Sect. 5). Self-assembled systems containing endohedral functionalities for coordination of metal complexes are beyond the scope of this article and have been reviewed elsewhere [7]. The use of coordination cages in catalytic tandem transformations ranging from preventing transition metal degradation in aqueous solution, over avoiding inhibition of enzymes by compartmentalizing the catalysts, to their use as orthogonal catalysts for stabilizing and converting high-energy intermediates for subsequent catalytic processes is reviewed in the last section of this chapter (Sect. 6).

2 Self-Assembled Coordination Cages and Organic Capsules

Non-covalent host molecules such as self-assembled coordination cages and organic capsules present several advantages over their covalent analogs since they can be prepared in a modular fashion from smaller building blocks. This approach allows facile adaption of the supramolecular assembly to the desired application without the arduous synthesis accompanied with the preparation of larger covalent hosts. Additionally, as self-assembled container molecules are held together by reversible non-covalent interactions, they exhibit a certain flexibility and dynamic behavior which facilitates not only guest encapsulation but also release of the product in catalytic applications. Scheme 1 illustrates the self-assembly of several non-covalent hosts which have been successfully used in catalytic transformations. Self-assembled coordination cages are prepared under thermodynamic reaction conditions from ligands and metal ions, as has been demonstrated by Raymond, Fujita, Ward and many more. Raymond and Bergman's coordination cage in Scheme 1a illustrates how small changes of the ligand structure can be used to form coordination capsules diastereoselectively. While the achiral ligand results in the formation of a racemic mixture of the tetrahedral coordination cages $\Delta\Delta\Delta\Delta$-**1a** and $\Lambda\Lambda\Lambda\Lambda$-**1a**, the enantiopure ligands enable the diastereoselective self-assembly of either $\Delta\Delta\Delta\Delta$-**1b** or $\Lambda\Lambda\Lambda\Lambda$-**1b**, thus leading to host-assemblies which provide in their cavity an asymmetric microenvironment and can be used for applications in enantioselective host-guest catalysis [8]. In an approach termed subcomponent self-assembly, the Nitschke group has taken advantage of the simultaneous formation of dynamic covalent imine bonds and coordination bonds to prepare coordination cages from even smaller fragments, as for example the coordination capsule $\Delta\Delta\Delta\Delta$-**3** (Scheme 1b) [9]. Since the bis-dicoordinate imine ligands are formed in situ, for example, from the diamine *SS*-**2** by reversible imine bond-formation with 2-pyridinecarboxaldehyde, control over guest binding strengths, selectivity, and dynamics can be exerted easily by subtle changes in the host framework resulting from substituent effects. In $\Delta\Delta\Delta\Delta$-**3**, the 12 glyceryl substituents do not only render this coordination capsule water-soluble but also allow it to form diastereoselectively, presumably by hydrogen bonding of the

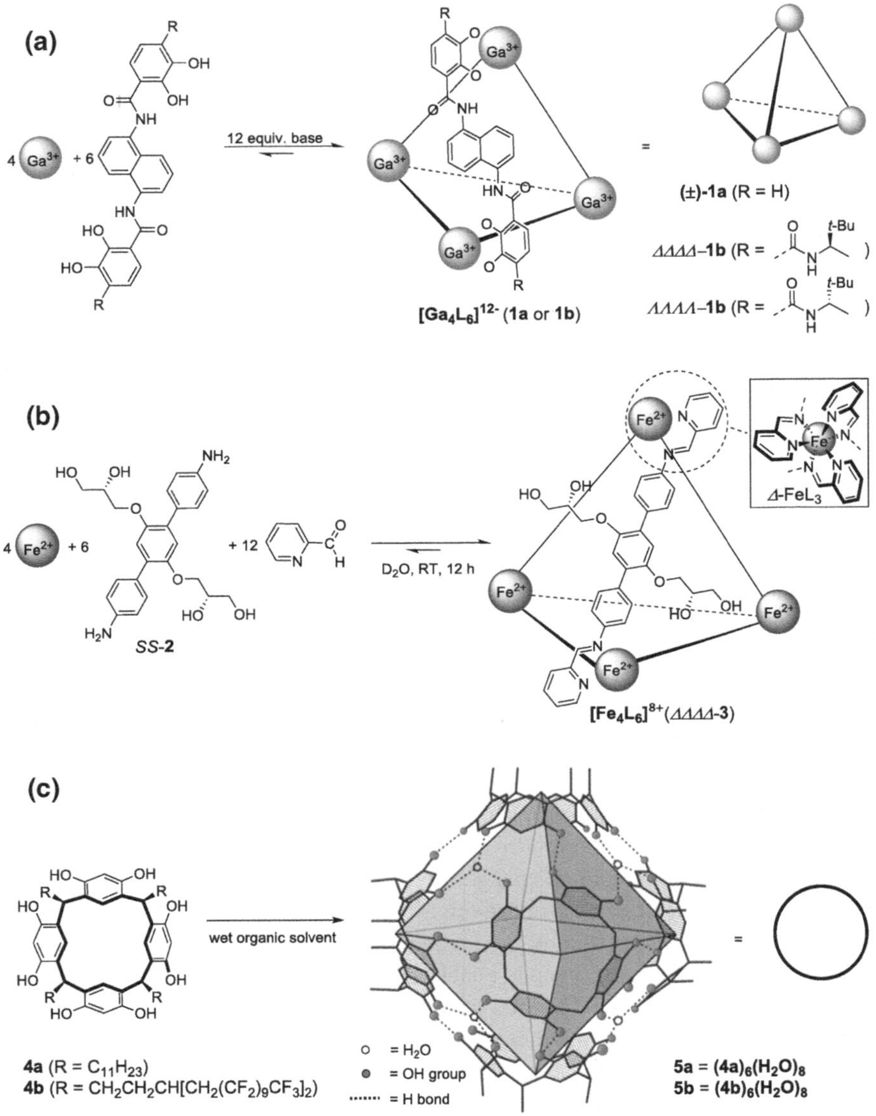

Scheme 1 Self-assembled capsules. **a** Example of self-assembled coordination cages; **b** subcomponent self-assembly of coordination cages; **c** hydrogen-bonded organic capsule (adapted from Chem. Commun., 2015, 51, 892–894; Published by The Royal Society of Chemistry)

glyceryl substituents on the faces of the tetrahedral host. Additionally, since these sp^3 hybridized substituents are very flexible, they allow rapid guest exchange and thereby a possible way to overcome product inhibition which is often encountered in supramolecular catalysis.

Next to coordination cages, also hydrogen-bonded organic capsules have been shown to provide binding cavities suitable for host-guest catalysis. Examples are Rebek's dimeric capsule (vide infra) or the hexameric resorcin[4]arene capsule (Scheme 1c), a chiral spherical molecular assembly held together by 60 hydrogen bonds which was initially described by MacGillivray and Atwood [10]. As shown in Scheme 1c, these hexameric hydrogen-bonded organic capsules (e.g. **5a**) form readily in wet organic solvents such as chlororform or benzene from six symmetrically substituted resorcin[4]arene units (e.g., **4a**) and incorporate eight water molecules in the hydrogen-bonded assembly (**5a**=(**4a**)$_6$(H_2O)$_8$).

3 Thermodynamics and Kinetics in Host-Guest Catalysis

3.1 Host-Guest Catalysis Compared to Enzyme Catalysis

Supramolecular chemists aim to copy four key features of enzyme catalysis when attempting to create artificial enzyme mimics. Biological systems make extensive use of tailored microenvironments to promote reactions by destabilizing the ground state and stabilizing certain transition state geometries. In analogy, an ideal self-assembled supramolecular catalyst would provide a (hydrophobic) binding pocket with strong affinity for the substrate, an even stronger stabilization of the transition state or intermediates of the reaction, and weak binding of the product to allow its release and ensure catalytic turnover. Nowadays, the design and synthesis of coordination cages which bind a desired substrate is aided by molecular modeling and generally does not present a problem. The challenges in host-guest catalysis most often are encountered when attempting to stabilize a certain transition state without increasing the affinity of the host for the reaction product. Therefore, while many groups have reported the use of coordination cages and organic capsules as supramolecular reaction flasks to increase the rate of a reaction or achieve unusual reactivity and selectivity [6, 11, 12], only a small proportion of these systems were demonstrated to be truly catalytic.

A simplified catalytic cycle for host-guest catalysis employing host H to convert substrate S to product P is displayed in Fig. 1a while Fig. 1c illustrates the relative free energy changes during the catalytic reaction. As generally the case, the catalyst —in this case the coordination cage or organic capsule—lowers the activation energy when compared with the uncatalyzed reaction ($\Delta G^{\ddagger}_{cat} < \Delta G^{\ddagger}$). The first step of the catalytic cycle involves the encapsulation of substrate S which is typically fast and reversible. Depending on the rates of the forward and reverse reaction (k_1 and k_{-1}), the host–guest complex [S$\subset$H] either is in fast or slow exchange with the empty host H and free substrate S. In order to achieve product formation in the second step, the host H must bind the transition state better than the ground state ($\Delta G^{\circ}_{TS} < \Delta G^{\circ}_{en}$). This rate-limiting step (k_2, also known as k_{cat} in enzymatic catalysis) yields the encapsulated product in form of [P$\subset$H]. For release of the product

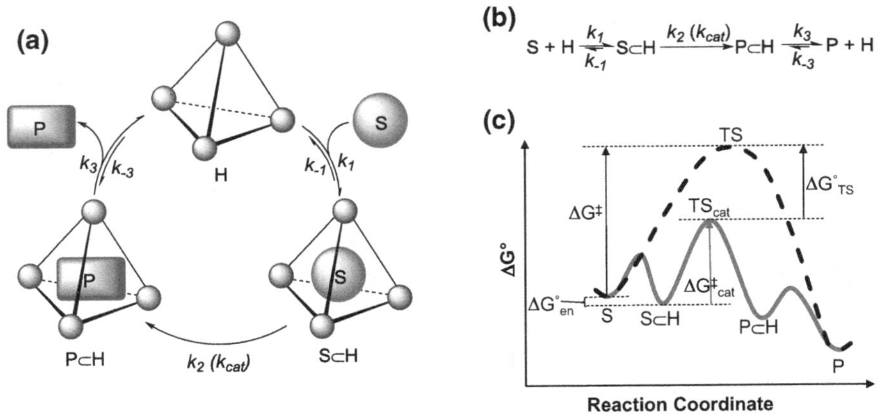

Fig. 1 Host-guest catalysis: same product with and without supramolecular catalyst. **a** Catalytic cycle; **b** rate constants; **c** typical reaction coordinate of the catalyzed reaction versus the uncatalyzed reaction

to take place, k_3 must be large compared to k_{-3}; additionally the product must be more weakly bound than the substrate otherwise product inhibition is likely to occur.

Occasionally, product inhibition does not pose a problem due to steric or electronic interactions resulting in a lower affinity of the product for the host when compared with the starting material [13, 14]. Otherwise, a potential strategy to ensure turnover and thereby closure of the catalytic cycle includes subsequent reactions of the catalytic product upon its release, e.g., the hydrolysis of an iminium cation to an aldehyde which shows significantly lower affinity for its host, thus making the product release irreversible [15]. Furthermore, a change in charge can facilitate catalytic turnover [16].

3.2 New Selectivity Due to Confinement

Encapsulation can significantly alter the properties of the guest which often results in unusual reactivity inside a supramolecular host. As seen in Fig. 2, confinement can lead to a stronger stabilization of an alternative transition state compared to the uncatalyzed reaction and thus explain unusual selectivity observed in the product obtained in cage-catalyzed reactions when compared to the reaction being carried out in bulk solution [17, 18].

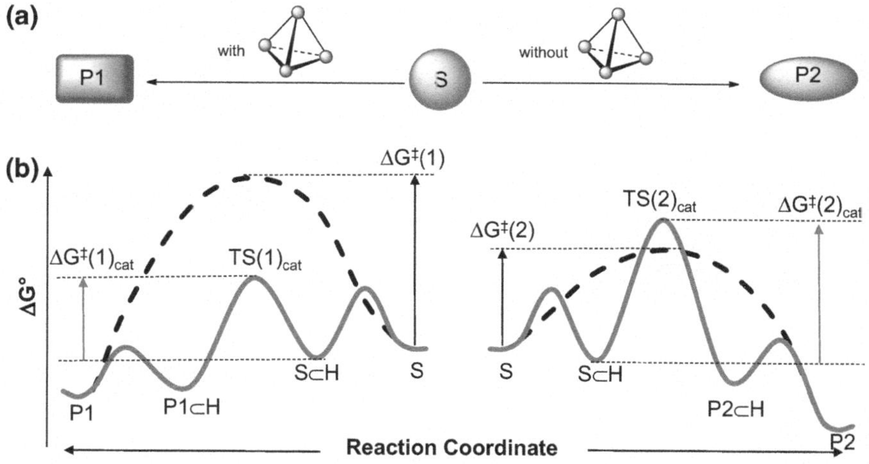

Fig. 2 Encapsulation can lead to new selectivity. **a** Confinement can change the outcome of a reaction; **b** stabilization of a different transition state inside a capsule

4 Host-Guest Catalysis in Self-Assembled Coordination Cages and Organic Capsules

4.1 Catalytic Hydrolysis by Self-Assembled Coordination Cages and Organic Capsules

Hydrolysis reactions can be significantly accelerated in negatively charged coordination cages which show a strong preference for positively charged guests and allow the protonation of encapsulated molecules even in basic media [19]. Likewise, positively charged cage molecules can facilitate hydrolysis by increasing the concentration of hydroxide in its vicinity and/or polarizing heteronuclear bonds of the encapsulated molecule, thus enabling a nucleophilic attack of hydroxide at neutral pH [9].

The author has recently demonstrated this concept in the catalytic hydrolysis of neurotoxic organophosphates by a $[Fe_4L_6]^{8+}$ coordination cage. The positively charged $[Fe_4L_6]^{8+}$ coordination cage *ΔΔΔΔ*-**3** (Scheme 1b) was found to encapsulate in water a wide range of neutral organic and organometallic compounds in its large hydrophobic cavity [9]. Water-soluble guests such as 1-adamantylmethanol were observed to undergo fast exchange on the NMR timescale with free molecules in solution while hydrophobic guests (e.g., dibenzyl) were in slow exchange with their free counterparts. Organophosphorus insecticides are one class of molecules which are encapsulated in *ΔΔΔΔ*-**3** (Scheme 2) and display the same behavior as previously observed with organic guest molecules: water-soluble insecticides (e.g., dichlorvos or demeton S methyl sulfon) were in fast exchange on the NMR timescale with free molecules in solution whereas the hydrophobic insecticides fenitrothion and parathion-methyl showed slow exchange on the NMR timescale.

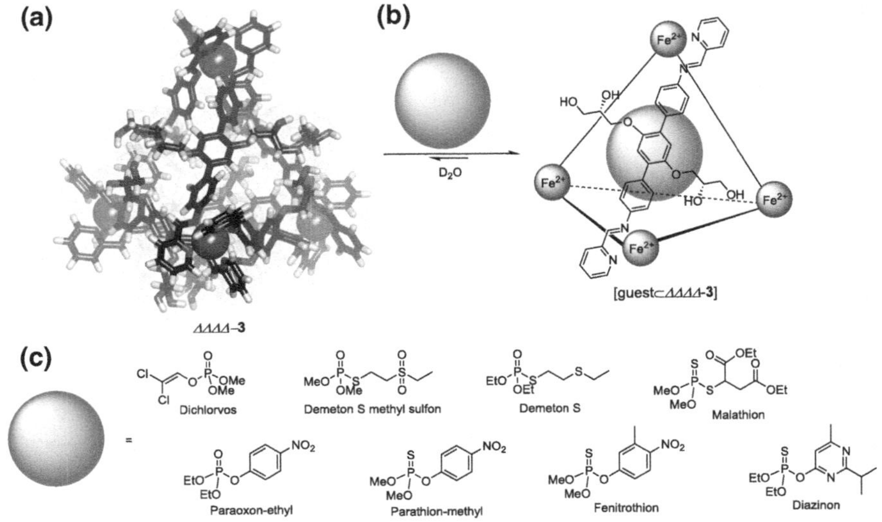

Scheme 2 Host-guest chemistry of *ΔΔΔΔ*-**3**. **a** Molecular model of *ΔΔΔΔ*-**3**; **b** encapsulation of guests in water; **c** examples of organophosphorus insecticides which are encapsulated

The Nitschke group has recently demonstrated that the organophosphate insecticide and chemical warefare agent (CWA) simulant dichlorvos can not only be encapsulated into *ΔΔΔΔ*-**3** but also that hydrolysis of the toxic compound occurs at an increased rate in the presence of *ΔΔΔΔ*-**3** (Fig. 3) [9, 20]. In the presence of 1 mol% of *ΔΔΔΔ*-**3** an increased rate of hydrolysis was measured at pH 7 at room temperature compared to a reference sample containing only buffer solution. *ΔΔΔΔ*-**3** is acting as a supramolecular catalyst in the hydrolysis of dichlorvos to dimethyl phosphate (DMP) and dichloroacetaldehyde (hydrate) or, alternatively, to dichlorovinylmethyl phosphate (DVMP) and methanol (Fig. 3a). As indicated in Fig. 3b, both pathways are accelerated in the presence of *ΔΔΔΔ*-**3**. Since the hydrolysis products are more water-soluble than the starting material, no product inhibition was observed in the hydrolysis of dichlorvos. The authors suggested as a possible mechanistic explanation for catalytic acceleration that the (reversible) encapsulation of the insecticide by the highly positively charged cage molecule leads to the polarization of the phosphorus-oxygen bonds, thus facilitating a nucleophilic attack at the phosphorus atom and leading to the observed higher rate of hydrolysis in the presence of the cage molecule.

Control reactions carried out in the presence of competing guests (e.g., 1-adamantylmethanol) and inhibiting guests (e.g., dibenzyl) supported the hypothesis that the encapsulation of dichlorvos into *ΔΔΔΔ*-**3** leads to faster hydrolysis. The hydrophobic guest dibenzyl inhibits the catalytic acceleration by forming a strong host–guest complex with *ΔΔΔΔ*-**3**, thereby preventing dichlorvos from binding and resulting in the same rate of reaction as the reference sample. On the other hand, the competing guest 1-adamantylmethanol was seen to slow the

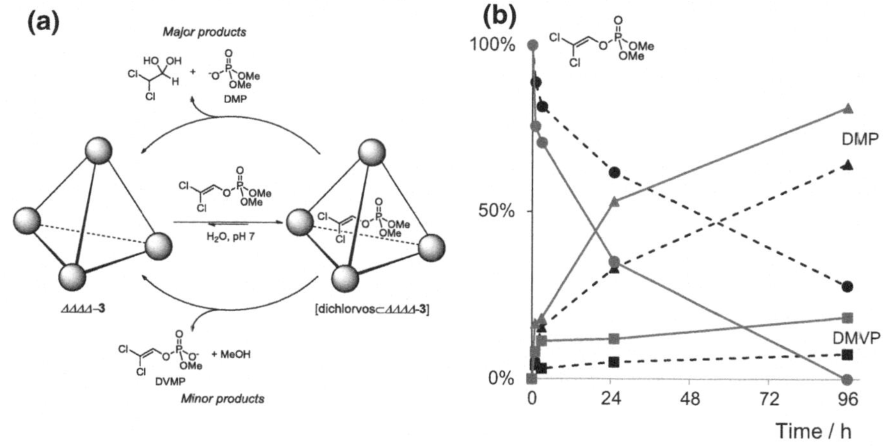

Fig. 3 Hydrolysis of dichlorvos in 0.1 M phosphate buffer at pH 7 and 298 K. **a** Encapsulation of dichlorvos in *ΔΔΔΔ*-**3** and hydrolysis to dimethyl phosphoric acid (DMP, major) and dichlorovinylmethyl phosphoric acid (DVMP, minor); **b** hydrolysis of dichlorvos in 0.1 M phosphate buffer at pH 7 and 298 K followed by NMR: *red line* in the presence of 1 mol% of *ΔΔΔΔ*-**3**, *dashed black line* reference reaction; *red circle* and *black circle* dichlorvos, *red triangle* and *black triangle* dimethyl phosphoric acid (DMP), *red square* and *black square* dichlorovinylmethyl phosphoric acid (DVMP) (Color figure online)

hydrolysis reaction to a rate somewhere between the reference sample and the sample in the presence of pure *ΔΔΔΔ*-**3**. Other control experiments involving the addition of equimolar amounts of subcomponents (12 mol% of 2-formylpyridine, 6 mol% of *SS*-**2**, 4 mol% of $FeSO_4$, or 4 mol% of a mononuclear iron complex formed from 2-formylpyridine and aniline) to the buffered solution at pH 7 showed no acceleration of the rate of hydrolysis of dichlorvos relative to the reference reaction, thus lending further support that the catalytic effect of *ΔΔΔΔ*-**3** can be assigned to encapsulation rather than some structural features present in the subcomponents.

Hydrolysis of the other organophosphate insecticides displayed in Scheme 2c either did not take place at all (e.g., fenitrothion is encapsulated in *ΔΔΔΔ*-**3** without being hydrolyzed) or no catalytic turnover occurred due to product inhibition as observed with diazinon.

During the past decade the Raymond and Bergman groups have published several examples of acid catalysis in basic solution employing the negatively charged $[Ga_4L_6]^{12-}$ coordination cage **1a** (Scheme 1a) as catalyst. This supramolecular host relies exclusively on electrostatic and hydrophobic interactions for the thermodynamic stabilization of protonated substrates as observed in the orthoformate hydrolysis promoted by **1a** (Scheme 3) [21, 22]. Orthoformates **6a–f** are small enough to undergo reversible encapsulation by **1a** and are readily hydrolyzed in basic solution while tripentyl orthoformate and triphenyl orthoformate are too large to be encapsulated and therefore not hydrolyzed under these

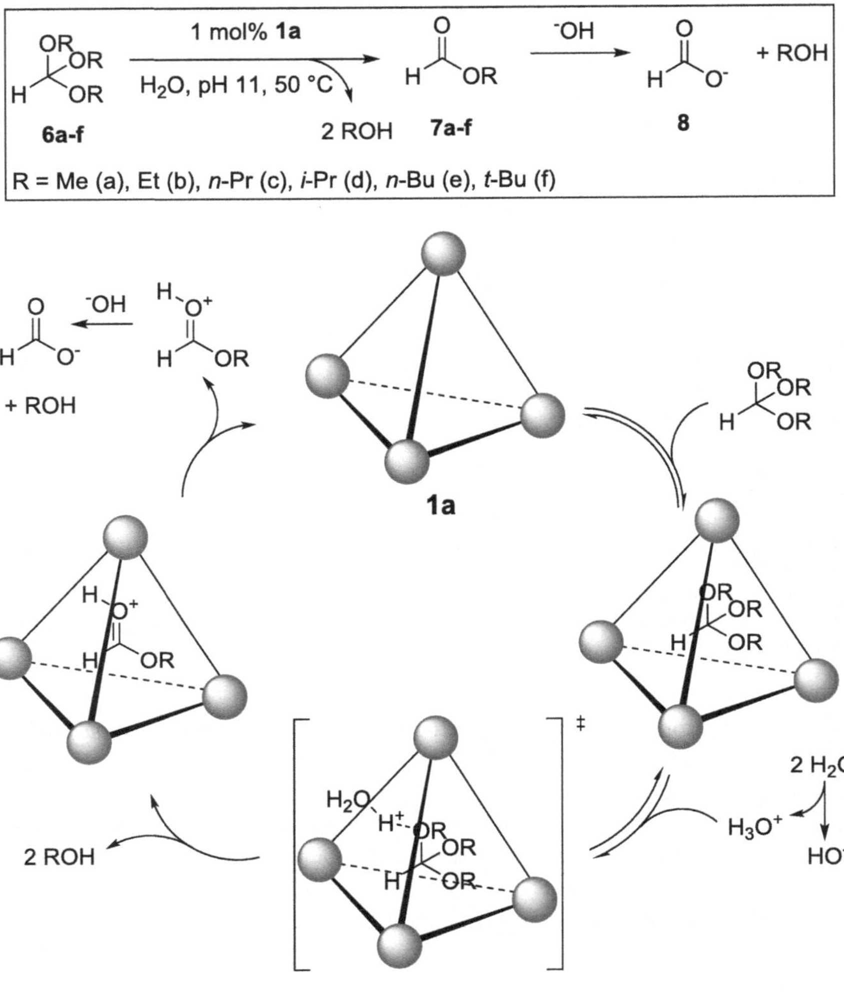

Scheme 3 Acid catalysis in basic solution: orthoformate hydrolysis catalyzed by the negatively charged $[Ga_4L_6]^{12-}$ coordination cage **1a**

reaction conditions. In analogy to enzymes, competitive inhibition is observed in presence of other strongly bound guests, thereby confirming that the cavity of **1a** was the active site for catalysis. The overall reaction mechanism obeys classical enzymatic Michaelis–Menten kinetics including a fast pre-equilibrium step involving the encapsulation of the orthoformate **6a–f** by **1a**, leading to the host–guest complexes [**6a–f**⊂**1a**] which are in equilibrium with the empty host **1a**. Analysis of the rate law indicated that the rate-limiting step involved the proton transfer from protonated water to the encapsulated orthoformate, thus enabling the two successive hydrolysis steps in the cavity which lead to the release of two equivalents of the corresponding alcohol. The resulting protonated formate esters

7a–f are released rapidly from **1a** and subsequently hydrolyzed by base in solution to give the third equivalent of alcohol and orthoformate.

The groups of Raymond and Bergman also investigated the acid catalyzed deprotection of acetals in basic solution inside the supramolecular host **1a** [23, 24]. Acetals are common protecting groups for aldehydes and ketones in organic synthesis due to their high stability under basic conditions. Their deprotection generally is carried out using a Brønsted or Lewis acid in the presence of water. Nevertheless, in the presence of 5 mol% of coordination cage **1a** a variety of smaller dimethoxy alkyl acetals and ketals (**9**) were readily hydrolyzed at pH 10 to the corresponding aldehydes or ketones (**10**) as shown in Scheme 4. Similarly to the above mentioned orthoformate hydrolysis, prospective substrates which are too large to fit into the cavity of **1a** (e.g. 2,2-dimethoxyundecane) do not undergo hydrolytic cleavage under basic conditions and remain unchanged in solution. An in depth analysis of the acid hydrolysis mechanism of acetals by the negatively charged host **1a** revealed that while the mechanism resembled that of orthoformate hydrolysis involving a pre-equilibrium of reversible host–guest complex formation [**9**$\subset$**1a**] driven by the hydrophobic effect, the subsequent protonation occurred fast, and the rate-limiting step was the attack of water on the protonated substrate (Scheme 4). Following the loss of two equivalents of methanol, the protonated carbonyl product **10** is released and immediately deprotonated in the basic solution. Product inhibition is not a problem since the hydrolysis product **10** is bound less strongly than the starting material **9** and is also generally less soluble in water which allows catalytic turnover and leads to the observed conversions of 87–95% in the presence of 5 mol% **1a**.

The Tiefenbacher group demonstrated that the hexameric resorcin[4]arene capsule **5a** (Scheme 1c) was capable of promoting acetal hydrolysis in wet organic solvents (Scheme 5a). Contrary to the coordination capsules of the Raymond group, this neutral hydrogen-bonded organic capsule is thought to act itself as reasonably strong Brønsted acid with a pK_a between 5.5 and 6 [25]. After protonation of the encapsulated diethyl acetal, the cationic transition state is thought to experience stabilization by cation-π interactions with the aromatic cavity. The rate of hydrolysis strongly depends on the substrate size with smaller diethyl acetals **11a** and **11b** being deprotected significantly faster than the larger acetals **11f** and **11g**, which is consistent with encapsulation being the key to catalytic hydrolysis. Additionally, adding a competing guest such as tetrabutylammonium occupies the catalytic cavity of **5a**, thus reducing the observed hydrolysis rate to the rate of the background reaction (Scheme 5b) which parallels enzymatic catalysis.

4.2 *Pre-Organization and Enzyme-Like Deprotonation of Encapsulated Substrates*

Enzyme-like pre-organization of reactants inside supramolecular cages can enable reactions at significantly milder conditions due to a higher effective molarity of the

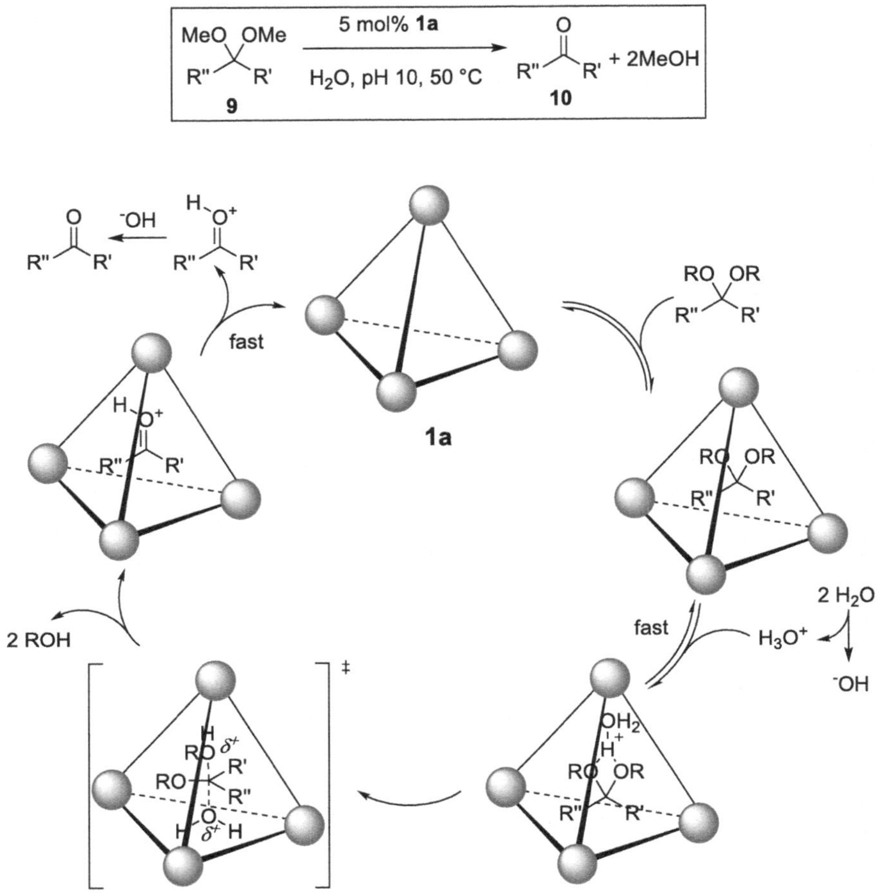

Scheme 4 Proposed mechanism of the acid catalyzed acetal deprotection in basic solution by cage **1a**

reacting molecules and their perfect alignment with respect to the transition state. Additionally, pre-organization can lead to regioselective deprotonation, thus allowing the (kinetic) formation of reaction products not observed in bulk solution.

The efficient catalysis of the Kemp elimination of benzisoxazole **14** with hydroxide to form 2-cyanophenolate **15** inside the cavity of the water-soluble $[Co_8L_{12}]^{16+}$ cage **13** (Fig. 4) has been assigned to enzyme-like pre-organization of the reactants [16]. As **13** binds neutral substrates strongly, benzisoxazole is readily encapsulated and forms a host–guest complex ([**14**$\subset$**13**]) where the CH bond is pointing towards the apertures on the faces of the cubic cage **13**. Additionally, ion pairing of the highly charged supramolecular host **13** with hydroxide leads to a high local concentration of partially desolvated hydroxide ions around the cavity. Due to the ideally situated hydrophobic guest molecule, deprotonation of the CH bond occurs at a pH as low as 8.5 (Fig. 4c) compared to a pH of 14 required in bulk

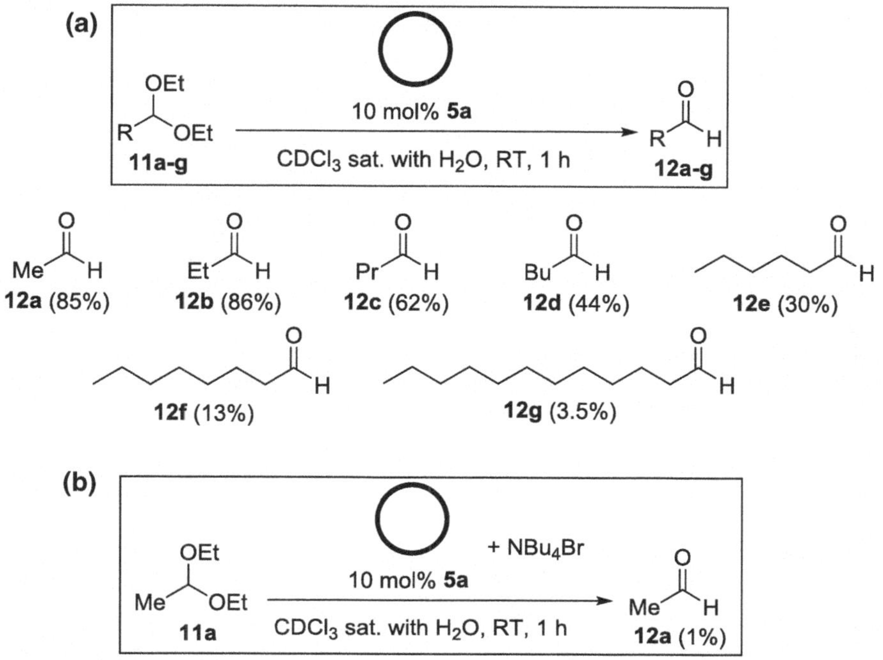

Scheme 5 Acid catalyzed acetal deprotection inside hexameric resorcin[4]arene capsule **5a**. **a** Size dependent rates of hydrolysis of various diethyl acetals; **b** cationic species, e.g. tetrabutylammonium salts, are inhibitors due to their encapsulation in **5a**

solution (Fig. 4b). Since **13** binds neutral guests much more strongly than either cationic or anionic molecules, the hydrophilic 2-cyanophenolate **15** is readily released, thus enabling catalytic turnover.

Similarly, enzyme-like control over the regioselectivity of carbocation deprotonation inside the coordination cage **1a** has been shown to lead to the kinetically favored products of the Nazarov cyclization rather than the most thermodynamically stable compounds [18]. This acid catalyzed reaction of pentadienols **16a–c** results under thermodynamic reaction conditions in bulk solution in the formation of cyclopentadiene **18** via a diallylic carbocation intermediate which undergoes a conrotatory electrocyclic ring closure as predicted by the Woodward-Hoffmann rules [26]. As previously seen in the orthoformate and acetal hydrolysis reactions [19], **1a** can act as an acid catalyst under neutral conditions which in the case of the Nazarov cyclization promotes the formation of the cyclized allyl cation and stabilizes this reactive intermediate inside the cavity of the negatively charged supramolecular host **1a** [27]. The cage-catalyzed Nazarov cyclization of pentadienols **16a** and **16b** leads to the formation of the kinetic product dihydrofulvene **17** as a result of the regioselective deprotonation of the intermediary *trans*-cyclopentadienyl cation by water through the apertures on the faces of the tetrahedral $[Ga_4L_6]^{12-}$ coordination cage **1a** (Scheme 6a). **16c** forms the *cis*-cyclopentadienyl cation which is deprotonated to give **18** (Scheme 6b).

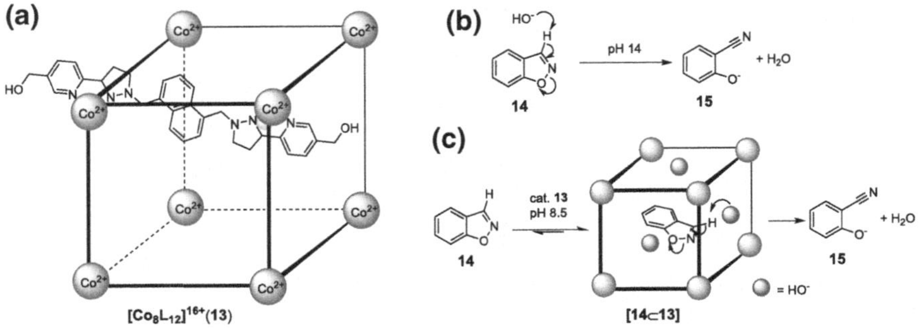

Fig. 4 Kemp elimination catalyzed by **13**. **a** A water-soluble cubic $[Co_8L_{12}]^{16+}$ coordination cage **13**; **b** Kemp elimination under strongly basic conditions; **c** Kemp elimination inside the supramolecular catalyst. **13** occurs at significantly lower pH

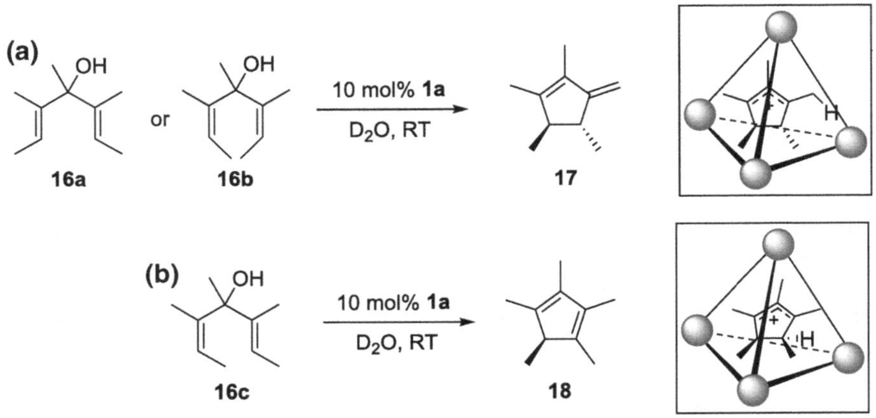

Scheme 6 Nazarov cyclization promoted by **1a** leads to the kinetic product by regioselective deprotonation of the intermediary carbocation stabilized by **1a** (*insets*)

4.3 Host-Guest Catalysis Involving Stabilization of Cationic Transition States in Unimolecular Rearrangement and Cyclization Reactions

When carried out in bulk solution, reactions involving highly reactive cationic intermediates often undergo undesired side reactions. Stabilizing these intermediates by non-covalent interactions and reducing their accessibility for nucleophilic attack by solvent molecules is expected to lead to more selective reactions. Catalysis of reactions proceeding via cationic transition states can therefore be expected to proceed more selectively and at a higher rate inside supramolecular assemblies which allow the stabilization of cationic intermediates and transition states by cation-π and cation dipole interactions, assuming product inhibition does not occur [4].

Aromatic supramolecular capsules, such as the hydrogen-bonded resorcin[4] arene hexamer **5a**, possess a hydrophobic cavity with a high affinity for positively charged species (e.g., ammonium cations) due to extensive cation-π interactions between the host and guest molecules. Taking advantage of the ability of **5a** to act as a Brønsted acid as well as stabilizing the resulting cationic transition state, the Tiefenbacher group demonstrated that the intramolecular hydroalkoxylation of the unsaturated alcohol **19a** to the corresponding cyclic ether **20a** could be carried out in 96% yield in the presence of 10 mol% of capsule **5a** (Scheme 7a) [28]. Adding a competing guest to a **5a**-catalyzed reaction, such as the strongly bound tetrabutylammonium cation, significantly reduced the observed yield (7%), while performing the reaction without **5a** gave no product at all, thus confirming that the reaction takes place inside the self-assembled enzyme-like organic capsule. Various differently substituted tetrahydopyranes and oxepanes were formed in moderate to excellent yield, giving the desired Markovnikov products as shown in Scheme 7b.

The same group also reported a biomimetic tail-to-head terpene cyclization catalyzed inside the cavity of the self-assembled capsule **5a** which is proposed to proceed via a series of cationic intermediates (Scheme 8) [29]. In analogy to the hydrophobic binding pocket of terpene cyclase, the hydrophobic cavity of the resorcin[4]arene hexamer **5a** is capable of stabilizing the allylic cation formed upon encapsulation of geranyl acetate **21** by cation-π interactions with the aromatic host (Scheme 8a). The catalytic cascade involves an isomerization step from the *transoid* to the *cisoid* cation (Scheme 8b), followed by a cation-olefin cyclization (Scheme 8c). A subsequent Wagner-Meerwein 1,2-H-shift (Scheme 8d) and selective deprotonation (Scheme 8e) leads to α-terpinene in good selectivity.

In a related monoterpene-like cyclization based on the Prins cyclization, the groups of Raymond, Bergman, and Toste explored the use of various racemic, homochiral, and enantiopure $[Ga_4L_6]^{12-}$ coordination cages as supramolecular catalysts [8, 30, 31]. As already seen above, in water the negatively charged host **1a** readily encapsulates cationic molecules (e.g., ammonium ions) or neutral compounds which undergo protonation inside the hydrophobic cavity of the coordination capsule and thereafter are stabilized by strong cation-π interactions between the charged guest and the naphthalene units of the cage. Thorough investigation of the proton-mediated Prins cyclization of the unsaturated aldehyde **23** to form preferably the *trans*-cyclization product **24** has led to the proposed catalytic cycle shown in Scheme 9. Although under the aqueous reaction conditions the aldehyde **23** is in equilibrium with its hydrate, only **23** is reversibly encapsulated by the coordination cage (K_1). This encapsulation is believed to be driven by the hydrophobic effect as it results in the release of bound solvent from **A**. Due to the confined space, the encapsulated substrate adopts a chair-like conformation (**B**), which is activated by protonation of the carbonyl group (K_2) for the intramolecular nucleophilic attack of the ideally situated alkene. The stabilized cationic intermediate in **C** is then irreversibly converted to the encapsulated alcohol **24** in **E** either stepwise via the cyclic carbocation in **D**, followed by rapid deprotonation (k_1, k_2) or

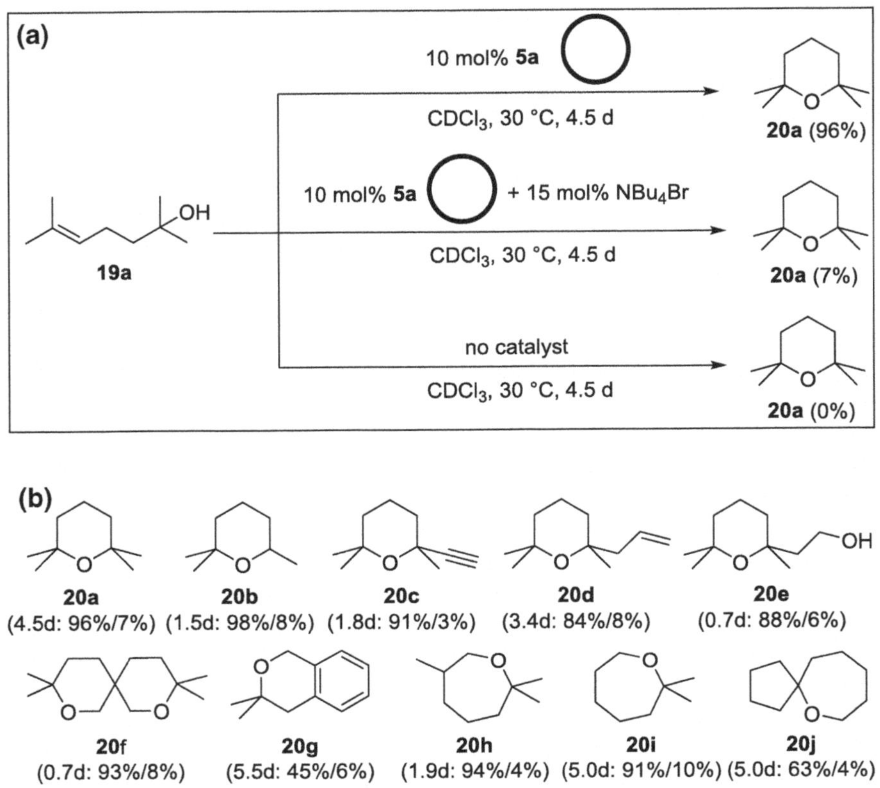

Scheme 7 Intramolecular hydroxyalkylation catalyzed inside **5a**. **a** Catalytic reaction and control experiments; **b** scope of the intramolecular hydroxylation (time: yield with 10 mol% of **5a** without/with 15 mol% NBu_4Br)

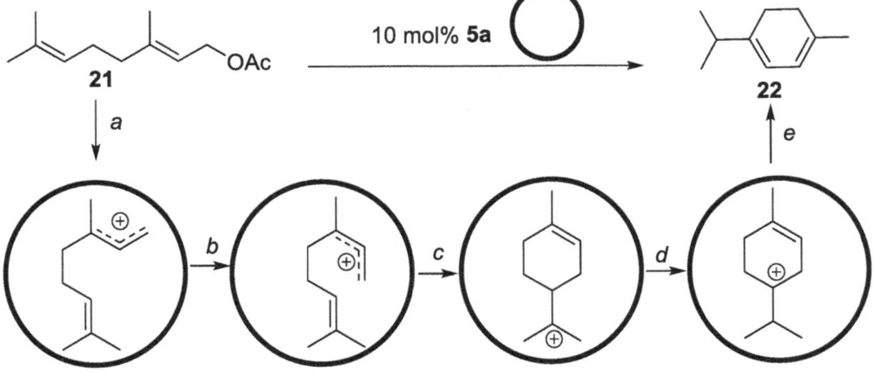

Scheme 8 Terpene cyclization catalyzed inside the cavity of the hexameric resorcin[4]arene capsule **5a**. The formation of α-terpinene **22** from geranyl acetate **21** is proposed to proceed via cationic intermediates (*a–e*)

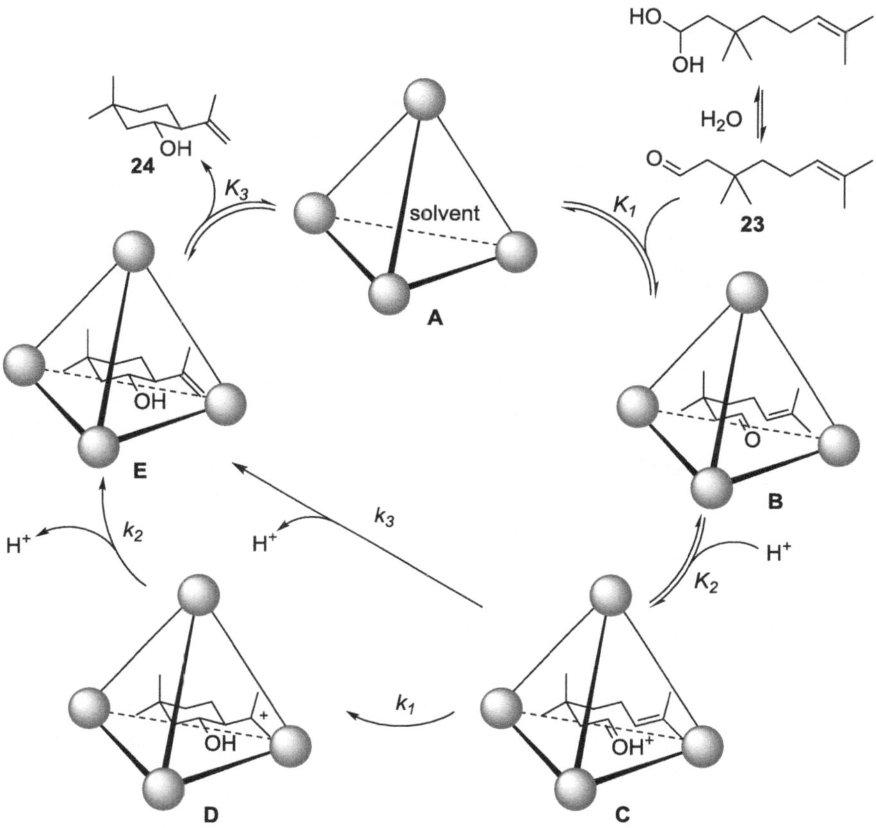

Scheme 9 Proposed mechanism of $[Ga_4L_6]^{12-}$ coordination cage catalyzed Prins cyclization. Both stepwise (k_1, k_2) or concerted mechanisms (k_3) are plausible

in a concerted fashion (k_3) and released from the supramolecular host (K_3). No product inhibition was observed in this case which the authors attribute to the higher solvation of the released alcoholic product **24** compared to the aldehyde **23**.

Since the coordination cage **1a** is chiral and exists in solution as a racemic mixture of homochiral cages (*ΔΔΔΔ*-**1a** and *ΛΛΛΛ*-**1a**), the authors pursued the preparation of the related enantiopure $[Ga_4L_6]^{12-}$ cages *ΔΔΔΔ*-**1b** and *ΛΛΛΛ*-**1b** (see Scheme 1b) which self-assemble diastereoselectively due to their enantiopure ligands. Similarly, the larger pyrene-based coordination cages **25a**, *ΔΔΔΔ*-**25b** and *ΛΛΛΛ*-**25c** (Fig. 5) have been prepared as catalysts to promote enantioselective reactions inside these chiral assemblies.

As could be demonstrated in the Prins cyclization of **23** catalyzed by a racemic mixture of homochiral cage **1a** (Schemes 9 and 10), the *trans*-isomer **24** (and its enantiomer) was obtained in high yield while the formation of the *cis*-isomer **26** was disfavored in the cage-catalyzed reaction. Employing the enantiopure *ΔΔΔΔ*-**1b** capsule did not change the diastereomeric ratio notably; however, one of the

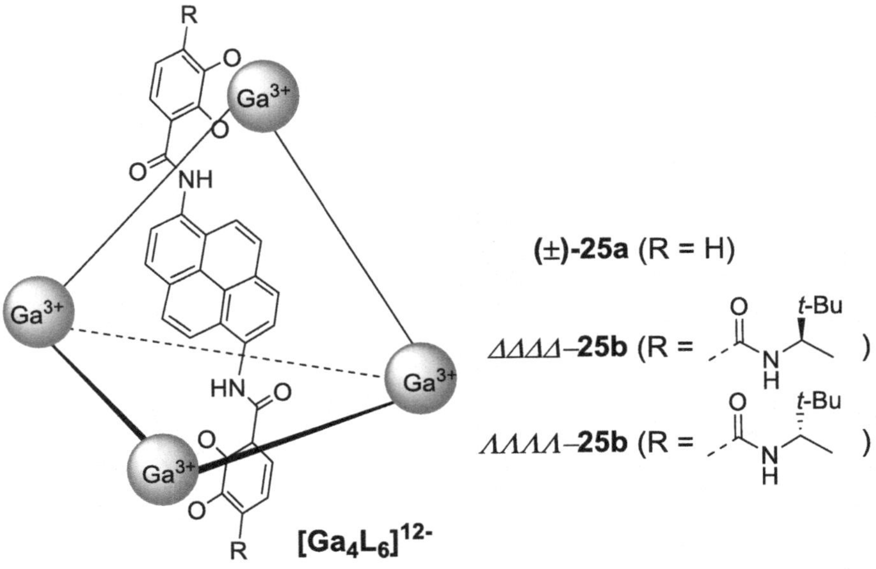

Fig. 5 Extended $[Ga_4L_6]^{12-}$ coordination cages. Diastereoselective self-assembly from enantiopure ligands gives *ΔΔΔΔ*-**25b** or *ΛΛΛΛ*-**25b**, respectively

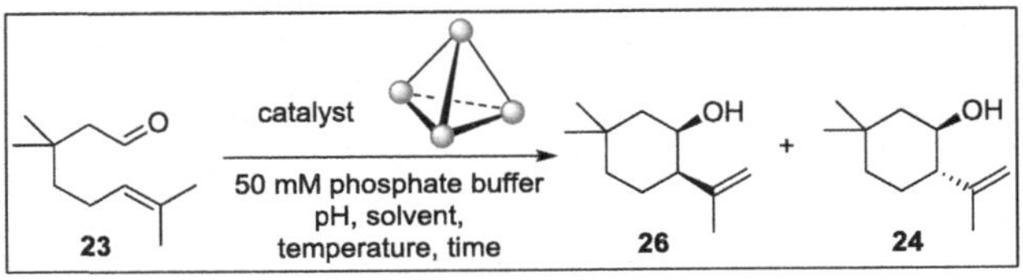

catalyst	mol%	temp/°C	pH (50 mM phosphate buffer in)	time/h	%conv.	% 26	% 24
(±)-1a	10	60	pH 7.50 (water)	28	91	14	86
ΔΔΔΔ-1b	2.5	25	pH 8.00 (water/methanol 1:1)	50	92	12	88 (61%ee)
ΛΛΛΛ-25b	2	40	pH 8.00 (water/methanol 1:1)	16	93	2	98 (33% ee)

Scheme 10 $[Ga_4L_6]^{12-}$ coordination cages as catalysts for the enantioselective Prins cyclization

trans-isomers was obtained in 61% ee. Changing to the larger coordination cage *ΛΛΛΛ*-**25b** led exclusively to the *trans*-isomer, albeit in the case of **24** with loss in enantioselectivity.

As one of the first unimolecular rearrangement reactions catalyzed by **1a** involving cationic species, the aza-Cope rearrangement of allylenammonium and propargylenammonium substrates was published collaboratively by the groups of Raymond and Bergman [15, 32–34]. Aza-Cope rearrangements involving allylenammonium substrates **27** have been carried out both in presence of a racemic mixture of homochiral cages **1a**[32] as well as in presence of enantiopure *ΔΔΔΔ*-**1a**

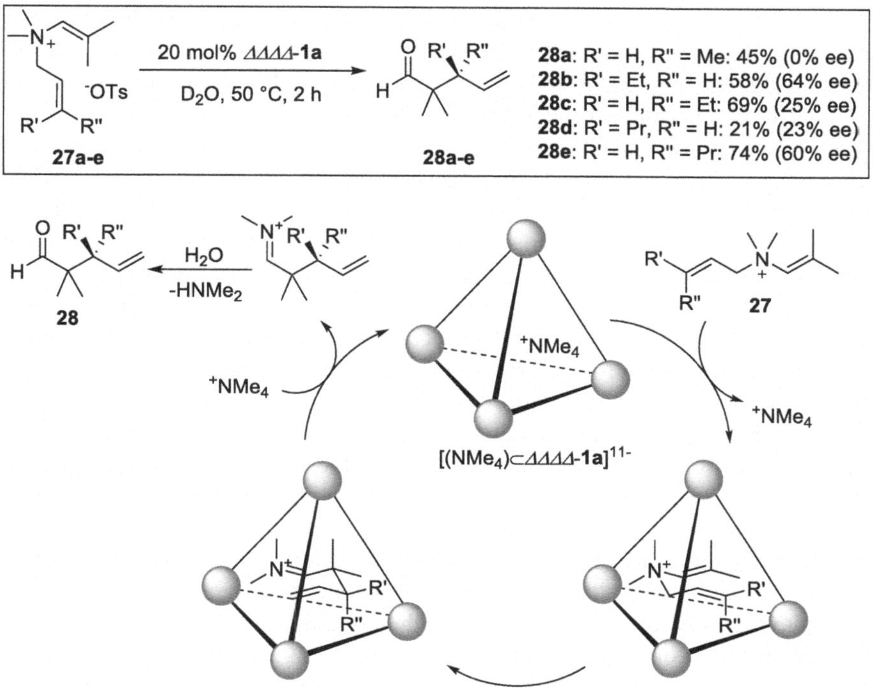

Scheme 11 Aza-Cope rearrangement of allylenammonium substrates **27a–e** catalyzed by the enantiopure $[Ga_4L_6]^{12-}$ coordination cage *ΔΔΔΔ*-**1a**

obtained by resolution from (±)-**1a** (Scheme 11) [34]. The observed enantioselectivities of up to 64% ee for the aldehyde products **28a–e** of the aza-Cope rearrangement reactions vary strongly with subtle changes in size and shape of the prochiral substrates **27a–e** when carried out with 20 mol% of *ΔΔΔΔ*-**1a** as catalyst. As indicated in the catalytic cycle in Scheme 11, these cage-catalyzed [3]-sigmatropic rearrangements proceed via a chair-like transition state which is enforced upon encapsulation of the substrate. The resulting iminium ion is readily hydrolyzed in solution to the corresponding aldehyde which is not encapsulated in the host and therefore does not inhibit the catalytic turnover.

4.4 Bimolecular Reactions Involving Nucleophilic Attack on Capsule-Stabilized Cationic Transition States and Intermediates

Bimolecular reactions inside the confined space of supramolecular assemblies include solvolysis and hydration reactions and can be promoted by activating, for example, the electrophilic substrate by protonation which facilitates the subsequent nucleophilic attack of the reaction partner.

In an exceptional case of nucleophilic substitution at sp^3 carbons, the groups of Raymond, Bergman, and Toste observed overall retention of absolute stereochemistry in reactions catalyzed by the water-soluble hosts **1a**, *ΔΔΔΔ*-**1b**, and *ΛΛΛΛ*-**1b** [17]. In contrast, nucleophilic substitutions at saturated carbon centers generally proceed via an S_N1 or an S_N2 pathway which dictates their stereochemical outcome: S_N1 reactions lead to racemized products while S_N2 reactions yield products with inversion of stereochemistry. This includes the phosphoric acid (**29**) catalyzed solvolysis of the secondary benzylic compound (*S*)-**30** which proceeds in bulk solution via an S_N2 mechanism to give (*R*)-**31** with 84% inversion of stereochemistry (Scheme 12a). The nucleophilic substitution reaction on the benzylic substrate (*S*)-**30** inside cage *ΔΔΔΔ*-**1b** proceeds with unprecedented 74% retention of stereochemistry found for the product (*R*)-**31** (Scheme 12b). As these $[Ga_4L_6]^{12-}$ cages are known to decrease the pK_a of encapsulated guests, protonation of the acid activated trichloroacetamide leaving group in (S)-**30** occurs upon encapsulation, followed by the loss of the leaving group. The authors proposed that the cavity of the host can stabilize the developing positively charged intermediate in the transition state through cation-π interactions with the naphthalene walls of the cage, thus blocking the back side of the carbocation from nucleophilic attack. This complexation also limits planarization and the stabilized carbocation is trapped by a nucleophile (in this case the solvent) faster than rotation can occur, resulting in the observed retention of stereochemistry.

Similarly, efficient isonitrile hydration takes place through encapsulation and activation by protonation within the hexameric resorcin[4]arene capsule **5a** (Scheme 13) [35]. No hydration takes place in bulk solution but neutral isonitriles **32a–e** are readily encapsulated and converted in the presence of water to the corresponding *N*-formylamides **33a–e** inside the organic capsule **5a**. Although the mechanism was not investigated experimentally, the catalytic reaction was proposed to proceed via protonation of the carbenic carbon atom followed by nucleophilic addition of water. Catalytic hydration was strongly dependent on size and electronic properties of isonitriles **32a–e** which gave the corresponding *N*-formylamides **33a–e** in variable yields.

4.5 Bimolecular Reactions Catalyzed by Coordination Cages and Organic Capsules

Bimolecular reactions catalyzed by self-assembled hosts either can proceed via encapsulation and activation of one of the substrates or by co-encapsulating two different molecules which react in the confines of the cavity with each other. Product inhibition is an often encountered obstacle; nevertheless, careful selection of reactants, host assemblies or reaction conditions have led to multiple examples of host-guest catalysis.

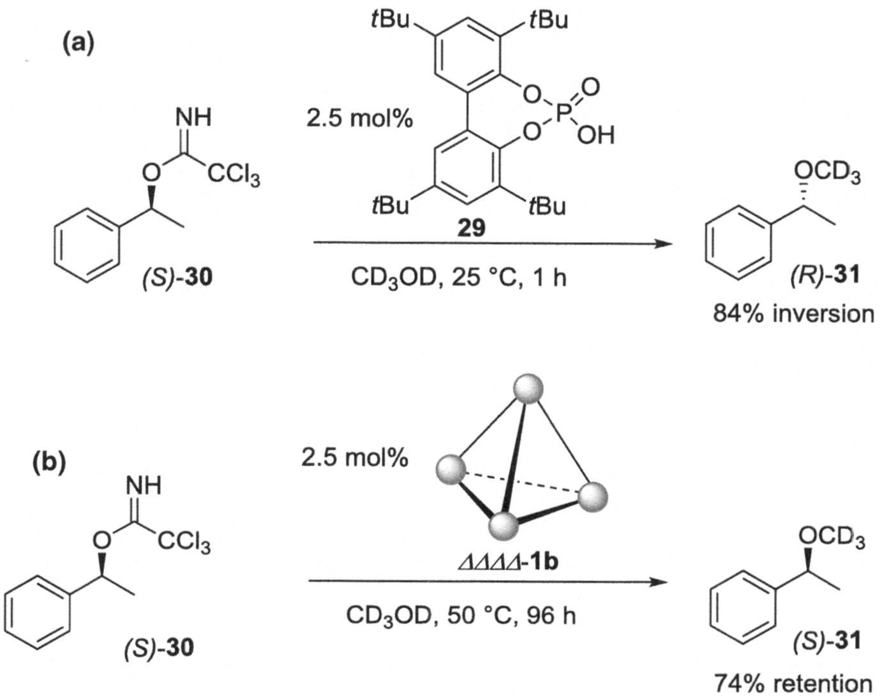

Scheme 12 Nucleophilic substitution and its stereochemical outcome. **a** The phosphoric acid catalyzed nucleophilic substitution in bulk solution leads to inversion of configuration as expected for an S_N2 reaction; **b** in contrast, the cage-catalyzed nucleophilic substitution reaction on a secondary benzylic carbon center gives the product with overall retention of the absolute configuration

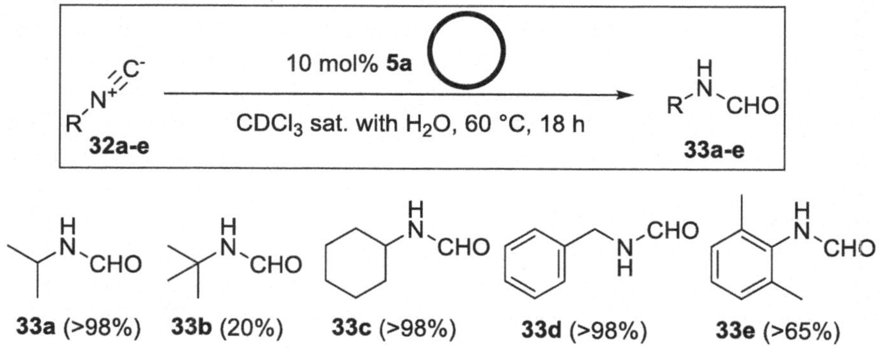

Scheme 13 Isonitrile hydration promoted by the hexameric resorcin[4]arene capsule **5a**

One of the first examples of catalytic hydrogen-bonded self-assembled molecular capsules was published by Rebek: dimerization of **34** in organic solvents leads to the "hydroxy softball" **35** which is held together by 16 hydrogen bonds

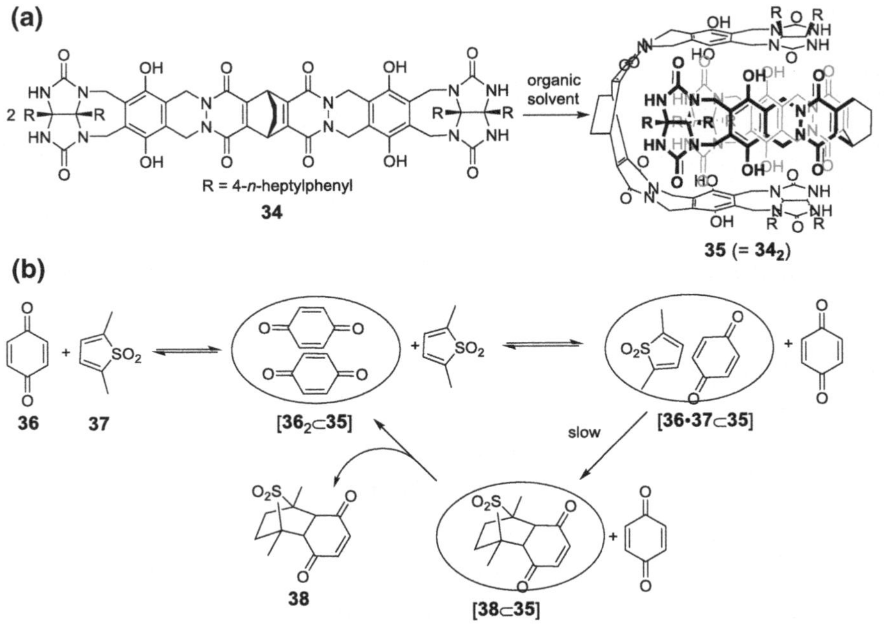

Scheme 14 **a** Dimerization of **34** in organic solvent results in the hydrogen-bonded capsule **35**. **b** A Diels-Alder reaction promoted by capsule **35**

(Scheme 14a). In a first example, this group observed a rate acceleration in the Diels-Alder reaction between *p*-quinone **36** and cyclohexadiene inside the organic capsule **35**, but this system suffered from product inhibition due to the high affinity of the Diels–Alder adduct for **35** [36]. Successful catalysis of the Diels–Alder reaction by the dimeric capsule **35** was subsequently achieved between **36** and the thiophene oxide derivative **37** [13]. The authors proposed a catalytic cycle (Scheme 14b) where the resting state of the capsule contains two molecules of *p*-quinone [**36**$_2$⊂**35**] which is in equilibrium with [**36**·**37**⊂**35**] containing the two co-complexed reactants. The rate-limiting cycloaddition takes place within the confines of the organic capsule, followed by a comparably fast displacement of the endo-adduct **38** from the complex [**38**⊂**35**] by two equivalents of quinone, thus completing the catalytic cycle.

Shimidzu and coworkers demonstrated the application and recycling of catalytic hydrogen-bonded organic capsules bearing fluorinated substituents which selectively dissolved in the fluorous layer of a biphasic system [37]. Due to the fluorophobic effect, the fluorinated version of the hexameric resorcin[4]arene capsule (**5b**) was observed to significantly accelerate the formation of the Diels-Alder adducts such as **41** from various dienes (e.g., **39**) and dienophiles (e.g., methyl acrylate **40**) when carried out in a fluorous biphasic system in comparison to the reactions carried out in the absence of **5b** (Scheme 15). Encapsulation was shown to increase the endo/exo ratio, particularly with larger substrates where only the

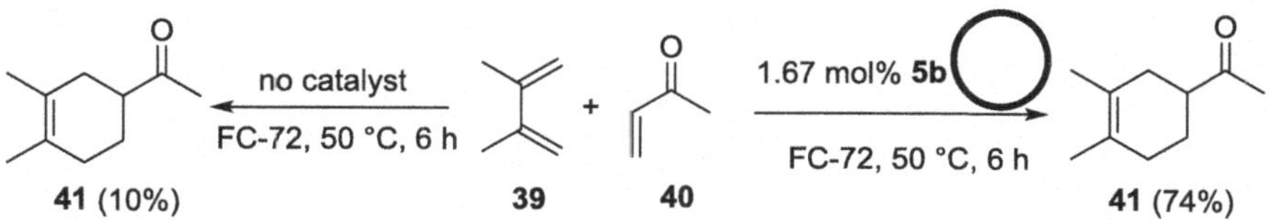

Scheme 15 Diels–Alder reaction promoted by the fluorinated capsule **5b**

endo product was formed. Furthermore, the authors demonstrated that **5b** could be recycled and reused for catalysis with no decrease in activity after five consecutive runs.

Fujita and coworkers reported the use of aqueous organopalladium cages **42** and **43a** (Fig. 6) as supramolecular reaction vessels for the Diels–Alder cycloaddition of anthracenes and maleimides (Scheme 16) [14]. The bowl-shaped host **42** efficiently catalyzes the Diels–Alder reaction between anthracene derivative **44** and maleimide **45** to yield **46** quantitatively (Scheme 16a). The catalytic cycle is thought to involve binding of the anthracene derivative which can stack onto the triazine ligands of the host, gaining stabilization from aromatic-aromatic or charge-transfer interactions. Likewise, maleimide encapsulation is favored due to stacking interactions and the reactant-like transition state experiences similar stabilization. However, upon formation of the adduct **46** most of these favorable aromatic stacking interactions are lost due to the bending of the former anthracene molecule which closes the catalytic cycle by release of **46** and encapsulation of a new pair of substrates. While this reaction is catalytic and does not suffer from product inhibition, the regioselectivity (addition at the 9,10-position of the anthracene) does not differ from reactions carried out in bulk solution. Although the Diels-Alder reaction carried out inside the octahedral coordination cage **43a** requires a stoichiometric amount of the

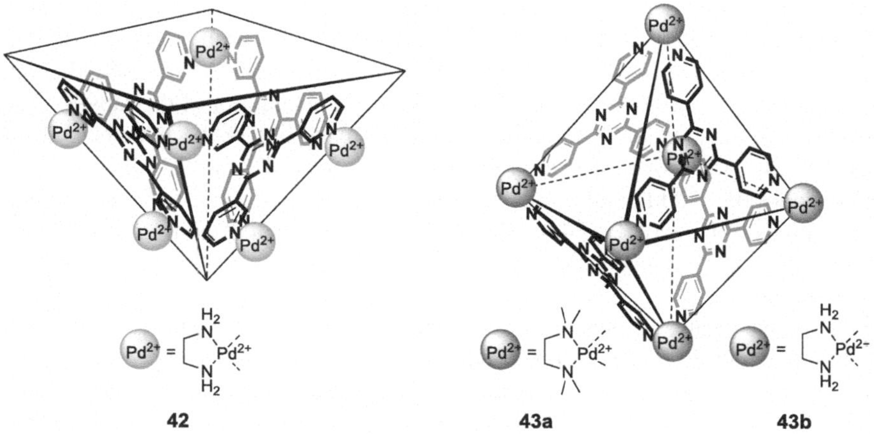

Fig. 6 Water-soluble organopalladium cages assembled from *cis*-end-capped Pd(II) salts and triazine-based tridentate ligands. **42** resembles an open basket while the palladium atoms in **43a** and **43b** are arranged in an octahedral fashion

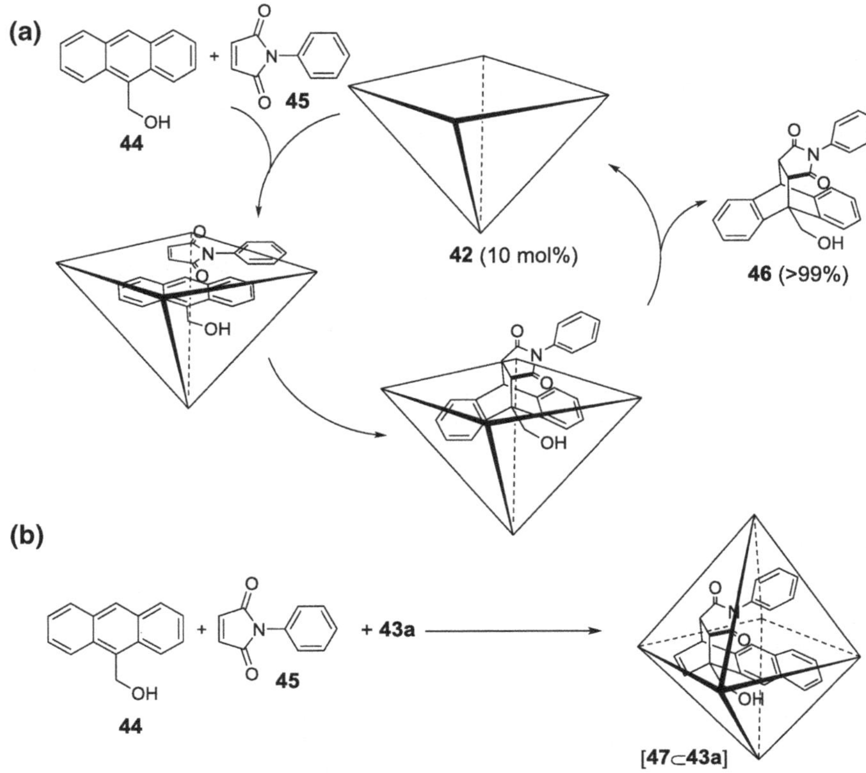

Scheme 16 **a** Diels–Alder reaction catalyzed by **42** proceeds with normal selectivity; **b** inside **43a** the DA reaction leads to an unusual product but is not catalytic (product inhibition)

organopalladium host, its unusual regioselectivity is worth mentioning (Scheme 16b). This palladium capsule forms rapidly and exclusively the desired host–guest complex [(**44**·**45**)⊂**35**] with the co-encapsulated substrates being held in a very defined position due to the restricted space in the interior of the capsule. The proximity of the dienophile **45** to the terminal aromatic ring of the anthracene in **44** leads to the exclusive formation of the 1,4-adduct **47** instead of usual the 9,10-adduct **46**. Additionally, only the exo-selective syn-adduct is formed within the confines of the cavity of **43a**. Unfortunately, the naphthalene ring of **47** forms strong π–π stacking interactions with the triazine ligand of **43a** which prevents catalytic turnover.

The cage-catalyzed Knoevenagel condensation between aldehyde **48** and Meldrum's acid **49** under neutral conditions in water was efficiently catalyzed by the previously problematic octahedral cage **43b** (Scheme 17) while the bowl-shaped organopalladium host **42** proved to be a poor catalyst, yielding the condensation product **50** even with stoichiometric quantities of **42** only in 17% yield [38]. The reaction catalyzed by **43b** involves the encapsulation of up to four equivalents of the aromatic aldehyde in the hydrophobic cavity of **43b**. Meldrum's

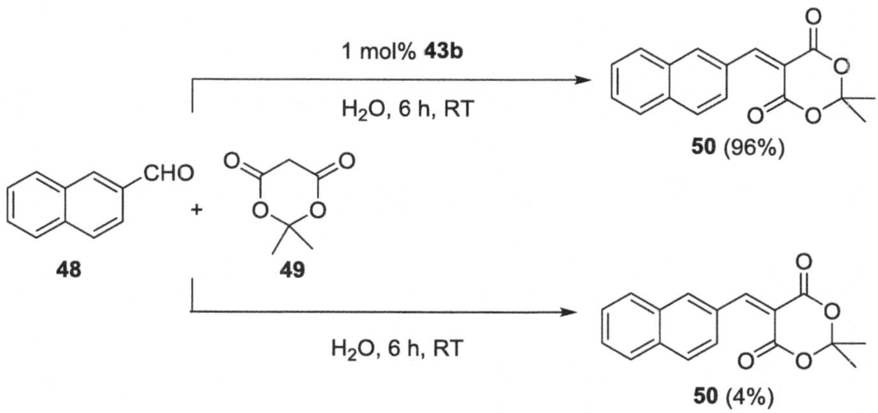

Scheme 17 Knoevenagel condensation under neutral conditions in water. **a** Catalyzed by the octahedral organopalladium capsule **43b**; **b** background reaction in the absence of cage

acid **49** is not encapsulated due to its high water-solubility and is under the reaction conditions in equilibrium with its enolate. Electrostatic interaction between the positively charged coordination cage and the monoanionic enolate is proposed to lead to the nucleophilic attack of the enolate on the encapsulated aldehyde which is facilitated by stabilization of the formed oxyanion intermediate by the cationic cage environment. Dehydration of this intermediate is possible inside the hydrophobic cavity of the coordination cage despite the aqueous solvent and leads to the desired condensation product **50**. As **50** is too large for the host it is spontaneously released, thereby allowing another molecule of substrate to undergo a catalytic Knoevenagel condensation inside **43b**.

As previous studies have shown, isonitriles are good guests for the organic capsule **5a** which presumably catalyzes the isonitrile hydration by protonation [35]. In the case of the catalytic preparation of substituted 1*H*-tetrazoles from isonitriles and TMS-azide there is no evidence for a protonation step being involved in the catalytic cycle [39]. Nevertheless, activation of the reaction between **51** and **52** clearly occurs by encapsulation into **5a** which leads to complete conversion to the desired 1*H*-tetrazole **53**, possibly assisted by the electron-rich interior surface of the cavity rather than hydrogen bonding with the hydroxyl groups of the separate resorcin[4]arene components. Control experiments in the presence of the capsule and the competing guest tetrabutylammonium significantly reduces the yield of **53** while reactions carried out in the absence of **5a** yielded no product at all (Scheme 18).

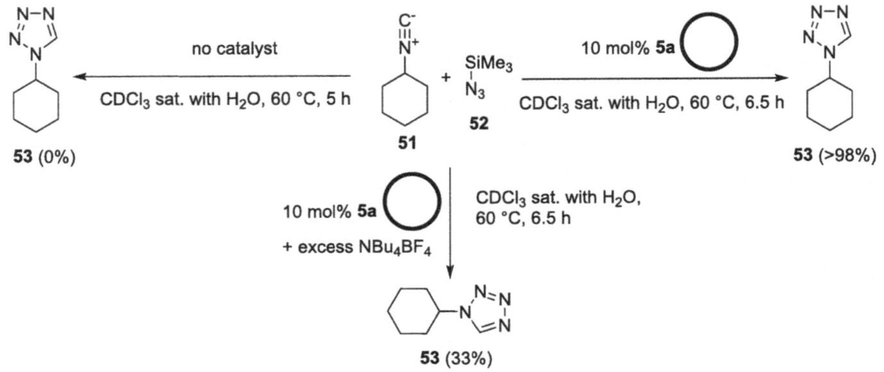

Scheme 18 1*H*-tetrazole formation promoted by hydrogen-bonded capsule **5a**

5 Encapsulated Metal Complexes

Confinement has been shown to be a viable method to modulate the activity and selectivity of transition metal catalysts. Metalloenzymes demonstrate beautifully how not only the directly coordinated ligands but also the second coordination sphere—in this case the binding pocket of the prosthetic group—play a crucial role in catalyzing highly selective reactions. Transition metals in confined spaces include covalently bound metal catalysts (see Chapter "Endohedral Functionalization of Molecular Cavities for Catalysis in Confined Spaces"), confined species obtained by the template-ligand approach, and metal complexes which are encapsulated into coordination cages and self-assembled organic capsules [7]. Encapsulation of metal catalysts has not only been employed to alter their reactivity and selectivity but also was found to enable their use under otherwise incompatible reaction conditions by preventing their degradation (vide infra).

The chemo- and regioselectivity of alkyne hydration is found to change upon encapsulation of the gold catalyst within the hexameric resorcin[4]arene host **5a** [40]. While the hydration of the terminal alkyne **55** catalyzed by the free (*i*-Pr-NHC)Au(OTf) complex **54** quantitatively yielded the expected Markovnikov adduct **56** (Scheme 19a) or formed selectively the 1,2-dihydronaphthalene **57** under anhydrous conditions, the cationic host–guest complex $[(i\text{-Pr-NHC})\text{Au}\subset\mathbf{5a}]^+$ was observed to lead to a different product distribution (Scheme 19b). Equimolar amounts of the Markovnikov product **56** and its dehydration product **57** were accompanied by the formation of a small amount of the anti-Markovnikov-adduct **58**. The origin the cyclic dehydration product and the linear aldehyde are not entirely clear although the hydrophobic interior of capsule **5a** is assumed to prevent efficient nucleophilic attack of water, thus changing the chemoselectivity of the reaction, while the shape of the cavity may further facilitate both intramolecular reaction leading to **57** and the nucleophilic attack on the terminal alkyne carbon giving **58** by enforcing a favorable geometry of the encapsulated gold-activated

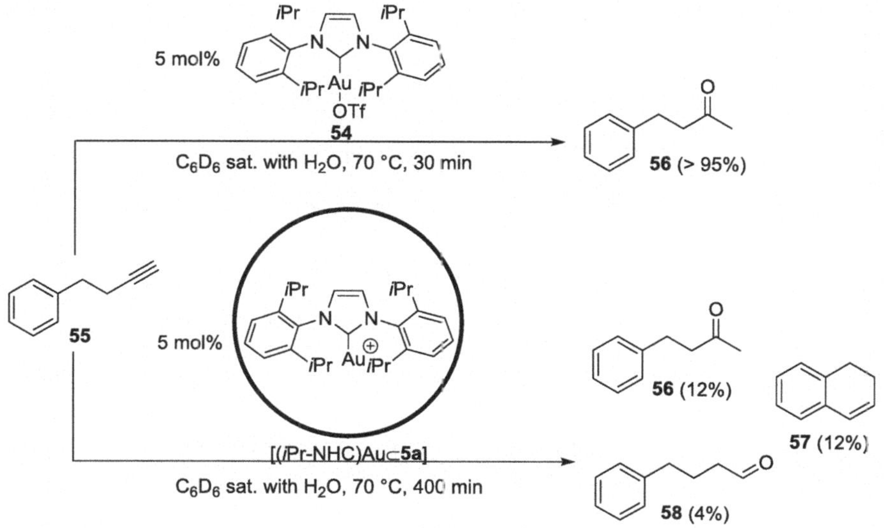

Scheme 19 Gold catalysis in hydrogen-bonded capsule

substrate. Investigation of the substrate selectivity in the alkyne hydration demonstrated a clear size-dependence for reactions mediated by the host–guest complex [(*i*-Pr-NHC)Au⊂**5a**]$^+$ with smaller 4-R-phenylacetelyenes reacting faster that larger substrates (for R=H, Me, *t*-Bu the relative rates were found to be 1.6/1.3/1.0) [41]. Reactions rates carried out in the presence of the free (*i*-Pr-NHC) Au(OTf) **54** complex were observed to proceed faster with more electron-rich alkynes (for R=H, Me, *t*-Bu the relative rates were found to be 1.0/1.4/1.5), thus showing a reversed order compared to the encapsulated gold catalyst.

High substrate specificity has been reported by the groups of Bergman and Raymond in the rhodium catalyzed allylic alcohol isomerization upon encapsulation of rhodium catalysts in the coordination cage **1a** [42]. The cationic hydrated bisphosphine rhodium(I) complex **59** does not distinguish in bulk solution between the allyl alcohols **60** and **62** and yields the corresponding isomerization products **61** and **62** quantitatively whereas the encapsulated catalyst [**59**⊂**1a**] isomerized **60** in quantitative yields to the aldehyde **61** but showed no reactivity in the isomerization of **62** (Scheme 20).

In a similar allylic alcohol isomerization, the water-sensitive ruthenium(II) catalyst was protected from decomposition by encapsulation into the self-assembled $[Ga_4L_6]^{12-}$ cage **1a** which permitted its use in aqueous solution (vide infra) [43]. Likewise, encapsulation of a gold catalyst led to an enhancement in catalytic activity in the hydroxyalkylation of allenes and furthermore increased its lifetime under the reaction conditions [44].

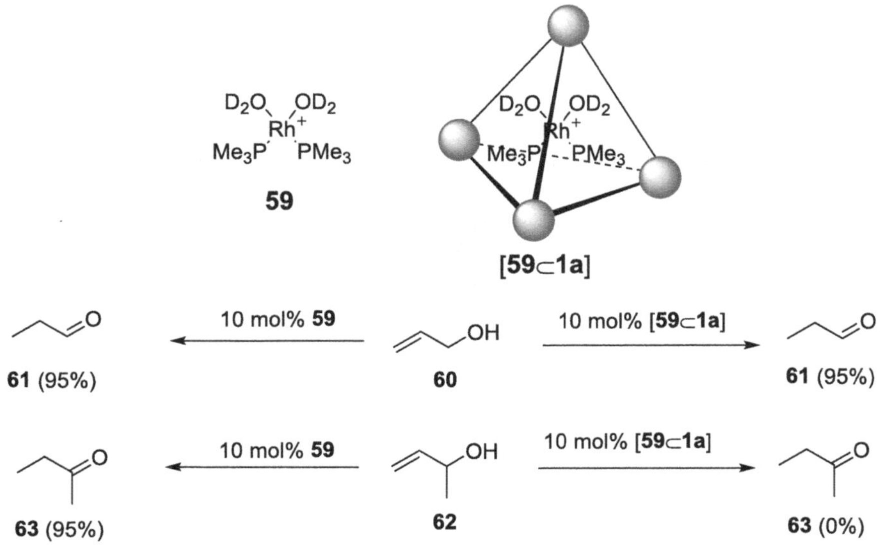

Scheme 20 Rh catalysis inside the coordination cage **1a** exhibits highly size and shape selective reactivity

6 Orthogonal Tandem Catalytic Reactions Using Self-assembled Coordination Cages

Tandem catalysis and catalytic cascade reactions allow the synthesis of complex structures while avoiding the isolation of any intermediate products. Orthogonal tandem catalysis describes one-pot reactions in which sequential catalytic processes occur through two or more distinct catalytic cycles [45]. The main challenge is to ensure orthogonal reactivity of all components and catalysts in this multistep process in order to obtain the desired product in good yield. Compartmentalization has been used to address catalyst incompatibilities by imposing physical barriers to avoid their direct contact, thus allowing the tandem reactions to be carried out in one pot.

Raymond, Bergman, and Toste recently demonstrated that a supramolecular approach allowed combining otherwise incompatible transition metal catalysts with enzymes by encapsulating the metal complexes into a $[Ga_4L_6]^{12-}$ coordination cage [46]. Protecting a ruthenium(II) complex inside a supramolecular host prevents its degradation in aqueous media and enables the transition metal catalyzed allylic isomerization of **60** to the aldehyde **61** under bioorthogonal conditions [43], which allows the following enzymatic reduction by alcohol dehydrogenase (ADH) to 1-propanol **65** to be carried out in the same pot without necessitating the isolation of the intermediate product (Scheme 21).

In a second example of a procedure combining enzymatic and transition metal catalysis the authors reported that encapsulation of the gold(I) catalyst was required

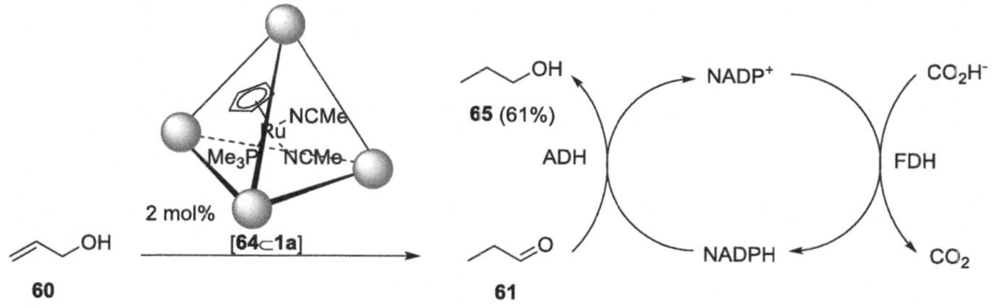

Scheme 21 A catalytic tandem reaction incorporating transition metal catalysis and enzymatic catalysis: the product of the ruthenium catalyzed allyl alcohol isomerization, **61**, is efficiently reduced by alcohol dehydrogenase (ADH) and formate dehydrogenase (FDH)

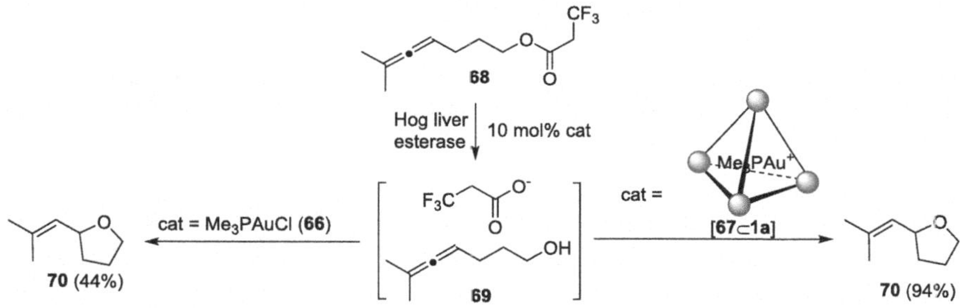

Scheme 22 A catalytic tandem reaction employing the enzyme hog liver esterase to generate in situ the hydroxy allene **69** which is transformed to the cyclic ether **70** by a gold catalyst

to achieve high yields in the sequential enzymatic ester hydrolysis of **68** to **69**, followed by the gold-catalyzed intramolecular hydroxyalkylation (Scheme 22) [44, 46]. Mechanistic investigations revealed that hog liver esterase is inhibited by the free gold catalyst which limits the yield of the intermediate product **69**, thus leading to low overall yield of **70** after the subsequent gold-catalyzed intramolecular hydroxyalkylation. Encapsulation (compartmentalization) of the gold complex prevents its inhibiting effect on the esterase and thereby allows quantitative conversion of the allene ester **68** to the tetrahydrofuran derivative **70** over two steps.

In a self-organizing chemical assembly line, the Nitschke group showed recently that coordination cages could not only be used as catalysts in multistep one-pot syntheses but also self-assembled in situ from its subcomponents [47]. The reported example consists of three noninterfering catalytic steps A, B, and C which transform furan (**72**) into 5-hydroxy-3-(nitromethyl)dihydrofuran-2(*3H*)-one **75** (Scheme 23). In the photooxidation cycle (A) singlet oxygen ($^1O_2$) is generated via absorption of visible light by the sensitizer methylene blue and undergoes a hetero-Diels-Alder reaction with furan **72** to give the high-energy intermediate **73**, an endoperoxide. The self-assembled coordination cage **71** encapsulates in the

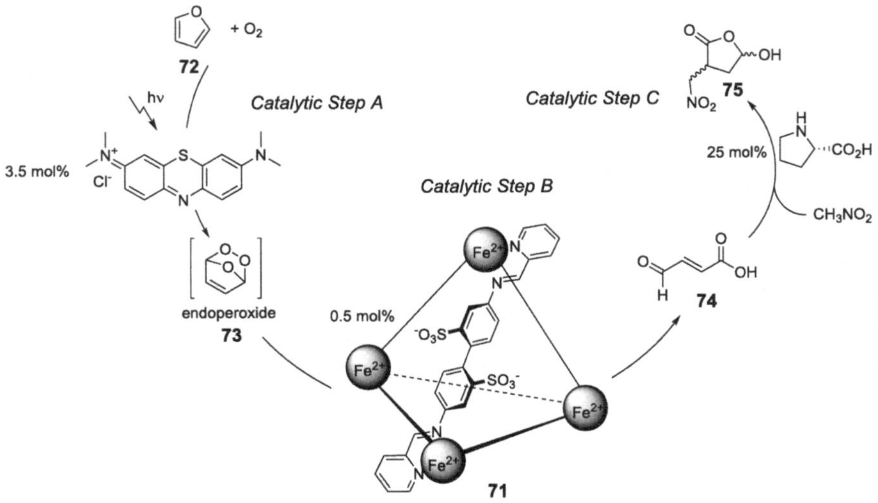

Scheme 23 One-pot conversion of furan **72** to 5-hydroxy-3-(nitromethyl)dihydrofuran-2(*3H*)-one **75** by three sequential catalytic steps: Photocatalysis (A), host-guest catalysis (B), and organocatalysis (C)

second catalytic cycle (B) selectively the endoperoxide **73** and transforms it into its linear isomer fumaraldehydic acid **74**. The L-proline catalyzed 1,4-addition of nitromethane to **74** followed by cyclization leads in the organocatalytic cycle (C) to the final product **75**.

References

1. Raynal M, Ballester P, Vidal-Ferran A, van Leeuwen PWNM (2014) Supramolecular catalysis. Part 1: non-covalent interactions as a tool for building and modifying homogeneous catalysts. Chem Soc Rev 43:1660–1733. doi:10.1039/C3CS60027K
2. Raynal M, Ballester P, Vidal-Ferran A, van Leeuwen PWNM (2014) Supramolecular catalysis. Part 2: artificial enzyme mimics. Chem Soc Rev 43:1734–1787. doi:10.1039/C3CS60037H
3. Brown CJ, Toste FD, Bergman RG, Raymond KN (2015) Supramolecular catalysis in metal-ligand cluster hosts. Chem Rev 115:3012–3035. doi:10.1021/cr4001226
4. Catti L, Zhang Q, Tiefenbacher K (2016) Self-assembled supramolecular structures as catalysts for reactions involving cationic transition states. Synthesis 48:313–328. doi:10.1055/s-0035-1560362
5. Gangemi CMA, Pappalardo A, Sfrazzetto GT (2015) Applications of supramolecular capsules derived from resorcin[4]arenes, calix[n]arenes and metallo-ligands: from biology to catalysis. RSC Adv 5:51919–51933. doi:10.1039/C5RA09364C
6. Vardhan H, Verpoort F (2015) Metal-organic polyhedra: catalysis and reactive intermediates. Adv Synth Catal 357:1351–1368. doi:10.1002/adsc.201400778
7. Leenders SHAM, Gramage-Doria R, de Bruin B, Reek JNH (2014) Transition metal catalysis in confined spaces. Chem Soc Rev 44:433–448. doi:10.1039/C4CS00192C

8. Zhao C, Sun Q-F, Hart-Cooper WM et al (2013) Chiral amide directed assembly of a diastereo- and enantiopure supramolecular host and its application to enantioselective catalysis of neutral substrates. J Am Chem Soc 135:18802–18805. doi:10.1021/ja411631v
9. Bolliger JL, Belenguer AM, Nitschke JR (2013) Enantiopure water-soluble [Fe4L6] cages: host-guest chemistry and catalytic activity. Angew Chem Int Ed 52:7958–7962. doi:10.1002/anie.201302136
10. MacGillivray LR, Atwood JL (1997) A chiral spherical molecular assembly held together by 60 hydrogen bonds. Nature 389:469–472. doi:10.1038/38985
11. Koblenz TS, Wassenaar J, Reek JNH (2008) Reactivity within a confined self-assembled nanospace. Chem Soc Rev 37:247–262. doi:10.1039/B614961H
12. van der Vlugt JI, Koblenz TS, Wassenaar J, Reek JNH (2010) Chemistry in self-assembled nanoreactors. In: Brinker UH, Mieusset J-L (eds) Mol. Encapsulation, Wiley, pp 145–174
13. Kang J, Santamaría J, Hilmersson G, Rebek J (1998) Self-assembled molecular capsule catalyzes a Diels–Alder reaction. J Am Chem Soc 120:7389–7390. doi:10.1021/ja980927n
14. Yoshizawa M, Tamura M, Fujita M (2006) Diels-Alder in aqueous molecular hosts: unusual regioselectivity and efficient catalysis. Science 312:251–254. doi:10.1126/science.1124985
15. Hastings CJ, Fiedler D, Bergman RG, Raymond KN (2008) Aza cope rearrangement of propargyl enammonium cations catalyzed by a self-assembled "nanozyme". J Am Chem Soc 130:10977–10983. doi:10.1021/ja8013055
16. Cullen W, Misuraca MC, Hunter CA et al (2016) Highly efficient catalysis of the Kemp elimination in the cavity of a cubic coordination cage. Nat Chem 8:231–236. doi:10.1038/nchem.2452
17. Zhao C, Toste FD, Raymond KN, Bergman RG (2014) Nucleophilic substitution catalyzed by a supramolecular cavity proceeds with retention of absolute stereochemistry. J Am Chem Soc 136:14409–14412. doi:10.1021/ja508799p
18. Hastings CJ, Backlund MP, Bergman RG, Raymond KN (2011) Enzyme-like control of carbocation deprotonation regioselectivity in supramolecular catalysis of the Nazarov cyclization. Angew Chem Int Ed 50:10570–10573. doi:10.1002/anie.201105325
19. Pluth MD, Bergman RG, Raymond KN (2009) Proton-mediated chemistry and catalysis in a self-assembled supramolecular host. Acc Chem Res 42:1650–1659. doi:10.1021/ar900118t
20. Bolliger JL (2014) [Fe4L6]8+ cages: encapsulation and catalytic degradation of an insecticide. Chim Int J Chem 68:204–207. doi:10.2533/chimia.2014.204
21. Pluth MD, Bergman RG, Raymond KN (2007) Acid catalysis in basic solution: a supramolecular host promotes orthoformate hydrolysis. Science 316:85–88. doi:10.1126/science.1138748
22. Pluth MD, Bergman RG, Raymond KN (2008) Supramolecular catalysis of orthoformate hydrolysis in basic solution: an enzyme-like mechanism. J Am Chem Soc 130:11423–11429. doi:10.1021/ja802839v
23. Pluth MD, Bergman RG, Raymond KN (2007) Catalytic deprotection of acetals in basic solution with a self-assembled supramolecular "nanozyme". Angew Chem Int Ed 46:8587–8589. doi:10.1002/anie.200703371
24. Pluth MD, Bergman RG, Raymond KN (2009) The acid hydrolysis mechanism of acetals catalyzed by a supramolecular assembly in basic solution. J Org Chem 74:58–63. doi:10.1021/jo802131v
25. Zhang Q, Tiefenbacher K (2013) Hexameric resorcinarene capsule is a brønsted acid: investigation and application to synthesis and catalysis. J Am Chem Soc 135:16213–16219. doi:10.1021/ja4080375
26. Woodward RB, Hoffmann R (1969) The conservation of orbital symmetry. Angew Chem Int Ed Engl 8:781–853. doi:10.1002/anie.196907811
27. Hastings CJ, Pluth MD, Bergman RG, Raymond KN (2010) Enzymelike catalysis of the Nazarov cyclization by supramolecular encapsulation. J Am Chem Soc 132:6938–6940. doi:10.1021/ja102633e

28. Catti L, Tiefenbacher K (2015) Intramolecular hydroalkoxylation catalyzed inside a self-assembled cavity of an enzyme-like host structure. Chem Commun 51:892–894. doi:10.1039/C4CC08211G
29. Zhang Q, Tiefenbacher K (2015) Terpene cyclization catalysed inside a self-assembled cavity. Nat Chem 7:197–202. doi:10.1038/nchem.2181
30. Hart-Cooper WM, Zhao C, Triano RM et al (2015) The effect of host structure on the selectivity and mechanism of supramolecular catalysis of Prins cyclizations. Chem Sci 6:1383–1393. doi:10.1039/C4SC02735C
31. Hart-Cooper WM, Clary KN, Toste FD et al (2012) Selective monoterpene-like cyclization reactions achieved by water exclusion from reactive intermediates in a supramolecular catalyst. J Am Chem Soc 134:17873–17876. doi:10.1021/ja308254k
32. Fiedler D, Bergman RG, Raymond KN (2004) Supramolecular Catalysis of a unimolecular transformation: Aza-Cope rearrangement within a self-assembled host. Angew Chem Int Ed 43:6748–6751. doi:10.1002/anie.200461776
33. Fiedler D, van Halbeek H, Bergman RG, Raymond KN (2006) Supramolecular catalysis of unimolecular rearrangements: substrate scope and mechanistic insights. J Am Chem Soc 128:10240–10252. doi:10.1021/ja062329b
34. Brown CJ, Bergman RG, Raymond KN (2009) Enantioselective catalysis of the Aza-Cope rearrangement by a chiral supramolecular assembly. J Am Chem Soc 131:17530–17531. doi:10.1021/ja906386w
35. Bianchini G, Sorella GL, Canever N et al (2013) Efficient isonitrile hydration through encapsulation within a hexameric self-assembled capsule and selective inhibition by a photo-controllable competitive guest. Chem Commun 49:5322–5324. doi:10.1039/C3CC42233J
36. Kang J, Rebek J (1997) Acceleration of a Diels-Alder reaction by a self-assembled molecular capsule. Nature 385:50–52. doi:10.1038/385050a0
37. Shimizu S, Usui A, Sugai M et al (2013) Hexameric capsule of a resorcinarene bearing fluorous feet as a self-assembled nanoreactor: a Diels-Alder reaction in a fluorous biphasic system. Eur J Org Chem 2013:4734–4737. doi:10.1002/ejoc.201300652
38. Murase T, Nishijima Y, Fujita M (2012) Cage-catalyzed Knoevenagel condensation under neutral conditions in water. J Am Chem Soc 134:162–164. doi:10.1021/ja210068f
39. Giust S, La Sorella G, Sperni L et al (2015) Supramolecular catalysis in the synthesis of substituted 1 H-Tetrazoles from isonitriles by a self-assembled hexameric capsule. Asian J Org Chem 4:217–220. doi:10.1002/ajoc.201402229
40. Cavarzan A, Scarso A, Sgarbossa P et al (2011) Supramolecular control on chemo- and regioselectivity via encapsulation of (NHC)-Au catalyst within a hexameric self-assembled host. J Am Chem Soc 133:2848–2851. doi:10.1021/ja111106x
41. Cavarzan A, Reek JNH, Trentin F et al (2013) Substrate selectivity in the alkyne hydration mediated by NHC–Au(I) controlled by encapsulation of the catalyst within a hydrogen bonded hexameric host. Catal Sci Technol 3:2898–2901. doi:10.1039/C3CY00300K
42. Leung DH, Bergman RG, Raymond KN (2007) Highly selective supramolecular catalyzed allylic alcohol isomerization. J Am Chem Soc 129:2746–2747. doi:10.1021/ja068688o
43. Brown CJ, Miller GM, Johnson MW et al (2011) High-turnover supramolecular catalysis by a protected ruthenium(II) complex in aqueous solution. J Am Chem Soc 133:11964–11966. doi:10.1021/ja205257x
44. Wang ZJ, Brown CJ, Bergman RG et al (2011) Hydroalkoxylation catalyzed by a gold(I) complex encapsulated in a supramolecular host. J Am Chem Soc 133:7358–7360. doi:10.1021/ja202055v
45. Lohr TL, Marks TJ (2015) Orthogonal tandem catalysis. Nat Chem 7:477–482. doi:10.1038/nchem.2262
46. Wang ZJ, Clary KN, Bergman RG et al (2013) A supramolecular approach to combining enzymatic and transition metal catalysis. Nat Chem 5:100–103. doi:10.1038/nchem.1531
47. Salles AG, Zarra S, Turner RM, Nitschke JR (2013) A self-organizing chemical assembly line. J Am Chem Soc 135:19143–19146. doi:10.1021/ja412235e

Artificial Metalloenzymes

Christian Trindler and Thomas R. Ward

Abstract While chemists are developing confined environments for catalysis, nature has evolved highly elaborate compartments to carry out reactions. Proteins offer such catalytic nano-environments that accept specific substrates to yield highly enantioenriched products. Metalloenzymes form a subclass that combines the functional diversity of proteins with the promiscuous activities of metals. In recent years, a variety of artificial metalloenzymes (ArMs) has been created upon incorporation of metal complexes into a protein scaffold. The following chapter discusses some of the protein scaffolds exploited for the creation of artificial metalloenzymes. Focus is laid on artificial metalloenzymes that catalyze abiotic and asymmetric reactions. Each subchapter presents the unique characteristics of a scaffold followed by a description of the reactions that were performed with it.

1 Introduction: The Choice of an Appropriate Scaffold for the Creation of Artificial Metalloenzymes

Current approaches to create metalloenzymes strongly focus on the modification of existing proteins. The living world offers a huge repertoire of enzymes that can be potentially modulated, but which ones will be the appropriate scaffold for asymmetric catalysis? A clear answer is not possible but several considerations may be useful:

1 Despite the broad range of protein functions, only a limited number of protein scaffolds are used by nature: a structural classification of ~38,000 protein crystal structures of the PDB revealed ~1200 different folds (SCOP survey 2009) [1]. Single scaffolds have obviously evolved to fulfill multiple functions.

C. Trindler · T.R. Ward (✉)
Department of Chemistry, University of Basel, Spitalstrasse 51, CH-4056 Basel, Switzerland
e-mail: thomas.ward@unibas.ch

R. Poli (ed.), *Effects of Nanoconfinement on Catalysis*,
Fundamental and Applied Catalysis, DOI 10.1007/978-3-319-50207-6_3

This suggests that engineers of artificial metalloenzymes should be able to take advantage of a single structural fold to repurpose it for multiple novel functions.

2 The artificial metal cofactors need to be anchored into the host protein. Covalent attachment, supramolecular assemblies, or dative anchoring are the most widely used strategies. Reactive amino acids such as cysteine, small molecule binding pockets or coordinating amino acids, respectively, are typically required to accommodate the artificial metal cofactors. More information on the attachment technologies can be found in recent reviews [2, 3].

3 A rigid fixation such as dual anchoring can help to reduce the motional degrees of freedom of a cofactor. Less motion may favor one specific chiral environment that promotes asymmetric induction [4–6].

4 The amino acids are responsible for a providing a chiral environment around the metal center. Accordingly, the cofactor should be close to the protein that forms the so-called second coordination sphere. While deep cavities or channels might provide a nearly complete shielding of the cofactor, they might also block access if cofactors or substrates are too large.

5 The native activity of an enzyme can be evolved according to desired needs. Thermal/pH/mechanical stability, high catalytic turnover, good stereoselectivity, and substrate scope are properties typically addressed with directed evolution strategies [7]. The same applies for artificial metalloenzymes: structure-based design and directed evolution haven proven efficient to identify improved mutants.

2 Hemoproteins P450 and Myoglobin

The enzymes P450 and Myoglobin have been thoroughly investigated and the huge amount of publications on these enzymes speaks for their robustness and evolvability [8]. While many enzyme scaffolds and their cofactors are synthetically assembled after purification, heme containing metalloenzmes such as P450 or Myoglobin are already naturally abundant in many organisms [9]. Several groups take advantage of the biological cofactor-protein assembly machinery that helps to avoid lengthy cofactor installation steps. However, when it comes to the replacement of the natural Fe-cofactors with other metal such as zinc, manganese, or cobalt, the purification of the apo-enzyme is typically necessary [10]. Fasan recently demonstrated an alternative route and introduced non-natural Mn/Co cofactors into Myoglobin in vivo. For this purpose, an outer membrane heme transporter was engineered into *Escherichia coli* that was then able to incorporate the Mn/Co-hemes in vivo [11]. Not only the natural in vivo biosynthesis of heme-proteins makes proteins such as P450 or Myoglobin very attractive targets for the design of artificial metalloenzymes, but also the demonstrated possibility to fine tune the electronic properties of the metal cofactors by genetic means [12]. In P450 and in Myoglobin, the native axial Fe-coordination sites are occupied by cysteine or

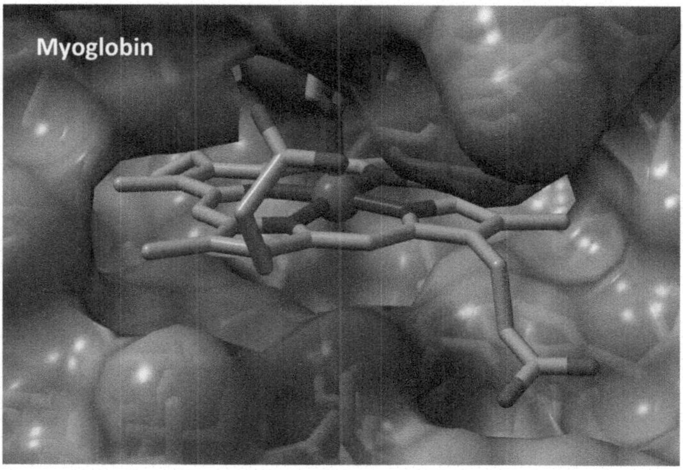

Fig. 1 Surface model of Myoglobin with the heme cofactor (pdb: 1A6 K). The porphyrin in Myoglobin is axially coordinated by a histidine (*stick color code* N = *blue*, C = *beige*, O = *red*, Fe = *brown sphere*). It is well exposed to the outer surface [28]. Several amino acids were modified to enlarge the access to the active site (shown in *red*) for cyclopropanation reactions [27] (color figure online)

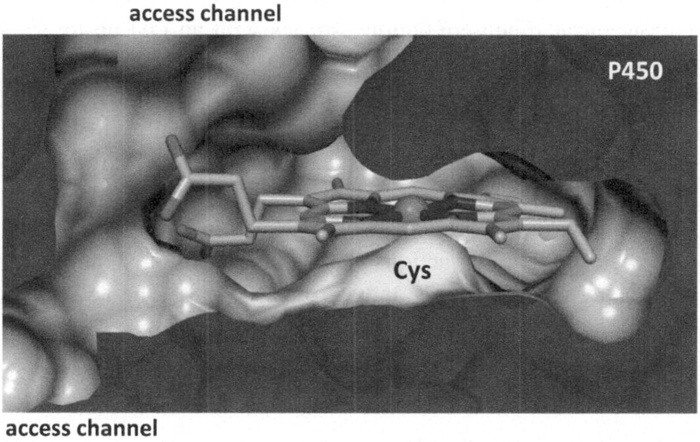

Fig. 2 Surface model of P450 with the heme cofactor (pdb: 4H24). [26] The iron atom is coordinated by cysteine (*yellow surface*) in the native P450 protein and was replaced by serine to enable carbenoid insertion reactions. The active site in P450 is deeply embedded within the protein so that substrates must transit through channels to reach the Fe-porphyrin [29, 30] (color figure online)

histidine ligands, respectively (Figs. 1 and 2). Replacement of the axial ligands alters the electronic state allowing to perform abiotic reactions such as nitrene [11, 13–19] and carbene insertions [20–25] (Fig. 3). Very high turnovers (>10,000)

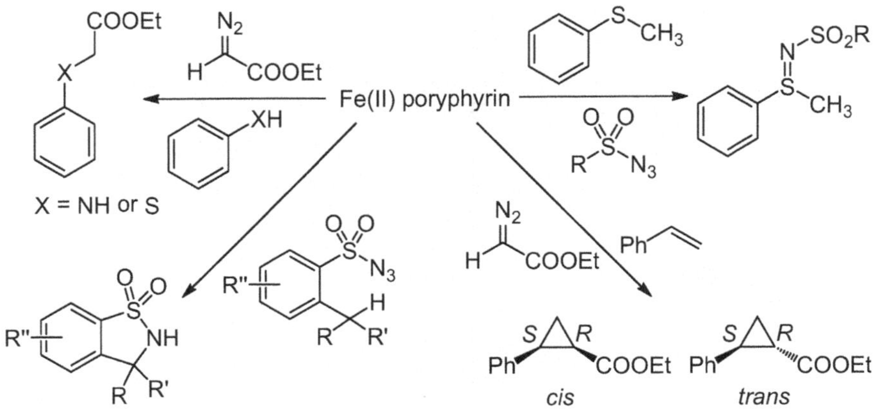

Fig. 3 A selection of carbene and nitrene insertion reactions catalyzed by myoglobin or P450 mutants

and high enantioselectivities have been obtained by Arnold and Fasan using P450 and Myoglobin mutants for cyclopropanation reactions [26, 27].

Obviously, the apolar heme binding sites can be engineered to offer a suitable environment to host a given transformation (Figs. 1 and 2). The engineered P450 were also shown to be active in whole cell/biomass conversions [17, 21, 22, 26, 31]. Arnold used concentrated cell-containing slurries with an OD up to 30.

In vitro assembly of hybrid Myoglobin catalysts allowed researchers to replace the natural cofactor by a variety of abiotic cofactors [32]. Cr(III)-salophen [5, 33], Fe/Mn-Porphycenes [10, 34] for oxidation and sulfoxidation reactions. Unfortunately, only modest enantiomeric excesses (ee) were reported for these oxidation reactions.

The original function of P450 was also fine-tuned in several evolution projects. Oxidation and hydroxylation reactions were performed on a number of apolar substrates such as terpene, alkaloids and steroids [35–38]. The ee's were typically quite high for these reactions.

3 Large Spherical Scaffolds

3.1 Ferritin

The 24 subunits of ferritin form a globular shell with an inner-diameter of about 80 Å (Fig. 4). The oxidation of Fe^{2+} to Fe^{3+} and controlled biomineralization of iron are the native functions of Ferritin. The cavity can accommodate about 4500 iron ions. Several groups refunctionalized apoferritin by introducing nanoparticles of Fe [39], Co [40], Ni [40], Au [41], Pt [42–45], Cu [46], Ce [47], Pd, Ag [41],

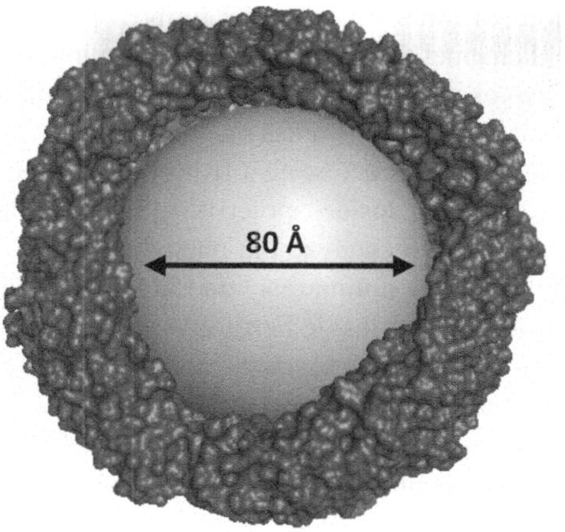

Fig. 4 Surface model of Ferritin (pdb: 1MFR). The subunits pointing toward the reader were omitted to allow view on the modeled nanoparticle (*gray sphere*) inside the cavity

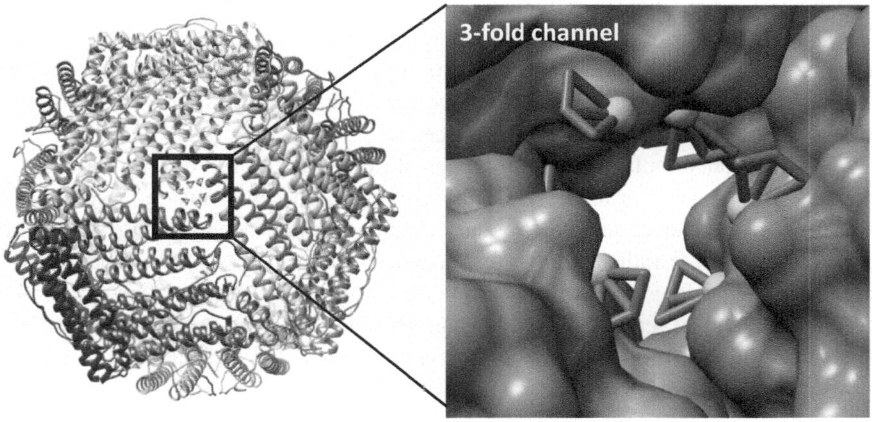

Fig. 5 Ribbon representation of horse liver Ferritin with magnification of a three-fold channel as surface model (pdb: 2ZG7). The cysteine residues inside the cavity and in the three-fold channels can bind metal complexes shown here as *green spheres* with *purple* allyl ligands. The limited diameter of the channels only allows transit of small molecules (color figure online)

CdSe [48, 49], or Y-V:Eu/Dy [50]. The metals were accumulated in the interior space in the form of metal nanoparticles by self-assembly or redox reactions. New functions include luminescence, superoxide dismutation, hydrogenation, enhanced ferroxidase activity, or the solubilization of supermagnetic particles.

The spatial constraint imposed by apoferritin not only helps to produce monodisperse nanoparticles but it also imposes a size exclusion barrier to its interior. The 24 shell monomers form two kinds of size-selective channels: one is located at the junctions of three subunits and forms a polar channel (Fig. 5). The other one is a fourfold channel that is lined by apolar amino acids. Building on this, Watanabe used the size exclusion for the selective hydrogenation of electron-poor olefins of low molecular weight by a Pd-nanocluster [51].

In addition to various nanoparticles, also $[Rh(nbd)Cl]_2$-and dinuclear $[Pd(allyl)Cl]_2$-complexes were anchored inside apoferritin [52–55]. The resulting artificial metalloenzymes were shown to catalyze the phenylacetylene (PA) polymerization [53, 56] and the Suzuki cross-coupling reaction [54], respectively. Both complexes were shown to attach at two sites inside Ferritin: at the threefold channel and at the so-called accumulation center. The crystal structures revealed cysteine and histidines as coordinating amino acids.

In line with the catalytic activity of the free $[Rh(nbd)Cl]_2$ complex, the metalloenzyme produced head-to-tail *cis/trans* PA-polymers within the protein cavity. The ferritin cage prevented the precipitation of the insoluble polymer and led to a narrow size distribution of the polymer. Despite a clear localization of the metal binding sites as revealed by X-ray crystallography, the authors speculate that the bound Rh-catalysts are merely precursors that migrate to an apolar site after coordination of the first monomer coordination [53].

The publications for PA polymerization and the Suzuki cross-coupling strongly focus on the crystal structures and the analysis of the cofactor binding. The relevant amino acids such as histidine and cysteine were substituted by alanine residues to study their relevance in catalysis. No extensive genetic engineering was reported to improve the metalloenzyme's performance.

The mentioned cofactor binding sites differ strongly by the way they are accommodated in the protein. While the accumulation center inside the cavity of Ferritin is clearly more related to a surface than a pocket, the binding site at the threefold channel is rather tightly surrounded by the protein scaffold. The latter is suggested to interfere with the penetration of substrates into the Ferritin cavity [54].

While the use of ferritin as host for nanoparticles has been widely exploited, no reports on the use of ferritin in asymmetric catalysis could be identified.

3.2 POP, Prolylpeptidase

The prolylpeptidases selected by Lewis originates from *Pyrococcus furiosus* (*Pfu*). This POP is known for its high thermal stability. The family of POPs shares a common α/β-hydrolase domain, which contains a Ser-Asp-His triad for the amide bond hydrolysis. The active site serine was substituted by Azidophenylalanine to covalently attach the cofactor by a strain promoted cycloaddition [6, 57]. Prior to this cycloaddition, the β-domain covering the "active" site was genetically engineered to widen up the entrance of the cavity for improved cofactor access. For this purpose, four large amino acids blocking the entrance were mutated to alanine residues. As no crystal structure was available, the modifications were engineered based on a homology model from porcine POP (pdb: 1QFS). The internal cavity of

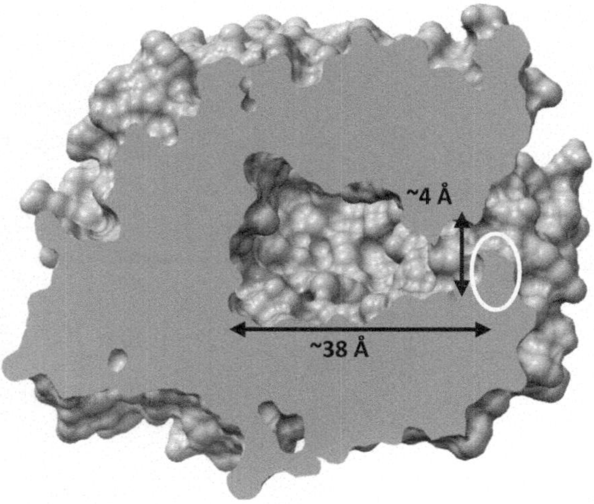

Fig. 6 View of the X-ray structure of porcine POP highlighting the large cavity and the narrow channel (pdb: 1QFS). For the assembly of the artificial metalloenzyme, the entrance (*white*) was widened. The artificial cofactor was covalently anchored into the active site (*red*) by [3 + 2] cycloaddition on an engineered azidophenylalanine (color figure online)

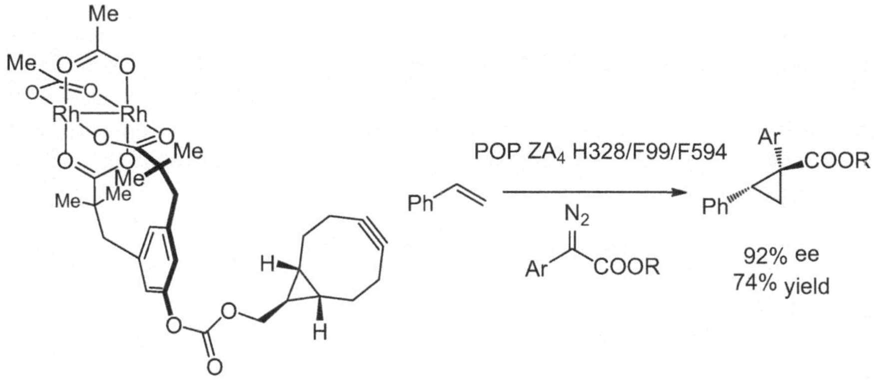

Fig. 7 Structure of the di-rhodium cofactor bearing an activated alkyne functionality for anchoring within POP bearing an azidophenylalanine. The resulting artificial metalloenzyme catalyzes the asymmetric cyclopropanation of styrene

porcine POP is large, but despite its size, the entrance is about 4 Å wide without the alanine modifications [58] (Fig. 6).

The WT protein initially afforded 19% yield and 11% ee for the cyclopropanation of styrene with a diazo compound (Fig. 7). Using salt-rich conditions and subsequent protein evolution, the yield increased to 74% and and 92% ee. A major contribution to the impressive improvement of the ee was traced back to the L328H

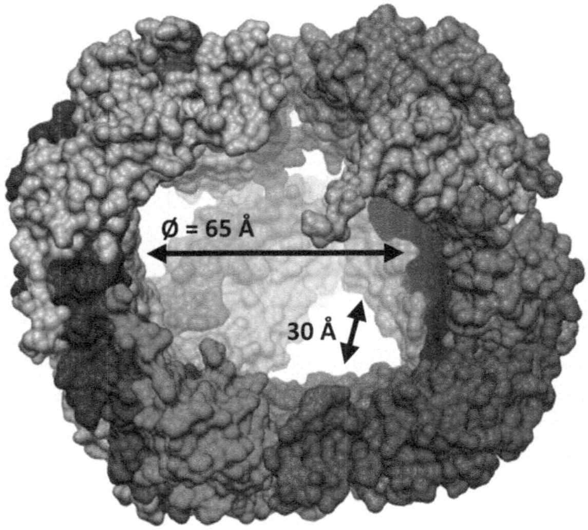

Fig. 8 Surface model of heat shock protein from *Methanocaldococcus jannaschii* (pdb: 1SHS). The heat shock protein consists of an assembly of 24 monomers that forms a cavity. The front monomers are not displayed. The interior can be accessed by several large pores which have a diameter of approximately 30 Å

mutation. The histidine likely binds to one Rh-atom, thereby reducing the cofactor mobility. Similar dual anchoring strategies for the improvement of catalytic performance of artificial metalloenzymes have been previously reported by Lu [5] and Ward [4] (Fig. 7).

In addition to the high ee obtained for the cyclopropanation, the evolution of the Rh-metalloenzyme also allowed to decrease the hydrolysis of the diazo reactant. The high diastereoselectivity, however, is not a result of the enzyme evolution but is dictated by the free cofactor. No reactions with acceptor/donor diazo substrates were performed so far by Arnold and Fasan in their P450 and Myoglobin systems, respectively.

3.3 MjHSP

Metathesis catalysts are sensitive against various functional groups. The internal environment of the heat shock protein from *Methanocaldococcus jannaschii* was scrutinized to determine if it may offer a suitable environment for an Hoveyda–Grubbs II catalyst. The monomer of the heat shock protein has a molecular weight of 16.5 kDa. It self-assembles into a 24-mer-shell that contains a large internal cavity (Fig. 8). In addition to the spacious cavity, the protein is heat stable up 70 °C and tolerates a broad pH range from 4 to 11. Hilvert anchored a metathesis catalyst at a cysteine inside the cavity [59]. The resulting metalloenzyme was active for the ring closing metathesis although less active than the free cofactors. The highest turnover numbers were achieved at pH 2 whereby the oligomeric protein is disassembled.

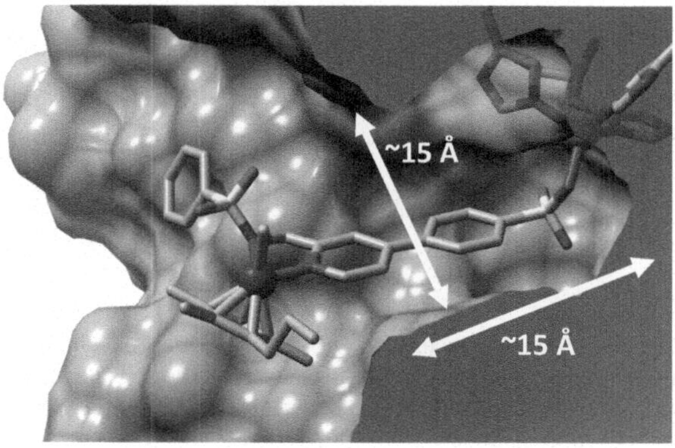

Fig. 9 Surface model of human carbonic anhydrase II with a bound pianostool cofactor (pdb: 3ZP9) [61]. Aryl-sulfonamides display high affinity for human carbonic anhydrase II, thus allowing to anchor organometallic catalysts within hCA. The piano stool moiety protrudes from the active site cavity allowing ready access of the substrate to the iridium ion (*blue sphere*) (color figure online)

4 Human Carbonic Anhydrase

Arylsulfonamides are versatile inhibitors of the native function of human carbonic anhydrase (hCA). These bind to the Zn-ion thus blocking the hydrophobic active site. Designed arylsulfonamide ligands bind to hCA with picomolar affinity. Such inhibitors may thus be used as anchor to localize a variety of abiotic organometallic catalysts [60–62]. Importantly, the cavity is about 15 Å deep and 15 Å wide at its mouth, thus allowing to localize both a cofactor and its substrate (Fig. 9) [63].

A series of bidentate ligands chelating the {Cp*Ir(III)Cl}-moiety were tested for their influence on the reduction of prochiral imines. The hCA WT and the hCA I91A mutants showed strong protein acceleration and modest ee's (70 and 68%, respectively). Interestingly, the mutant I91A did not show any substrate inhibition compared to the WT [61]. Based on computational design, the second coordination sphere was tailored around the cofactor. The mutants obtained showed higher affinity for the cofactor than the WT and gave better turnovers and excellent ee's. The authors reckoned that a higher TON may relate to the reduced weight of nonproductive conformations of the cofactor–protein assembly [64].

Linking an arylsulfonamide with the NHC ligand of a Hoveyda–Grubbs II catalyst enabled the formation of an artificial metathesase that operated under physiological conditions (Fig. 10) [65].

An alternative strategy to an artificial CA-metalloenzyme was the direct substitution of the zinc ion by rhodium or manganese ions [66]. In presence of hydrogen peroxide, the manganese–metalloenzymes catalyzed the epoxidation of styrene derivatives with moderate ee's [67, 68]. The specific replacement of the Zinc by rhodium ion was more laborious than for manganese: to direct specific binding of rhodium into the active site of the carbonic anhydrase, all superficial

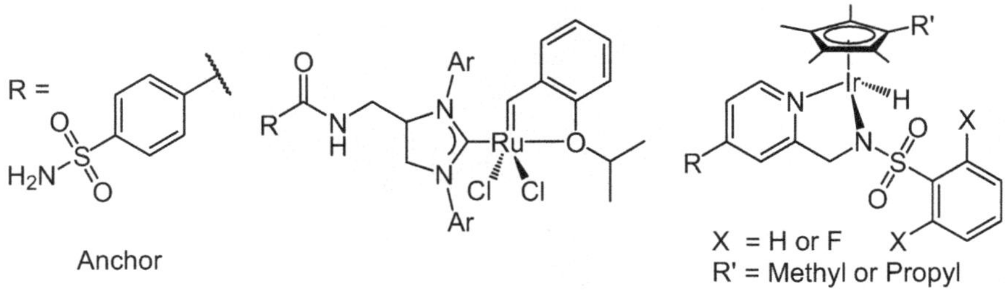

Fig. 10 A selection of artificial cofactors for hCA II: A Hoveyda–Grubbs II catalyst and a piano-stool iridium complex were anchored to human carbonic anhydrase II thanks to a sulfonamide anchor. The resulting artificial metalloenzymes catalyze the ring-closing metathesis and the asymmetric reduction of imines

histidines had to be removed. *Cis*-stilbene could be selectively reduced by the artificial Rh-metalloenzyme but the reaction was slightly slower compared to the free catalyst [66]. The same Rh-metalloenzyme was also able to hydroformylate styrene with good regioselectivity. The regioselectivity was increased when the metalloenzyme was treated with diethylpyrocarbonate to block all accessible imidazole functional groups [69].

5 Streptavidin

The streptavidin (SAv) molecule is well known for its extremely high affinity for biotin ($K_a \approx 10^{15}\ M^{-1}$) [70]. It forms a homotetramer that is remarkably thermostable [71] and robust against organic solvents [72]. Inspired by Whitesides [73], Ward established a library of artificial metalloenzymes with biotinylated cofactors (Fig. 11).

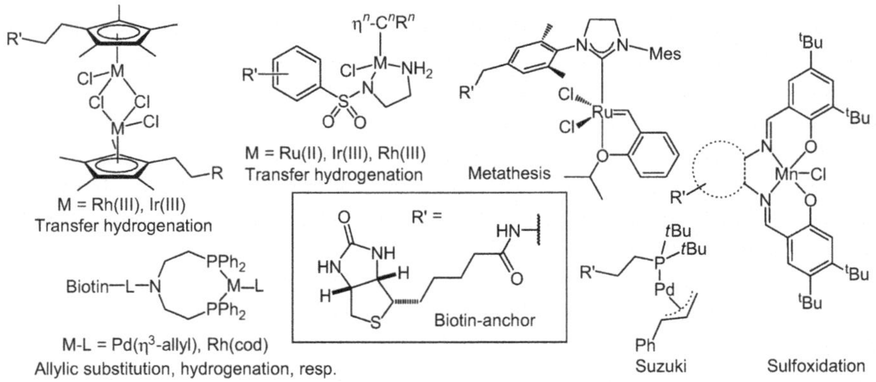

Fig. 11 A selection of biotinylated cofactors and reactions catalyzed by the corresponding metalloenzymes upon incorporation in streptavidin

The metalloenzymes were able to catalyze regio-, diastereo-, and enantioselective reactions including Suzuki cross-couplings [74], allylic alkylations [75], sulfoxidations [76, 77], or the hydrogenation of olefins, [72, 78–83] the transfer hydrogenations of ketones [84–87], enones, and imines [4, 88–92]. Stereocontrol is the major advantage of artificial metalloenzymes but the acceleration of catalysis, catalyst protection from poisoning substances or the increase of cofactor solubility are other benefits the artificial metalloenzymes can offer [88, 91]. Accordingly, the SAv scaffold was also chosen as support for olefin metathesis [93]. While the synthesis of cofactors is a first major task when novel metalloenzymes are created, the improvement of the secondary ligand sphere around the active site is another big challenge. The combination of chemical and genetic optimization was shown to be successful for the design of several artificial streptavidin-metalloenzymes [78, 86, 94]. Rational design and site directed saturation mutagenesis provided stepwise improvements [85, 95, 96]. The inversion of stereoselectivity could be obtained in most of the studies. The biotin anchor is deeply buried in the streptavidin cavity while the cofactor metals are moderately exposed to the surrounding medium at a cleft that is formed by two monomers of the tetramer (Fig. 12).

Especially for stereoselective catalysis, localization of the active site metal within a well defined chiral environment is believed to play an essential role. For this purpose, dual anchoring can be ensured by coordination of the strong coordinating ligands such as histidine or by apolar/polar interactions which preferentially interact with apolar/polar parts of the cofactor, respectively [4].

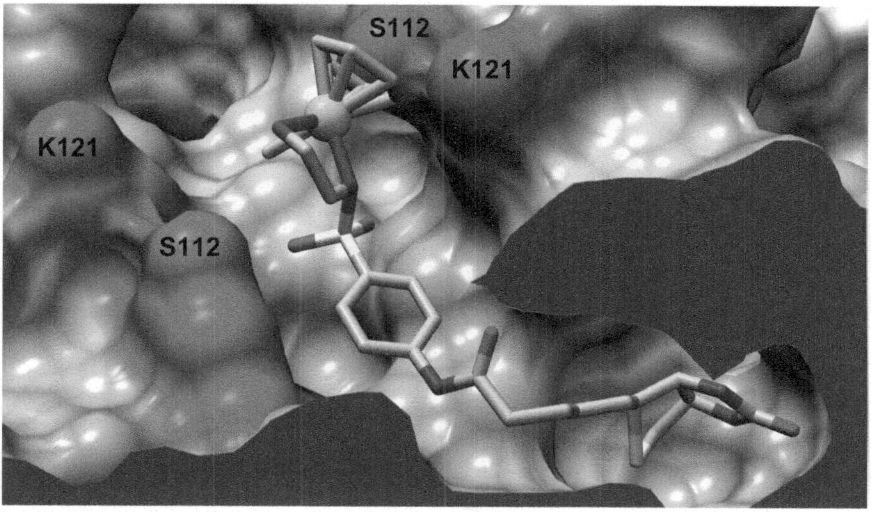

Fig. 12 Structure of [η^6-(arene)RuCl(Biot-*p*-L)] bound to SAv (surface model, pdb: 2QCB, iridium = *turquoise sphere*). Two monomers of the tetrameric protein are highlighted in *gray* and *beige*, respectively. The *red* surfaces belong to the amino acids S112 and K121 of the two monomers. These sites are very close to the precious metal and their mutations have contributed to significant catalytic improvement in most cases (color figure online)

In contrast to the biotin anchoring, SAv has been loaded with metals that had no high-affinity anchor attached: SAv treated with VO^{2+} catalyzes the sulfoxidation of prochiral sulfides with high enantioselectivity [77] and OsO_4 conjugates dihydroxylates olefins in ee's up to 95% [97]. A similar approach of a dative anchoring based on $K_2OsO_2(OH)_4$ in Laccase lead to a very high enantiomeric excess for the dihydroxylation of olefins as well [98].

6 Albumins

Serum Albumins are the most abundant serum proteins in mammals. They are well soluble in water (>585 mg/mL for BSA) [99] and are known to bind a variety of hydrophobic metabolites, drugs and small molecules. Serum albumins transport unesterified fatty acids [100] (Fig. 13). Large complexes, that could be potentially active in stereoselective catalysis, such as metal–salen complexes [101, 102], porphyrins, phthalocyanines [103] or corroles [104, 105] were also shown to bind albumin.

Despite the high affinity of albumins for various large organic molecules the first albumin metallohybrid was created by dative cofactor anchoring of a small inorganic compound: BSA-osmate esters were shown to stereoselectively dihydroxylate olefins up to an ee of 68% [106]. The osmate was believed to be bound by primary

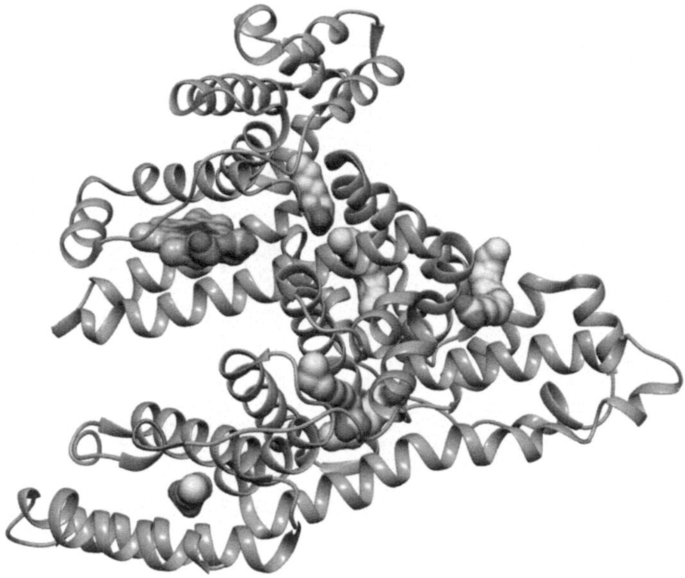

Fig. 13 Structure of Hemin (*gold-surface display*) bound to HSA (*cartoon display*). This binding site as well as six additional sites can bind fatty acids (*green-surface display*) (pdb: 1O9X) [104] (color figure online)

amino groups provided by lysines. In comparison: streptavidin-OsO_4 conjugates gave ee's up to 95% [97]. $Rh(CO)_2(acac)$ complexes in the presence of BSA were shown to catalyze the hydroformylation of olefins [107]. The hydroformylation of styrene preferentially yields the branched product in a 9:1 ratio. On the one hand, it appears that the regioselectivity is dictated by the rhodium moiety, as shown by Kazlauskas: the free rhodium catalysts preferentially hydroformylate styrene to the branched aldehyde [69]. On the other hand, regioselectivity depends on the nature of the olefin [108]. The same Rh-HSA hybrid and its $[Ir(COD)Cl]_2$ analog also catalyzed the reduction of carbonyl/alkene functionalities, although with no stereoselectivity [109].

The high affinity of porphyrinoids for albumins was exploited by Reetz for Diels Alder reactions [103] and by Gross for the stereoselective sulfoxidation of organic sulfides [110]. Screenings with human, porcine, bovine, sheep and rabbit albumins were performed. Sulfoxidation with Fe/Mn corroles gave the best results in combination with BSA (ee of 74%) [110]. The sensitivity of the BSA metalloenzymes toward cofactor variation became obvious with the substitution of the corroles by Mn-salen cofactors that lead to a complete loss of enantioselectivity [101]. In contrast, the Diels–Alder reactions gave again an excellent enantiomeric excess up to 93% with an *endo*:*exo*-ratio of 96:4 in the presence Cu–Phthalocyanine as cofactor and azachalcone/cyclopentadiene as substrates. In the absence of the phthalocyanine no stereoselectivity was observed [103].

A major difficulty in the design of artificial albumin hybrid enzymes remains the limited structural information. In addition, recombinant expression has only been developed for HSA [111], but no genetic optimization, that improved the catalytic performance of artificial HSA-metalloenzymes, have been reported so far [112]. The next generation HSA-metalloenzymes require extensive genetic optimization.

7 Papain

The endoprotease Papain offers a single reactive cysteine site for modification by maleinimide or by activated haloacetamides. The active site, e.g., the reactive cysteine, is located in a hydrophobic pocket with a cleft of about 15 Å and a groove of about 25 Å [113].

Several metal catalysts have been anchored to this groove via linkers. Unfortunately, the ee's remain modest in all cases reported so far [114]. The wide Papain pocket might be too spacious to exert a significant second coordination sphere influence on the metal cofactor. Perhaps a dual anchoring strategy—as exploited by Ward [4], Lu [5] and Lewis [6]—may allow to improve the selectivity of such hybrids relying on papain as host.

$[(\eta^5\text{-Cp*})Rh(N\text{^}N)Cl]^+$ complexes have successfully applied for the reduction of NAD^+ and various acetophenone derivatives [115–117]. The rhodium complexes clearly outperformed the Ru(arene) in terms of activity, but enantioselectivity remained modest.

Similarly, the reduction of methyl 2-acetamidoacrylate by Rh(COD)BF_4 ligated to the Papain via a monodentate phosphite ligand gave good yields but no ee's [118].

Gratifyingly, a Diels-Alder cycloaddition reaction between cyclopentadiene and acrolein was accelerated 100-fold by the protein using a $[(\eta^6\text{–arene})Ru(N^\wedge N)Cl]^+$ complex when compared to the free catalyst. [119] No enantiomeric excess was noticed.

8 Chymotrypsin

The chymotrypsin hydrolyses peptides at hydrophobic residues. Hirota took advantage of a known inhibitor to covalently install an artificial metathesis cofactor into the hydrophobic pocket. The resulting hybrid was shown to catalyze ring closing metathesis. Best results, were obtained with the water soluble glucose derivative GlcDAA bearing two allyl functionalities [120] (Fig. 14). Solubilization of the apolar Hoveyda–Grubbs II catalyst was the only obvious benefit that resulted from the incorporation of the catalyst into chymotrypsin.

9 Lipases

CALB and Cutinase catalyze the hydrolysis of hydrophobic esters. In contrast to the classical lipases, both enzymes do not possess a hydrophobic lid that must be displaced to allow access to the active site [121]. In the case of CALB, it was

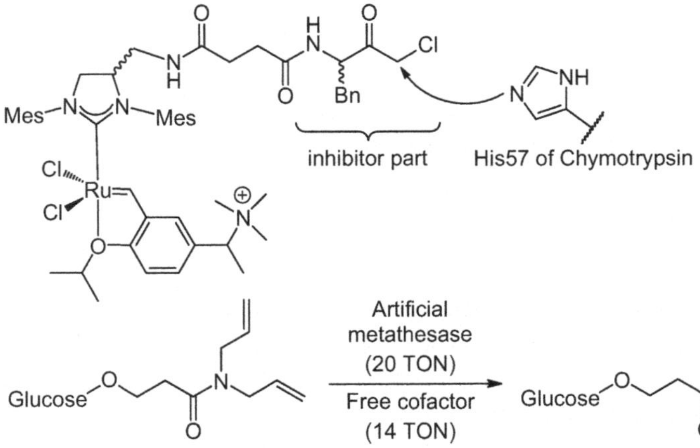

Fig. 14 An artificial metathesase was created by covalently modifying the active site of Chymotrypsin with an inhibitor that reacts irreversibly with the His57. The best substrate for RCM is a glucosylated water soluble diallylamine

suggested that the interfacial interaction still might play a role in the hydrolysis of big substrates [122]. Despite this structural regulation, lipases can accept a broad variety of substrates. The broad substrate acceptance, good stereoselectivities, and a very high solvent tolerance render lipases attractive for industrial applications but also for the design of artificial metalloenzymes [123].

As phosphonate ester selectively reacts with the active site serine residue, they are well suited to anchor artificial cofactors into the active site of lipases. Reetz applied this active site hybridization technique to covalently immobilize phosphonate ligands. Unexpectedly, the selected di-*para*-nitrophenyl-phosphonate hydrolysed within one day, and the activity of the original lipase was recovered [114] (Fig. 15). Van Koten and Gebbink modified the reactive phosphonate by substituting one of the destabilizing *para*-nitrophenyl functionalities by an ethyl group. The novel hybrid proteins harbored platinum- and palladium pincer complexes and were shown to be stable for at least 48 h at pH 8 and even up to pH 11 [124–126]. No catalytic results were published for these hybrids, however.

Cutinase was selected as scaffold for a metathesase due to its reactivity toward hydrophobic substrates, its relatively exposed active site, and its capability to induce steric hindrance in the artificial metalloenzymes. However, the desired steric hindrance strongly reduced the catalytic efficiency of an anchored Hoveyda–Grubbs II catalyst. By lengthening the linker, the turnover number was improved to a level that is comparable to the free catalysts [127].

The active site cavity of Candida Antarctica lipase B is deeper than that of cutinase. It is believed to improve the chemoselectivity that was observed for the

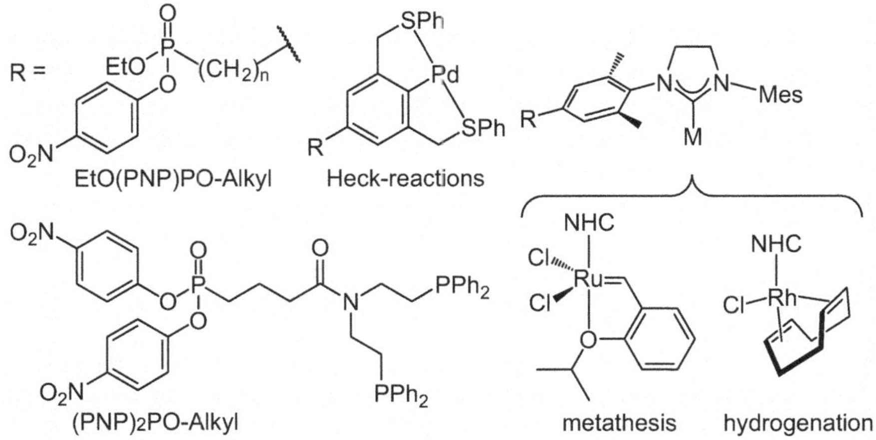

Fig. 15 Organometallic cofactors bearing a phosphonate for the anchoring within lipases rely on the nucleophilic active site serine residue. The di-*para*-nitrophenyl-phosphonate $(PNP)_2PO$-Alkyl was found to be unsuitable for anchoring a diphosphine ligand as the resulting phosphonate hydrolyzed. The substitution of one *para*nitrophenyl functionality by an ethyl group (EtO(PNP)PO-Alkyl) allows the formation of stable phosphonates. A Pd-pincer-complex and *N*-heterocyclic-carbene-complexes of rhodium and ruthenium were successfully incorporated into the lipases

cutinase-NHC(Rh)(cod)-metalloenzyme [128] (Fig. 15). The Cutinase hybrid displayed no enantioselectivity for the reduction of acetophenone or acetamidoacrylate, but a moderate preference for the reduction of the olefin over the ketone. A strong improvement of chemoselectivity was observed for the CALB metalloenzyme in single substrate-conversion experiments. Unfortunately, no competition experiments were performed to confirm this hypothesis for substrate mixtures [128]. An immobilized CALB was selected by Palomo for the anchoring of Pd-pincer complexes. The immobilized metalloenzyme catalyzed Heck asymmetric reactions with excellent ee of 97%. Remarkably, the hybrid protein was deployed at 120 °C in 75% DMF [129].

While most artificial metalloenzymes are developed to improve some kind of selectivity, Gebbink immobilized a $Ph_4CpRuCl_3$ catalyst on CALB for the racemization of alcohols in toluene. The racemization was combined with subsequent enantioselective acylation. This kinetic resolution catalyzed by the native lipase and by the artificial metalloenzyme gave excellent enantioselectivities of the acylated product (>99%) but the native enzyme was markedly faster [130].

So far, no genetic optimization has been performed on the artificial cutinase and CALB-metalloenzymes. Most recently, Bäckvall reported on the directed evolution of CALB using *P. pastoris* [131, 132]. It will be interesting to see if this organism may allow to improve the performance of artificial metalloenzymes relying on CALB as host.

10 Channels/Pores

Deep cavities are attractive sites for the localization of artificial metalloenzymes as they might help to surround the active site by a nearly complete secondary protein sphere. They often consist of β–barrels. In the following section, three proteins with β-barrel scaffolds are presented. Some form channels that offer access to the catalytic site from two directions.

10.1 tHisF

The thermostable α,β-barrel protein tHisF originally constitutes the synthase subunit of a glutaminase–synthase bi-enzyme complex that catalyzes the formation of imidazole glycerol phosphate in the histidine biosynthesis of *Thermotoga maritima*. The barrel forms a 25 Å long channel (Fig. 16) that has a large and a small entrance [57, 133].

Lewis ligated Mn- and Cu-terpyridine complexes and a dirhodium tetraacetate cofactor to a central moiety using a strain-promoted azide–alkyne cycloaddition analogous to the functionalization of a prolyl oligopeptidase [6]. The resulting dirhodium ArM catalyzed two similar carbene reactions with diazo compounds: an

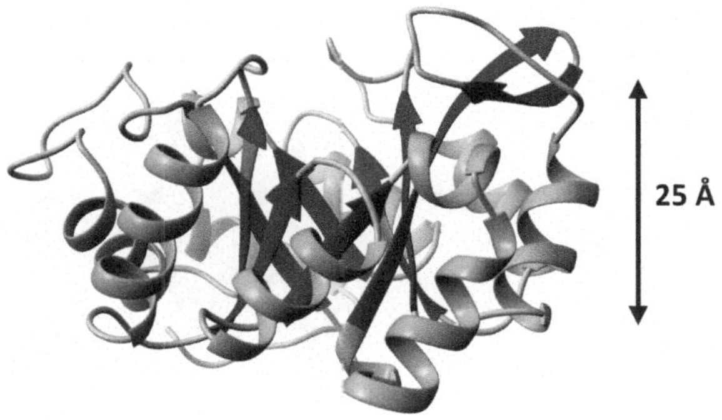

Fig. 16 The protein tHisF forms a channel that is open toward one side while the access from the other side is blocked by a salt bridge. Various cofactors were attached to an engineered azido-amino acid by a cupper-free cycloaddition. The three azido-modified positions are highlighted in *red* (color figure online)

insertion into a silane Si-H bond and a cyclopropanation with an olefin. Even if no protein acceleration was observed, the construct is a nice example for the application of "click" chemistry for the incorporation of an artificial cofactor. The yields of the carbene insertion were dependent on how deeply inside the cavity the cofactor was attached: attachment at the larger mouth gave 81% yield, in the middle 69% and at the smaller exit 60% [57].

Reetz selected the broad channel side for the attachment of an artificial Cu cofactor as it was considered ideal for the introduction of a 2-His-1-carboxylate motif (His/His/Asp) [134, 135]. A potentially coordinating cysteine was removed to avoid multiple Cu-binding. The resulting Cu–metalloenzyme not only accelerated the desired Diels Alder reactions when compared to free catalyst, it also provided the product with an ee of 46%.

10.2 FhuA

FhuA is a large channel that was selected for the design of an artificial metalloenzyme. The native cork domain was removed to obtain a passive mass-transfer channel fhuA Δ1-159 [136]. The channel of fhuA is about 11–14 Å wide (Fig. 17). A cysteine moiety was engineered inside the channel at a central position having about 30 and 15 Å distance to the upper and lower rim, respectively. In addition, two mutations N548V and E501F were introduced to improve accessibility of the engineered cysteine and to avoid additional catalyst coordination by the glutamic acid, respectively. A TEV cleavage site was also included to facilitate the analysis of the covalent anchoring position.

The engineered FhuA ΔCVF^{tev} was stable over a broad pH (5–11) and temperature range (4–60 °C) and tolerated organic solvents. This allowed the study of

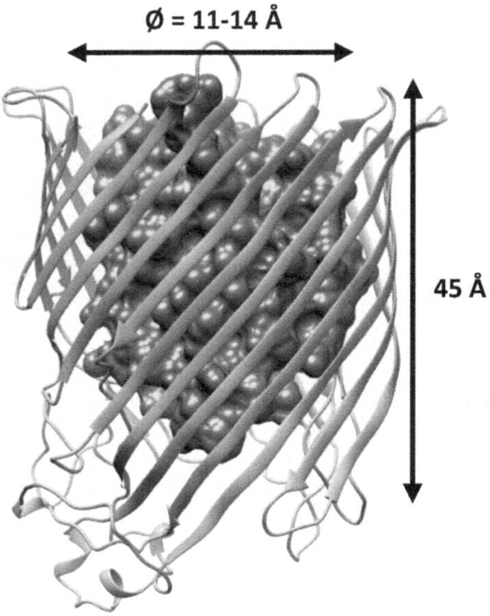

Fig. 17 FhuA as a host for the covalent anchoring abiotic cofactors (pdb 1BY3). A plug consisting of 159 amino acids (*surface display*) was removed to generate an open channel FhuA Δ1-159. The channel is large: about 45 Å long and 11–14 Å wide. A cysteine was engineered inside the channel with a distance of 15 and 30 Å to the channel edges (not shown)

reactions with up to 40% (v/v) THF or 10% (v/v) EtOH. Despite chemical and genetic engineering, the artificial metathesase displayed less turnover than the free catalyst; however, a small preference for the formation of *cis*-polymer was demonstrated [136].

10.3 lmrR

lmrR is a multidrug ABC transporter in *Lactococcus lactis*. LmrR is a dimeric helices-protein that forms a large flat hydrophobic pore. As the pore can bind various planar drugs it was considered to be large enough to incorporate a metal complex while still leaving enough space for substrate binding. Bidentate-Cu ligands were covalently attached at engineered cysteines. The final Cu-metalloenzymes catalyzed Diels–Alder reactions with very high ee's and the addition of water with good ee [137, 138] (Fig. 18).

In an alternative approach, the bipyridylalanine aminoacid was genetically engineered in lmrR using the amber codon strategy. The resulting bipyridine Cu(II) complex catalyzed the (Fig. 19) Friedels-Craft alkylation with promising ee's [139].

In contrast to the cofactors X_1–X_3 (Figs. 18 and 19), the incorporation of phenanthroline (X_4) into lmrR does not require any kind of anchoring (Fig. 19). The dissociation constants in the lower micromolar range ensure nearly quantitative localization under catalytic conditions. Excellent ee's and turnovers were obtained for hydration reaction. By site directed mutagenesis, two tryptophan residues W96

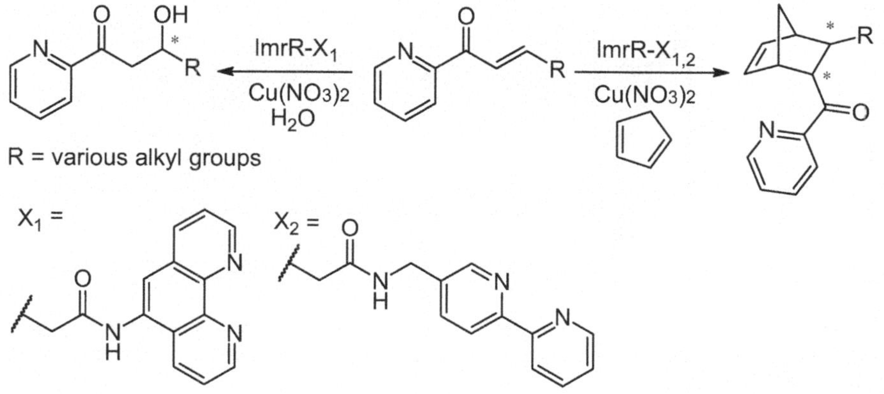

Fig. 18 lmrR was covalently modified with bidentate ligands X_1 and X_2. After $Cu(NO_3)_2$ addition, the resulting metalloenzymes catalyzed stereoselective Diels–Alder reactions and the hydration of α,β-unsaturated ketones

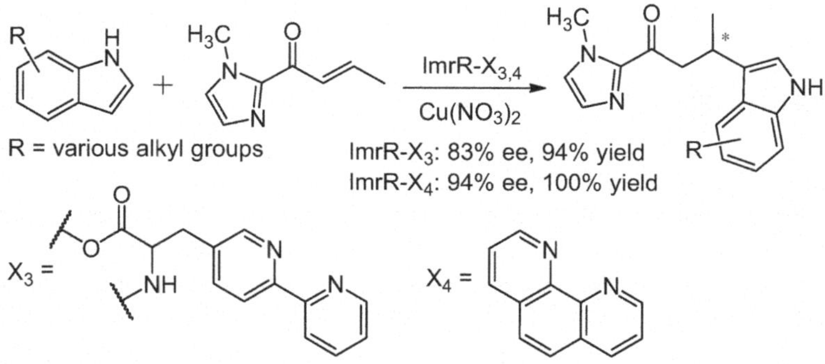

Fig. 19 Friedels-Craft alkylations were performed by bidentate Cu-complexes. Ligand X_3 was incorporated into the peptide chain of lmrR in vivo while the phenanthroline X_4 was able to bind lmrR in a supramolecular fashion

and W96′ (one in each subunit) were identified as key residues influencing catalysis outcome. Presumably, they may be involved in π-stacking interactions with the Cu (II) phenanthroline complex [140].

10.4 SCP-2L

Roelfes designed lmrR with a Cu(II)-phenanthroline cofactor for Diels–Alder reactions (Fig. 20) [138]. The same complex was installed by Kamer inside a channel of the sterol carrier protein type II (SCP–2L) by a cysteine–maleimide linkage [141]. In contrast to lmrR, SCP-2L possesses a rather cylindrical channel

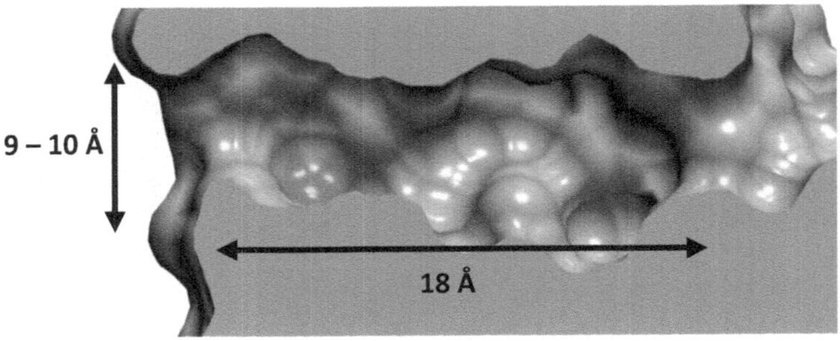

Fig. 20 The channel of SCP-2L (pdb: 1IKT). Cu(II)-phenanthroline was attached to a cysteine (*red surface*) that was engineered in either of two positions. Both positions are close to the channel exits (color figure online)

that is known to accommodate various hydrophobic molecules (Fig. 20). The resulting Diels–Alderase catalyzed the reaction with moderate enantioselectivity and yield.

11 Antibodies

While the design of the majority of metalloenzymes is based on the incorporation of a cofactor into a known protein scaffold, antibodies are evolved by living organisms to tightly bind to a molecule of choice via non-covalent interactions. Typically, a protein is modified with the molecule of choice and then injected into an organism that produces the antibody against this antigen. Based on this technique, Keinan obtained an antibody, SN37.4, for Ru- and Sn-porphyrins that catalyzed the sulfoxidation of thioanisole with an ee of 43% [142]. In the similar spirit, Mahy raised antibodies against a Fe–porphyrin–peptide conjugate, MP8 (microperoxidase 8). MP8 carries a histidine that occupies the proximal site of the iron. One antibody raised against MP8, 3A3-MP8, oxidized thioanisole with 45% enantiomeric excess [143]. It also catalyzed regioselectively the nitration of phenol in a 24:76 ratio in favor of 2–nitrophenol with low turnover [144]. In an alternative "Trojan" horse strategy, Mahy tested metalloporphyrin-estradiol conjugates with an anti-oestradiol antibody for their sulfoxidation activity. Again in this case, the enantiomeric excess remained low (10%) [145, 146].

In contrast to the low ee's obtained with porphyrins, Yamaguchi developed a monoclonal antibody against a rhodium diphosphane that stereoselectively reduced 2-acetamidoacrylic acid with very high ee (98%) [147]. Yamaguchi's result demonstrates the feasibility to perform asymmetric catalysis with antibodies elicited with metallic cofactors [148].

12 Lactoglobulin

β-Lactoglobulin is the major protein in cow's milk. In contrast to its function, its structure is well understood. Two anchoring strategies were applied to generate artificial metalloenzymes: (1) supramolecular anchoring based on fatty acids that bind into the hydrophobic pocket and (2) covalent modification of a cysteine residue by maleimide. The first method allowed Salmain to install a catalyst for the transfer hydrogenation of ketones. Fatty acid-cofactor conjugates bound into a pocket that only opens up above pH 7 in a process called Tanford transition [149]. The loop, that allows entry of the cofactor, was judged to be the most suited site for future mutations improving the stereoselectivity of the reactions.

The reduction of ketones by piano stool Rh/Ru complexes gave moderate enantiomeric excess (up to ee of 32%) [150–152].

Banse used a maleimide anchor for the covalent attachment of a pentadentate iron chelate in a pocket other than the fatty acid binding site. The resulting metalloenzyme catalyzed the enantioselective sulfoxidation of thioansiole in modest ee's (up to 20%) [153].

13 Xylanase

Xylanases catalyze the degradation of hemicellulose. They find extensive applications in the paper industry and in food technologies. Expression of Xylanase in *E. coli* lead to the discovery that Xylanase binds Fe-porphyrins [154].

As the binding moiety is a positively charged cleft, Mahy, and Ricoux synthesized various anionic porphyrins and tested the catalytic efficiency of the resulting hybrids [155]. Xylanase-Fe-porphyrin (Fe(T*p*CPP)) catalyzed the sulfoxidation of thioanisole with an ee of 40% [156]. The substitution of the Fe-porphyrin by a manganese analog yielded a metalloenzyme that enantioselectively epoxidized styrene in presence of oxone with a good ee of 80% [157].

14 Phytase

Phytases are phosphatases that catalyze the hydrolysis of phosphate esters. Sheldon took advantage of their affinity for phosphate and loaded the active site with isomorphous vanadate [158]. The resulting peroxidase was able to oxidize thioanisole stereoselectively up to an ee of 68% with excellent turnover numbers [159–161]. Other isostructural anions such as MoO_4^{2-}, ReO_4^{-}, SeO_4^{2-}, and WO_4^{2-} lead to lower enantiomeric excesses [161].

15 Nitrobindin

The β-barrel of nitrobindin typically harbors a heme molecule. The protein was selected as a host for a Rh(Cp)(cod) cofactor as the apo-protein keeps its rigid beta-fold despite the removal of the heme moiety [162]. Its hydrophobic moiety was considered ideal for reshaping the active site. A cysteine was introduced for covalent modification at the entrance of the pocket. While the pocket mouth is about 1–2 nm wide, the influence of the pocket size on the *cis/trans* ratio was tested by mutating small amino acids to large ones and vice versa within a distance of 6 Å around the Rh metal. Interestingly, the larger cavities promoted the *cis*-selectivity up to 82%, probably by enabling a precise positioning of the rhodium complex [163].

Nitrobindin was also explored for its suitability on oxidation reactions based on a manganese terpyridine complex. High yields were obtained for the epoxidation of various olefins and for the oxidation of benzylic and α-ethereal C–H carbons [164].

16 Ribonuclease S

The design of this artificial RNAse-metalloenzmye resembles the Trojan horse strategy. The Trojan horse strategy typically comprises the use of small molecules that bind to a protein with high affinity. In this approach the small molecule was replaced by a peptide of 20 amino acids.

The peptide can be prepared by cleaving it from the *N*-terminus of RNAse A. The resulting so-called *S*-peptide still displays high affinity for the mother protein. Ward identified two histidines on the *S*-peptide that were able to bind iridium complexes for transfer hydrogenation. Upon assembly with the mother protein, the product salsolidine was formed in higher yields but with a lower enantiomeric excess. The best stereocontrol was found with an unbound *S*-peptide [165].

17 Additional Scaffolds

The recombinant **Photoactive Yellow Protein** (POP) displays a free cysteine for modification. It is located in a hydrophobic pocket that is typically occupied by a chromophore, *p*-hydroxycinnamic acid. Kamer attached several mono- and di-phosphine ligands with the aim of creating a metalloenzyme for the nucleophilic substitution of allylacetates catalyzed by an immobilized Pd(allyl)Cl moiety [166]. $[Rh(cod)(MeCN)_2]BF_4$ was also attached to the same ligands for the stereoselective reduction of an α,β-unsaturated ester [167]. However, no enantiomeric excess was detected for the allylic substitution and for the reduction. The authors reckon that the protein denatures upon addition of the required organic solvents. The modification of another protein "**Adipocyte lipid binding protein**" (ALBP) was also tested with the phosphine ligands but the activated esters had too little chemoselectivity for cysteine [167]. Distefano circumvented this problem by applying activated iodoacetamide linkers to attach phenanthroline ligands at a cysteine inside the large cavity of ALBP. The addition of Cu(II) lead to a metalloenzyme that hydrolyzed racemic amino acid esters with enantioselectivity (up to 86%) at low turnover numbers [168]. The enantiomeric excess was even increased to 94% by the application of a related protein: **Intestinal Fatty Acid Binding Protein** [169].

Based on the Trojan horse strategy, Mahy, and Ricoux attached a Fe(III) poryphyrin-testosterone conjugate to **Neocarcinostatin-3.24**. The metalloenzyme catalyzed the sulfoxidation of thionanisole with low enantiomeric excess [170]. Molecular modeling studies suggest that the porphyrin is well sandwiched by the protein but the iron atom remains exposed to solvent. No genetic optimization was attempted.

18 Summary of Reactions

The following table (Table 1) summarizes the literature that reports asymmetric catalysis by an artificial metalloenzyme. While the protein scaffolds have been the major focus of this chapter, this summary focusses on the reaction types. Literature that did not report an enantiomeric/diastereoisomeric excess or regioselectivity is not included. Identical cofactors and reaction types were merged into a single entry. Only the best result concerning stereocontrol is listed.

The list contains 26 reaction types with an enantiomeric excess above 80%. Fifteen of these (i.e., 58%) were performed by metalloenzymes that were genetically modified to improve catalysis. 32 reactions were reported to have a lower enantiomeric excess. Only 7 (22%) of these metalloenzymes were genetically engineered to improve the stereocontrol of the reactions. This result highlights the importance of genetic optimization following the cofactor anchoring within a host protein. Of course, it is questionable whether a protein such as P450 shall appear several times in this statistics, even if the reactions are completely different. The statistical result gives a rough estimate on the importance of genetic optimization. Obviously, a wise choice of protein scaffold represents an important step toward a good stereocontrol and might significantly reduce the screening efforts for a good mutant.

19 Perspectives

Most of the artificial metalloenzymes presented require laborious protein purification prior to cofactor installation and catalysis. One way to circumvent this problem is the establishment of in vivo catalysis thus avoiding the purification procedure.

Table 1 Summary of asymmetric reactions including the ones with regioselectivity

Enantiomeric/diastereoisomeric excess	≥90%	≥80%	≥50%	≥10%	<10%
Regioselectivity	≥95:5	≥9:1	≥8:2	≥6:4	<6:4
TON	≥10k	≥1k	≥100	≥10	<10
Yield	≥90%	≥80%	≥50%	≥10%	<10%

Reaction type	Cofactor	Scaffold	ee	r.s.	d.e.	Yield	TON	Genetic modification	
Allylic substitution	Pd(η^3-allyl)-diphosphine	Streptavidin						✓	[75]
	Pd(η^3-allyl)Cl-phosphine	PYP							[166]
Amination	Fe-heme	P450						✓	[15]
Aziridination	Fe-heme	P450						✓	[16]
Cyclopropanation	Fe-heme	P450						✓	[21, 23, 26]
	Fe-heme	BSA							[23]
	Fe-heme	HSA							[23]
	Fe-heme	Myoglobin						✓	[27]
	$Rh_2(OAc)_4$-BCN	tHisF							[57]
	$Rh_2(OAc)_4$-BCN	POP							[6]
Diels–Alder	Cu(II)-phenanthroline	LmrR						✓	[137]
	Cu(II) His/His/Asp	tHisF						✓	[134]
	Cu(II) phthalocyanine	BSA							[103]
	Cu(II) phenanthroline	SCP-2L							[141]
	Ru(II) phenanthroline	Papain							[119]
Dihydroxylation	OsO_4	Streptavidin						✓	[97]
	$K_2OsO_2(OH)_4$	Laccase							[98]
Friedels–Craft alkylation	Cu(II)-bipyridinyl alanine	LmrR						✓	[139]
	Cu(II)-phenanthroline	LmrR							[140]
Heck coupling	Pd-pncer	CALB							[129]
Hydro hydroxylation	Cu(II)-phenanthroline	lmrR						✓	[138]
Hydroformylation	$Rh(CO)_2(acac)$	hCA II						✓	[69]
	$Rh(CO)_2(acac)$	BSA/HSA							[107]
Hydrogenation	$Rh(cod)_2BF_4$	hCA II						✓	[66]
	Rh(COD)-diphosphine	Streptavidin						✓	[78, 80, 82, 95, 96]
	Rh(COD)-diphosphine	Avidin							[81]
	Rh(COD)-diphosphine	Antibody 1G8							[147]
	$RhCl_3$ dipyridyl	Papain							[114]
	Rh(COD)-diphosphine	Papain							[118]
Hydrolysis	Cu(II) phenanthroline	ADBP							[168]
	Cu(II) phenanthroline	IFABP							[169]
Hydroxylation	Mn-porphycene	Myoglobin							[34]
	Fe-heme	P450						✓	[38]
Metathesis/ROMP	Hoveyda–Grubbs II	FhuA							[136]
Nitration	Fe-porphyrin	Antibody 3A3-MP8							[144]
Oxidation	Fe-heme	P450						✓	[35]
	Fe-heme	P450							[37]
	Mn-terpyridine	Nitrobindin							[164]
	Fe-heme	P450							[36]
	$MnCl_2$	hCA II							[67]
	$MnOAc_2$	hCA II							[68]
	Mn porphyrin	Xylanase							[157]

(continued)

Reaction type	Cofactor	Scaffold	ee	r.s.	d.e.	Yield	TON	Genetic modification	
	Mn ^{3}salen	Papain							[114]
Polymerization	Rh(I)(Cp)(cod)	Nitrobindin						✓	[162, 163]
S-H carbenoid insertion	Fe-heme	Myoglobin							[24]
Sulfimidation	Fe-heme	P450						✓	[19]
Sulfonamidation	Fe-heme	P450						✓	[17, 22]
	Fe/Co/Mn-heme	Myoglobin						✓	[11]
	Fe-heme	P450						✓	[14, 18]
Sulfoxidation	Fe-porphycene	Myoglobin							[34]
	Cr(III)-salenophen	Myoglobin						✓	[33]
	Cr(III)	BSA							[33]
	VO^{2+}	Streptavidin							[77]
	Mn-salen	Streptavidin							[76]
	Fe/Mn corrol	Albumins							[108]
	Fe-porphyrin	Antibody 3A3-MP8							[143]
	Ru/Sn-porphyrin	Antibody SN37.4							[142]
	Fe-Porphyrin	Antibody 7A3							[148]
	Fe(II) pentaamino-ligand	Lactoglobulin							[153]
	Fe porphyrin	Xylanase							[156]
	VO_4^{3-}	Phytase							[158–161]
	Fe porphyrin	Neocarcinostation							[170]
Suzuki–Miyaura	PdCl(cinnamyl)	Streptavidin							[74]
Transfer hydrogenation	[(η^5-Cp*)Ir(pico)Cl]	hCA II						✓	[64]
	[(η^5-Cp*)RhCl$_2$]$_2$	Streptavidin						✓	[4]
	[η^6-(arene)Ru(Biot-p-L)Cl]	Streptavidin						✓	[84–87, 94]
	[Cp*Ir-(Biot-p-L)Cl]	Streptavidin						✓	[79, 83, 90]
	Cp*RhCl dipyridyl	Lactoglobulin							[150–152]
	[Cp*IrCl$_2$]$_2$	Ribonuclease S						✓	[165]
	Cp*RhCl N^N ligands	Papain							[116, 117]

ee = enantiomeric excess, r.s. = regioselectivity, d.e. = diastereoisomeric excess, TON = turnover number

Accordingly, the screening for better mutants will be significantly accelerated. While some proteins such as hemoproteins are naturally formed, most of the other artificial metalloenzyme require smart assembling strategies. Covalent attachment strategies are likely to fail in vivo as they typically do not offer enough selectivity for the selected biomolecules. Supramolecular recognition of a cofactor based on a Trojan horse strategy may thus be most promising.

Not all metals are biocompatible. The natural abundance of Fe or Mn transition metals in native proteins underlines their biocompatibility. In contrast, iridium, rhodium, ruthenium, etc., are metals that might show non-specific high affinities for biomolecules. Thiols or nucleobases typically have a high affinity for these metals. Ward found a way to drastically improve catalysis in cell lysates by pretreating the cell mass with diamide that oxidize thiols to disulfides [88]. The technology strongly increased the performance of a Noyori-type iridium catalyst in cell lysates. Alternative strategies should be pursued to allow the genetic optimization of artificial metalloenzymes in vivo.

References

1. Murzin AG, Brenner SE, Hubbard T, Chothia C (1995) SCOP: a structural classification of proteins database for the investigation of sequences and structures. J Mol Biol 247:536–540. doi:10.1016/S0022-2836(05)80134-2
2. Pàmies O, Diéguez M, Bäckvall J-E (2015) Artificial metalloenzymes in asymmetric catalysis: Key developments and future directions. Adv Synth Catal 357:1567–1586. doi:10.1002/adsc.201500290
3. Steinreiber J, Ward TR (2008) Artificial metalloenzymes as selective catalysts in aqueous media. Coord Chem Rev 252:751–766. doi:10.1016/j.ccr.2007.09.016
4. Zimbron JM, Heinisch T, Schmid M et al (2013) A dual anchoring strategy for the localization and activation of artificial metalloenzymes based on the biotin-streptavidin technology. J Am Chem Soc 135:5384–5388. doi:10.1021/ja309974s
5. Carey JR, Ma SK, Pfister TD et al (2004) A site-selective dual anchoring strategy for artificial metalloprotein design. J Am Chem Soc 126:10812–10813. doi:10.1021/ja046908x
6. Srivastava P, Yang H, Ellis-Guardiola K, Lewis JC (2015) Engineering a dirhodium artificial metalloenzyme for selective olefin cyclopropanation. Nat Commun 6:7789. doi:10.1038/ncomms8789
7. Denard C, Ren H, Zhao H (2015) ScienceDirect improving and repurposing biocatalysts via directed evolution. Curr Opin Chem Biol 25:55–64. doi:10.1016/j.cbpa.2014.12.036
8. Jung ST, Lauchli R, Arnold FH (2011) Cytochrome P450: taming a wild type enzyme. Curr Opin Biotechnol 22:809–817. doi:10.1016/j.copbio.2011.02.008
9. Munro AW, Leys DG, McLean KJ et al (2002) P450 BM3: the very model of a modern flavocytochrome. Trends Biochem Sci 27:250–257. doi:10.1016/S0968-0004(02)02086-8
10. Hayashi T, Murata D, Makino M et al (2006) Crystal structure and peroxidase activity of myoglobin reconstituted with iron porphycene. Inorg Chem 45:10530–10536. doi:10.1021/ic061130x
11. Bordeaux M, Singh R, Fasan R (2014) Intramolecular C(sp3)H amination of arylsulfonyl azides with engineered and artificial myoglobin-based catalysts. Bioorg Med Chem 22:5697–5704. doi:10.1016/j.bmc.2014.05.015
12. Yu Y, Cui C, Liu X et al (2015) A designed metalloenzyme achieving the catalytic rate of a native enzyme. J Am Chem Soc 137:11570–11573. doi:10.1021/jacs.5b07119
13. Svastits EW, Dawson JH, Breslow R, Gellman SH (1985) Functionalized nitrogen atom transfer catalyzed by cytochrome P-450. J Am Chem Soc 107:6427–6428. doi:10.1021/ja00308a064
14. Singh R, Bordeaux M, Fasan R (2014) P450-catalyzed intramolecular sp3 C-H amination with arylsulfonyl azide substrates. ACS Catal 4:546–552. doi:10.1021/cs400893n
15. Singh R, Kolev JN, Sutera P, Fasan R (2015) Enzymatic C(sp3)-H amination: P450-catalyzed conversion of carbonazidates into oxazolidinones. ACS Catal 5:1685–1691. doi:10.1021/cs5018612
16. Farwell CC, Zhang RK, McIntosh J a., et al. (2015) Enantioselective enzyme-catalyzed aziridination enabled by active-site evolution of a cytochrome P450. ACS Cent Sci 1:89–93. doi:10.1021/acscentsci.5b00056
17. McIntosh J, Coelho PS, Farwell CC et al (2013) Enantioselective Intramolecular C-H amination catalyzed by engineered cytochrome P450 enzymes in vitro and in vivo. Angew Chemie Int Ed 52:9309–9312. doi:10.1002/anie.201304401
18. Hyster TK, Farwell CC, Buller AR et al (2014) Enzyme-controlled nitrogen-atom transfer enables regiodivergent C-H amination. J Am Chem Soc 136:15505–15508. doi:10.1021/ja509308v
19. Farwell CC, McIntosh J, Hyster TK et al (2014) Enantioselective imidation of sulfides via enzyme-catalyzed intermolecular nitrogen-atom transfer. J Am Chem Soc 136:8766–8771. doi:10.1021/ja503593n

20. Sreenilayam G, Fasan R (2015) Myoglobin-catalyzed intermolecular carbene N-H insertion with arylamine substrates. Chem Commun 51:1532–1534. doi:10.1039/C4CC08753D
21. Wang ZJ, Renata H, Peck NE et al (2014) Improved cyclopropanation activity of histidine-ligated cytochrome P450 enables the enantioselective formal synthesis of levomilnacipran. Angew Chemie Int Ed 53:6810–6813. doi:10.1002/anie.201402809
22. Renata H, Wang ZJ, Kitto RZ, Arnold FH (2014) P450-catalyzed asymmetric cyclopropanation of electron-deficient olefins under aerobic conditions. Catal Sci Technol 4:3640–3643. doi:10.1039/C4CY00633J
23. Heel T, McIntosh J, Dodani SC et al (2014) Non-natural olefin cyclopropanation catalyzed by diverse cytochrome P450s and other hemoproteins. ChemBioChem 15:2556–2562. doi:10.1002/cbic.201402286
24. Tyagi V, Bonn RB, Fasan R (2015) Intermolecular carbene S-H insertion catalysed by engineered myoglobin-based catalysts. Chem Sci 6:2488–2494. doi:10.1039/C5SC00080G
25. Wang ZJ, Peck NE, Renata H, Arnold FH (2014) Cytochrome P450-catalyzed insertion of carbenoids into N-H bonds. Chem Sci 5:598–601. doi:10.1039/C3SC52535J
26. Coelho PS, Wang ZJ, Ener ME et al (2013) A serine-substituted P450 catalyzes highly efficient carbene transfer to olefins in vivo. Nat Chem Biol 9:485–487. doi:10.1038/nchembio.1278
27. Bordeaux M, Tyagi V, Fasan R (2015) Highly diastereoselective and enantioselective olefin cyclopropanation using engineered myoglobin-based catalysts. Angew Chemie Int Ed 54:1744–1748. doi:10.1002/anie.201409928
28. Vojtěchovský J, Chu K, Berendzen J et al (1999) Crystal structures of myoglobin-ligand complexes at near-atomic resolution. Biophys J 77:2153–2174. doi:10.1016/S0006-3495(99)77056-6
29. Urban P, Truan G, Pompon D (2015) Access channels to the buried active site control substrate specificity in CYP1A P450 enzymes. Biochim Biophys Acta Gen Subj 1850:696–707. doi:10.1016/j.bbagen.2014.12.015
30. Yu X, Cojocaru V, Wade RC (2013) Conformational diversity and ligand tunnels of mammalian cytochrome P450s. Biotechnol Appl Biochem 60:134–145. doi:10.1002/bab.1074
31. Wallace S, Balskus EP (2015) Interfacing microbial styrene production with a biocompatible cyclopropanation reaction. Angew Chemie Int Ed. doi:10.1002/anie.201502185
32. Bacchi M, Berggren G, Niklas J et al (2014) Cobaloxime-based artificial hydrogenases. Inorg Chem 53:8071–8082. doi:10.1021/ic501014c
33. Ohashi M, Koshiyama T, Ueno T et al (2003) Preparation of artificial metalloenzymes by insertion of chromium(III) Schiff base complexes into apomyoglobin mutants. Angew Chemie Int Ed 42:1005–1008. doi:10.1002/anie.200390256
34. Oohora K, Kihira Y, Mizohata E et al (2013) C(sp 3)–H bond hydroxylation catalyzed by myoglobin reconstituted with manganese porphycene. J Am Chem Soc 135:17282–17285. doi:10.1021/ja409404k
35. Lewis JC, Mantovani SM, Fu Y et al (2010) Combinatorial alanine substitution enables rapid optimization of cytochrome P450BM3 for selective hydroxylation of large substrates. ChemBioChem 11:2502–2505. doi:10.1002/cbic.201000565
36. Tang WL, Li Z, Zhao H (2010) Inverting the enantioselectivity of P450pyr monooxygenase by directed evolution. Chem Commun 46:5461. doi:10.1039/c0cc00735h
37. Kim D-H, Ahn T, Jung H-C et al (2009) Generation of the human metabolite piceatannol from the anticancer-preventive agent resveratrol by bacterial cytochrome P450 BM3. Drug Metab Dispos 37:932–936. doi:10.1124/dmd.108.026484
38. Seifert A, Vomund S, Grohmann K et al (2009) Rational design of a minimal and highly enriched CYP102A1 mutant library with improved regio-, stereo- and chemoselectivity. ChemBioChem 10:853–861. doi:10.1002/cbic.200800799
39. Li M, Wong KKW, Mann S (1999) Organization of inorganic nanoparticles using biotin–streptavidin connectors. Chem Mater 11:23–26. doi:10.1021/cm980610m

40. Gálvez N, Sánchez P, Domínguez-Vera JM et al (2006) Apoferritin-encapsulated Ni and Co superparamagnetic nanoparticles. J Mater Chem 16:2757. doi:10.1039/b604860a
41. Sennuga A, van Marwijk J, Whiteley CG (2013) Multiple fold increase in activity of ferroxidase–apoferritin complex by silver and gold nanoparticles. Nanomedicine Nanotechnology Biol Med 9:185–193. doi:10.1016/j.nano.2012.05.020
42. Fan J, Yin J-J, Ning B et al (2011) Direct evidence for catalase and peroxidase activities of ferritin–platinum nanoparticles. Biomaterials 32:1611–1618. doi:10.1016/j.biomaterials.2010.11.004
43. Zhang L, Laug L, Münchgesang W et al (2010) Reducing Stress on Cells with Apoferritin-Encapsulated Platinum Nanoparticles. Nano Lett 10:219–223. doi:10.1021/nl903313r
44. Carmona U, Zhang L, Li L et al (2014) Tuning, inhibiting and restoring the enzyme mimetic activities of Pt–apoferritin. Chem Commun 50:701–703. doi:10.1039/C3CC48000C
45. Li L, Zhang L, Carmona U, Knez M (2014) Semi-artificial and bioactive ferroxidase with nanoparticles as the active sites. Chem Commun 50:8021. doi:10.1039/c4cc03477e
46. Wang M, Yin H, Fu Z et al (2014) A label-free electrochemical biosensor for microRNA detection based on apoferritin-encapsulated Cu nanoparticles. J Solid State Electrochem 18:2829–2835. doi:10.1007/s10008-014-2531-y
47. Liu X, Wei W, Yuan Q et al (2012) Apoferritin–CeO_2 nano-truffle that has excellent artificial redox enzyme activity. Chem Commun 48:3155–3157. doi:10.1039/C1CC15815E
48. Iwahori K, Yamane M, Fujita S, Yamashita I (2015) Synthesizing CdSe nanoparticles by using a low concentration of cadmium ions and the apoferritin protein cage of marine pennate diatoms. Mater Lett 160:154–157. doi:10.1016/j.matlet.2015.07.117
49. Wong KKW, Mann S (1996) Biomimetic synthesis of cadmium sulfide-ferritin nanocomposites. Adv Mater 8:928–932. doi:10.1002/adma.19960081114
50. Harada T, Yoshimura H (2014) Synthesis of rare earth doped yttrium–vanadate nanoparticles encapsulated within apoferritin. Phys Chem Chem Phys 16:14947. doi:10.1039/c4cp02131b
51. Ueno T, Suzuki M, Goto T et al (2004) Size-selective olefin hydrogenation by a Pd nanocluster provided in an apo-ferritin cage. Angew Chemie Int Ed 43:2527–2530. doi:10.1002/anie.200353436
52. Abe S, Hikage T, Watanabe Y et al (2010) Mechanism of accumulation and incorporation of organometallic Pd complexes into the protein nanocage of apo-ferritin. Inorg Chem 49:6967–6973. doi:10.1021/ic1003758
53. Abe S, Hirata K, Ueno T et al (2009) Polymerization of phenylacetylene by rhodium complexes within a discrete space of apo-ferritin. J Am Chem Soc 131:6958–6960. doi:10.1021/ja901234j
54. Abe S, Niemeyer J, Abe M et al (2008) Control of the coordination structure of organometallic palladium complexes in an apo-ferritin cage. J Am Chem Soc 130:10512–10514. doi:10.1021/ja802463a
55. Ueno T, Abe M, Hirata K et al (2009) Process of accumulation of metal ions on the interior surface of apo-ferritin: crystal structures of a series of apo-ferritins containing variable quantities of Pd(II) Ions. J Am Chem Soc 131:5094–5100. doi:10.1021/ja806688s
56. Ke Z, Abe S, Ueno T, Morokuma K (2012) Catalytic mechanism in artificial metalloenzyme: QM/MM study of phenylacetylene polymerization by rhodium complex encapsulated in apo-ferritin. J Am Chem Soc 134:15418–15429. doi:10.1021/ja305453w
57. Yang H, Srivastava P, Zhang C, Lewis JC (2014) A general method for artificial metalloenzyme formation through strain-promoted azide-alkyne cycloaddition. ChemBioChem 15:223–227. doi:10.1002/cbic.201300661
58. Harris MN (2001) Kinetic and mechanistic studies of prolyl oligopeptidase from the hyperthermophile *Pyrococcus furiosus*. J Biol Chem 276:19310–19317. doi:10.1074/jbc.M010489200
59. Mayer C, Gillingham DG, Ward TR, Hilvert D (2011) An artificial metalloenzyme for olefin metathesis. Chem Commun 47:12068. doi:10.1039/c1cc15005g

60. Grzybowski B, Ishchenko AV, Kim C-Y et al (2002) Combinatorial computational method gives new picomolar ligands for a known enzyme. Proc Natl Acad Sci 99:1270–1273. doi:10.1073/pnas.032673399
61. Monnard FW, Nogueira ES, Heinisch T et al (2013) Human carbonic anhydrase II as host protein for the creation of artificial metalloenzymes: the asymmetric transfer hydrogenation of imines. Chem Sci 4:3269. doi:10.1039/c3sc51065d
62. Monnard FW, Heinisch T, Nogueira ES et al (2011) Human carbonic anhydrase II as a host for piano-stool complexes bearing a sulfonamide anchor. Chem Commun 47:8238. doi:10.1039/c1cc10345h
63. Jude KM, Banerjee AL, Haldar MK et al (2006) Ultrahigh resolution crystal structures of human carbonic anhydrases I and II complexed with "two-prong" inhibitors reveal the molecular basis of high affinity. J Am Chem Soc 128:3011–3018. doi:10.1021/ja057257n
64. Heinisch T, Pellizzoni M, Dürrenberger M et al (2015) Improving the catalytic performance of an artificial metalloenzyme by computational design. J Am Chem Soc 137:10414–10419. doi:10.1021/jacs.5b06622
65. Zhao J, Kajetanowicz A, Ward TR (2015) Carbonic anhydrase II as host protein for the creation of a biocompatible artificial metathesase. Org Biomol Chem 13:5652–5655. doi:10.1039/C5OB00428D
66. Jing Q, Okrasa K, Kazlauskas RJ (2009) Stereoselective hydrogenation of olefins using rhodium-substituted carbonic anhydrase—a new reductase. Chem - A Eur J 15:1370–1376. doi:10.1002/chem.200801673
67. Okrasa K, Kazlauskas RJ (2006) Manganese-substituted carbonic anhydrase as a new peroxidase. Chem - A Eur J 12:1587–1596. doi:10.1002/chem.200501413
68. Fernández-Gacio A, Codina A, Fastrez J et al (2006) Transforming carbonic anhydrase into epoxide synthase by metal exchange. ChemBioChem 7:1013–1016. doi:10.1002/cbic.200600127
69. Jing Q, Kazlauskas RJ (2010) Regioselective hydroformylation of styrene using rhodium-substituted carbonic anhydrase. ChemCatChem 2:953–957. doi:10.1002/cctc.201000159
70. Weber P, Ohlendorf D, Wendoloski J, Salemme F (1989) Structural origins of high-affinity biotin binding to streptavidin. Science 80:243:85–88. doi:10.1126/science.2911722
71. González M, Argaraña CE, Fidelio GD (1999) Extremely high thermal stability of streptavidin and avidin upon biotin binding. Biomol Eng 16:67–72. doi:10.1016/S1050-3862(99)00041-8
72. Skander M, Malan C, Ivanova A, Ward TR (2005) Chemical optimization of artificial metalloenzymes based on the biotin-avidin technology: (S)-selective and solvent-tolerant hydrogenation catalysts via the introduction of chiral amino acid spacers. Chem Commun 4815. doi:10.1039/b509015f
73. Wilson ME, Whitesides GM (1978) Conversion of a protein to a homogeneous asymmetric hydrogenation catalyst by site-specific modification with a diphosphinerhodium(I) moiety. J Am Chem Soc 100:306–307. doi:10.1021/ja00469a064
74. Chatterjee A, Mallin H, Klehr J et al (2015) An enantioselective artificial suzukiase based on the biotin-streptavidin technology. Chem Sci. doi:10.1039/C5SC03116H
75. Pierron J, Malan C, Creus M et al (2008) Artificial metalloenzymes for asymmetric allylic alkylation on the basis of the biotin-avidin technology. Angew Chemie Int Ed 47:701–705. doi:10.1002/anie.200703159
76. Pordea, Mathis D, Ward TR (2009) Incorporation of biotinylated manganese-salen complexes into streptavidin: new artificial metalloenzymes for enantioselective sulfoxidation. J Organomet Chem 694:930–936. doi:10.1016/j.jorganchem.2008.11.023
77. Pordea A, Creus M, Panek J et al (2008) Artificial metalloenzyme for enantioselective sulfoxidation based on vanadyl-loaded streptavidin. J Am Chem Soc 130:8085–8088. doi:10.1021/ja8017219
78. Rusbandi UE, Lo C, Skander M et al (2007) Second generation artificial hydrogenases based on the biotin-avidin technology: Improving activity, stability and selectivity by introduction

of enantiopure amino acid spacers. Adv Synth Catal 349:1923–1930. doi:10.1002/adsc.200700022
79. Heinisch T, Langowska K, Tanner P et al (2013) Fluorescence-based assay for the optimization of the activity of artificial transfer hydrogenase within a biocompatible compartment. ChemCatChem 5:720–723. doi:10.1002/cctc.201200834
80. Skander M, Humbert N, Gradinaru J et al (2004) Artificial metalloenzymes: (strept)avidin as host for enantioselective hydrogenation by achiral biotinylated rhodium–diphosphine complexes by the 2001 Nobel Prize in chemistry, which was shared by. J Am Chem Soc 126:14411–14418
81. Lin CC, Lin CW, Chan ASC (1999) Catalytic hydrogenation of itaconic acid in a biotinylated Pyrphos–rhodium(I) system in a protein cavity. Tetrahedron: Asymmetry 10:1887–1893. doi:10.1016/S0957-4166(99)00193-7
82. Collot J, Gradinaru J, Humbert N et al (2003) Artificial metalloenzymes for enantioselective catalysis based on biotin-avidin. J Am Chem Soc 125:9030–9031. doi:10.1007/978-3-540-87757-8_5
83. Quinto T, Häussinger D, Köhler V, Ward TR (2015) Artificial metalloenzymes for the diastereoselective reduction of NAD+ to NAD_2H. Org Biomol Chem 13:357–360. doi:10.1039/C4OB02071E
84. Pordea, Creus M, Letondor C et al (2010) Improving the enantioselectivity of artificial transfer hydrogenases based on the biotin-streptavidin technology by combinations of point mutations. Inorganica Chim Acta 363:601–604. doi:10.1016/j.ica.2009.02.001
85. Creus M, Pordea A, Rossel T et al (2008) X-ray structure and designed evolution of an artificial transfer hydrogenase. Angew Chemie Int Ed 47:1400–1404. doi:10.1002/anie.200704865
86. Letondor C, Pordea A, Humbert N et al (2006) Artificial transfer hydrogenases based on the biotin-(strept)avidin technology: fine tuning the selectivity by saturation mutagenesis of the host protein. J Am Chem Soc 128:8320–8328. doi:10.1021/ja061580o
87. Letondor C, Humbert N, Ward TR (2005) Artificial metalloenzymes based on biotin-avidin technology for the enantioselective reduction of ketones by transfer hydrogenation. Proc Natl Acad Sci U S A 102:4683–4687
88. Wilson YM, Dürrenberger M, Nogueira ES, Ward TR (2014) Neutralizing the detrimental effect of glutathione on precious metal catalysts. J Am Chem Soc 136:8928–8932. doi:10.1021/ja500613n
89. Köhler V, Wilson YM, Dürrenberger M et al (2013) Synthetic cascades are enabled by combining biocatalysts with artificial metalloenzymes. Nat Chem 5:93–99. doi:10.1038/nchem.1498
90. Dürrenberger M, Heinisch T, Wilson YM et al (2011) Artificial transfer hydrogenases for the enantioselective reduction of cyclic imines. Angew Chemie Int Ed 50:3026–3029. doi:10.1002/anie.201007820
91. Schwizer F, Ko V, Du M et al (2013) Genetic optimization of the catalytic efficiency of artificial imine reductases based on biotin–streptavidin technology. ACS Catal 3:1752–1755
92. Robles VM, Dürrenberger M, Heinisch T et al (2014) Structural, kinetic, and docking studies of artificial imine reductases based on biotin-streptavidin technology: an induced lock-and-key hypothesis. J Am Chem Soc 136:15676–15683. doi:10.1021/ja508258t
93. Lo C, Ringenberg MR, Gnandt D et al (2011) Artificial metalloenzymes for olefin metathesis based on the biotin-(strept)avidin technology. Chem Commun (Camb) 47:12065–12067. doi:10.1039/c1cc15004a
94. Pordea A, Ward TR (2008) Chemogenetic protein engineering: an efficient tool for the optimization of artificial metalloenzymes. Chem Commun (Camb) 4239–4249. doi:10.1039/b806652c
95. Reetz MT, Rentzsch M, Pletsch A et al (2007) Directed evolution of enantioselective hybrid catalysts: a novel concept in asymmetric catalysis. Tetrahedron 63:6404–6414. doi:10.1016/j.tet.2007.03.177

96. Reetz MT, Peyralans JJ-P, Maichele A, et al. (2006) Directed evolution of hybrid enzymes: evolving enantioselectivity of an achiral Rh-complex anchored to a protein. Chem Commun 4318. doi:10.1039/b610461d
97. Köhler V, Mao J, Heinisch T et al (2011) OsO_4·streptavidin: a tunable hybrid catalyst for the enantioselective cis-dihydroxylation of olefins. Angew Chemie Int Ed 50:10863–10866. doi:10.1002/anie.201103632
98. Konieczny S, Leurs M, Tiller JC (2015) Polymer enzyme conjugates as chiral ligands for sharpless dihydroxylation of alkenes in organic solvents. ChemBioChem 16:83–90. doi:10.1002/cbic.201402339
99. Yoshikawa H, Hirano A, Arakawa T, Shiraki K (2012) Effects of alcohol on the solubility and structure of native and disulfide-modified bovine serum albumin. Int J Biol Macromol 50:1286–1291. doi:10.1016/j.ijbiomac.2012.03.014
100. Curry S, Mandelkow H, Brick P, Franks N (1998) Crystal structure of human serum albumin complexed with fatty acid reveals an asymmetric distribution of binding sites. Nat Struct Biol 5:827–835. doi:10.1038/1869
101. Rousselot-Pailley P, Bochot C, Marchi-Delapierre C et al (2009) The protein environment drives selectivity for sulfide oxidation by an artificial metalloenzyme. ChemBioChem 10:545–552. doi:10.1002/cbic.200800595
102. Li G, Zhang H, Wang R et al (2013) Preparation and antioxidant activity of albumin binding Salen Schiff-base metal complexes. Chinese Sci Bull 58:2956–2963. doi:10.1007/s11434-013-5787-1
103. Reetz MT, Jiao N (2006) Copper-Phthalocyanine conjugates of serum albumins as enantioselective catalysts in Diels-Alder reactions. Angew Chemie Int Ed 45:2416–2419. doi:10.1002/anie.200504561
104. Zunszain P, Ghuman J, Komatsu T et al (2003) Crystal structural analysis of human serum albumin complexed with hemin and fatty acid. BMC Struct Biol 3:6. doi:10.1186/1472-6807-3-6
105. Mahammed A, Gray HB, Weaver JJ et al (2004) Amphiphilic corroles bind tightly to human serum albumin. Bioconjug Chem 15:738–746. doi:10.1021/bc034179p
106. Kokubo T, Sugimoto T, Uchida T et al (1983) The bovine serum albumin-2-phenylpropane-1,2-diolatodioxo-osmium(VI) complex as an enantioselective catalyst for cis-hydroxylation of alkenes. J Chem Soc Chem Commun 769–770
107. Bertucci C, Botteghi C, Giunta D et al (2002) Aqueous biphasic hydroformylation catalysed by protein-rhodium complexes. Adv Synth Catal 344:556. doi:10.1002/1615-4169(200207)344:5<556:AID-ADSC556>3.0.CO;2-E
108. Marchetti M, Mangano G, Paganelli S, Botteghi C (2000) A protein-rhodium complex as an efficient catalyst for two-phase olefin hydroformylation. Tetrahedron Lett 41:3717–3720. doi:10.1016/S0040-4039(00)00473-1
109. Marchetti M, Minello F, Paganelli S, Piccolo O (2010) Aqueous biphasic hydrogenations catalyzed by rhodium and iridium complexes modified with human serum albumin. Appl Catal A Gen 373:76–80. doi:10.1016/j.apcata.2009.10.042
110. Mahammed A, Gross Z (2005) Albumin-conjugated corrole metal complexes: extremely simple yet very efficient biomimetic oxidation systems. J Am Chem Soc 127:2883–2887. doi:10.1021/ja045372c
111. Kobayashi K, Ohmura T (1998) Physicochemical and immunochemical properties of recombinant human serum albumin from *Pichia pastoris*. Anal Biochem 256:56–62. doi:10.1006/abio.1997.2480
112. Albanese DCM, Gaggero N (2015) Albumin as a promiscuous biocatalyst in organic synthesis. RSC Adv 5:10588–10598. doi:10.1039/C4RA11206G
113. McGrath ME (1999) The lyosomal cysteine proteases. Annu Rev Biophys Biomol Struct 28:181–204. doi:10.1146/annurev.biophys.28.1.181
114. Reetz MT, Rentzsch M, Pletsch A, Maywald M (2002) Towards the directed evolution of hybrid catalysts. Chim Int J Chem 56:721–723. doi:10.2533/000942902777679920

115. Haquette P, Talbi B, Barilleau L et al (2011) Chemically engineered papain as artificial formate dehydrogenase for NAD(P)H regeneration. Org Biomol Chem 9:5720. doi:10.1039/c1ob05482a
116. Madern N, Talbi B, Salmain M (2013) Aqueous phase transfer hydrogenation of aryl ketones catalysed by achiral ruthenium(II) and rhodium(III) complexes and their papain conjugates. Appl Organomet Chem 27:6–12. doi:10.1002/aoc.2929
117. Madern N, Queyriaux N, Chevalley A et al (2015) Piano-stool d6-rhodium(III) complexes of chelating pyridine-based ligands and their papain bioconjugates for the catalysis of transfer hydrogenation of aryl ketones in aqueous medium. J Mol Catal B Enzym 122:314–322. doi:10.1016/j.molcatb.2015.10.007
118. Panella L, Broos J, Jin J et al (2005) Merging homogeneous catalysis with biocatalysis; papain as hydrogenation catalyst. Chem Commun 23:5656. doi:10.1039/b512138h
119. Talbi B, Haquette P, Martel A et al (2010) (η6-Arene) ruthenium(ii) complexes and metallo-papain hybrid as Lewis acid catalysts of Diels-Alder reaction in water. Dalt Trans 39:5605. doi:10.1039/c001630f
120. Matsuo T, Imai C, Yoshida T et al (2012) Creation of an artificial metalloprotein with a Hoveyda-Grubbs catalyst moiety through the intrinsic inhibition mechanism of α-chymotrypsin. Chem Commun 48:1662. doi:10.1039/c2cc16898g
121. Martinez C, De Geus P et al (1992) Fusarium solani cutinase is a lipolytic enzyme with a catalytic serine accessible to solvent. Nature 356:615–618. doi:10.1038/356615a0
122. Zisis T, Freddolino PL, Turunen P et al (2015) Interfacial activation of *Candida antarctica* lipase B: combined evidence from experiment and simulation. Biochemistry 54:5969–5979. doi:10.1021/acs.biochem.5b00586
123. Stepankova V, Damborsky J, Chaloupkova R (2015) Hydrolases in non-conventional media: implications for industrial biocatalysis. In: Grunwald P (ed) Industrial biocatalysis. Pan Stanford Publishing Pte Ltd, Germany, pp 583–630
124. Kruithof C., Casado M, Guillena G et al (2005) Lipase active-site-directed anchoring of organometallics: metallopincer/protein hybrids. Chem Eur J 11:6869–6877. doi:10.1002/chem.200500671
125. Rutten L, Wieczorek B, Mannie J-PBA et al (2009) Solid-state structural characterization of cutinase-ECE-pincer-metal hybrids. Chem Eur J 15:4270–4280. doi:10.1002/chem.200801995
126. Wieczorek B, Snelders DJM, Dijkstra HP et al (2012) Coordination chemistry in water of a free and a lipase-embedded cationic NCN-pincer platinum center with neutral and ionic triarylphosphines. Organometallics 31:2810–2820. doi:10.1021/om2010832
127. Basauri-Molina M, Verhoeven DG, van Schaik AJ et al (2015) Ring-closing and cross-metathesis with artificial metalloenzymes created by covalent active site-directed hybridization of a lipase. Chem Eur J. doi:10.1002/chem.201502381
128. Basauri-Molina M, Riemersma CF, Würdemann M et al (2015) Lipase active site covalent anchoring of Rh(NHC) catalysts: towards chemoselective artificial metalloenzymes. Chem Commun 51:6792–6795. doi:10.1039/C4CC09700A
129. Filice M, Romero O, Aires A et al (2015) Preparation of an immobilized lipase-palladium artificial metalloenzyme as catalyst in the Heck reaction: role of the solid phase. Adv Synth Catal 357:2687–2696. doi:10.1002/adsc.201500014
130. Wieczorek B, Träff A, Krumlinde P et al (2011) Covalent anchoring of a racemization catalyst to CALB-beads: towards dual immobilization of DKR catalysts. Tetrahedron Lett 52:1601–1604. doi:10.1016/j.tetlet.2011.01.106
131. Wikmark Y, Svedendahl Humble M, Bäckvall J-E (2015) Combinatorial library based engineering of *Candida antarctica* lipase A for enantioselective transacylation of sec-alcohols in organic solvent. Angew Chemie Int Ed 54:4284–4288. doi:10.1002/anie.201410675
132. Engström K, Nyhlén J, Sandström AG, Bäckvall JE (2010) Directed evolution of an enantioselective lipase with broad substrate scope for hydrolysis of α-substituted esters. J Am Chem Soc 132:7038–7042. doi:10.1021/ja100593j

133. Sterner R, Höcker B (2005) Catalytic versatility, stability, and evolution of the $(\beta\alpha)_8$-barrel enzyme fold. Chem Rev 105:4038–4055. doi:10.1021/cr030191z
134. Podtetenieff J, Taglieber A, Bill E et al (2010) An artificial metalloenzyme: creation of a designed copper binding site in a thermostable protein. Angew Chemie Int Ed 49:5151–5155. doi:10.1002/anie.201002106
135. Reetz MT, Rentzsch M, Pletsch A et al (2008) A robust protein host for anchoring chelating ligands and organocatalysts. ChemBioChem 9:552–564. doi:10.1002/cbic.200700413
136. Philippart F, Arlt M, Gotzen S et al (2013) A hybrid ring-opening metathesis polymerization catalyst based on an engineered variant of the β-barrel protein FhuA. Chemistry 19:13865–13871. doi:10.1002/chem.201301515
137. Bos J, García-Herraiz A, Roelfes G (2013) An enantioselective artificial metallo-hydratase. Chem Sci 4:3578. doi:10.1039/c3sc51449h
138. Bos J, Fusetti F, Driessen AJM, Roelfes G (2012) Enantioselective artificial metalloenzymes by creation of a novel active site at the protein dimer interface. Angew Chemie Int Ed 51:7472–7475. doi:10.1002/anie.201202070
139. Drienovská I, Rioz-Martínez A, Draksharapu A, Roelfes G (2015) Novel artificial metalloenzymes by in vivo incorporation of metal-binding unnatural amino acids. Chem Sci 6:770–776. doi:10.1039/C4SC01525H
140. Bos J, Browne WR, Driessen AJM, Roelfes G (2015) Supramolecular assembly of artificial metalloenzymes based on the dimeric protein LmrR as promiscuous scaffold. J Am Chem Soc 137:9796–9799. doi:10.1021/jacs.5b05790
141. Deuss PJ, Popa G, Slawin AMZ et al (2013) Artificial copper enzymes for asymmetric Diels-Alder reactions. ChemCatChem 5:1184–1191. doi:10.1002/cctc.201200671
142. Nimri S, Keinan E (1999) Antibody–metalloporphyrin catalytic assembly mimics natural oxidation enzymes. J Am Chem Soc 121:8978–8982. doi:10.1021/ja990314q
143. Ricoux R, Lukowska E, Pezzotti F, Mahy J-P (2004) New activities of a catalytic antibody with a peroxidase activity. Eur J Biochem 271:1277–1283. doi:10.1111/j.1432-1033.2004.04032.x
144. Ricoux R, Girgenti E, Sauriat-Dorizon H et al (2002) Regioselective nitration of phenol induced by catalytic antibodies. J Protein Chem 21:473–477. doi:10.1023/A:1021351120772
145. Raffy Q, Ricoux R, Sansiaume E et al (2010) Coordination chemistry studies and peroxidase activity of a new artificial metalloenzyme built by the "Trojan horse" strategy. J Mol Catal A: Chem 317:19–26. doi:10.1016/j.molcata.2009.10.016
146. Raffy Q, Ricoux R, Mahy J-P (2008) Synthesis of a new estradiol–iron metalloporphyrin conjugate used to build up a new hybrid biocatalyst for selective oxidations by the "Trojan horse" strategy. Tetrahedron Lett 49:1865–1869. doi:10.1016/j.tetlet.2008.01.022
147. Yamaguchi H, Hirano T, Kiminami H et al (2006) Asymmetric hydrogenation with antibody-achiral rhodium complex. Org Biomol Chem 4:3571. doi:10.1039/b609242j
148. Sansiaume E, Ricoux R, Gori D, Mahy J-P (2010) Oxidation of organic molecules in homogeneous aqueous solution catalyzed by hybrid biocatalysts (based on the Trojan Horse strategy). Tetrahedron Asymmetry 21:1593–1600. doi:10.1016/j.tetasy.2010.03.050
149. Qin BY, Bewley MC, Creamer LK et al (1998) Structural basis of the Tanford transition of bovine β-lactoglobulin. Biochemistry 37:14014–14023. doi:10.1021/bi981016t
150. Cherrier MV, Engilberge S, Amara P et al (2013) Structural basis for enantioselectivity in the transfer hydrogenation of a ketone catalyzed by an artificial metalloenzyme. Eur J Inorg Chem 2013:3596–3600. doi:10.1002/ejic.201300592
151. Chevalley A, Cherrier MV, Fontecilla-Camps JC et al (2014) Artificial metalloenzymes derived from bovine β-lactoglobulin for the asymmetric transfer hydrogenation of an aryl ketone – synthesis, characterization and catalytic activity. Dalt Trans 43:5482. doi:10.1039/c3dt53253d
152. Chevalley A, Salmain M (2012) Enantioselective transfer hydrogenation of ketone catalysed by artificial metalloenzymes derived from bovine β-lactoglobulin. Chem Commun 48:11984. doi:10.1039/c2cc36980j

153. Buron C, Sénéchal-David K, Ricoux R et al (2015) An artificial enzyme made by covalent grafting of an Fe II complex into β-lactoglobulin: molecular chemistry, oxidation catalysis, and reaction-intermediate monitoring in a protein. Chem Eur J 21:12188–12193. doi:10.1002/chem.201501755
154. Komatsu T, Ishihara S, Tsuchida E et al (2005) Heat-resistant oxygen-carrying hemoproteins consist of recombinant xylanases and synthetic iron(II) porphyrin. Biomacromolecules 6:1489–1494. doi:10.1021/bm049255l
155. Ricoux R, Dubuc R, Dupont C et al (2008) Hemozymes peroxidase activity of artificial hemoproteins constructed from the *Streptomyces lividans* xylanase A and iron(III)-carboxy-substituted porphyrins. Bioconjug Chem 19:899–910. doi:10.1021/bc700435a
156. Ricoux R, Allard M, Dubuc R et al (2009) Selective oxidation of aromatic sulfide catalyzed by an artificial metalloenzyme: new activity of hemozymes. Org Biomol Chem 7:3208. doi:10.1039/b907534h
157. Allard M, Dupont C, Muñoz Robles V et al (2012) Incorporation of manganese complexes into xylanase: new artificial metalloenzymes for enantioselective epoxidation. ChemBioChem 13:240–251. doi:10.1002/cbic.201100659
158. van de Velde F, Könemann L (1998) Enantioselective sulfoxidation mediated by vanadium-incorporated phytase: a hydrolase acting as a peroxidase. Chem Commun 1891–1892. doi:10.1039/a804702b
159. Van De Velde F, Arends IWCE, Sheldon R (2000) Biocatalytic and biomimetic oxidations with vanadium. J Inorg Biochem 80:81–89. doi:10.1016/S0162-0134(00)00043-X
160. van de Velde F, Könemann L, van Rantwijk F, Sheldon R (2000) The rational design of semisynthetic peroxidases. Biotechnol Bioeng 67:87–96. doi:10.1002/(SICI)1097-0290(20000105)67:1<87::AID-BIT10>3.0.CO;2-8
161. van de Velde F, Arends IWCE, Sheldon R (2000) Vanadium-catalysed enantioselective sulfoxidations: rational design of biocatalytic and biomimetic systems. Top Catal 13:259–265. doi:10.1023/A:1009094619249
162. Onoda A, Fukumoto K, Arlt M et al (2012) A rhodium complex-linked β-barrel protein as a hybrid biocatalyst for phenylacetylene polymerization. Chem Commun 48:9756. doi:10.1039/c2cc35165j
163. Fukumoto K, Onoda A, Mizohata E et al (2014) Rhodium-complex-linked hybrid biocatalyst: stereo-controlled phenylacetylene polymerization within an engineered protein cavity. ChemCatChem. doi:10.1002/cctc.201301055
164. Zhang C, Srivastava P, Ellis-Guardiola K, Lewis JC (2014) Manganese terpyridine artificial metalloenzymes for benzylic oxygenation and olefin epoxidation. Tetrahedron 70:4245–4249. doi:10.1016/j.tet.2014.03.008
165. Genz M, Köhler V, Krauss M et al (2014) An artificial imine reductase based on the ribonuclease S Scaffold. ChemCatChem 6:736–740. doi:10.1002/cctc.201300995
166. Laan W, Muñoz BK, den Heeten R, Kamer PCJ (2010) Artificial metalloenzymes through cysteine-selective conjugation of phosphines to photoactive yellow protein. ChemBioChem 11:1236–1239. doi:10.1002/cbic.201000159
167. den Heeten R, Muñoz BK, Popa G et al (2010) Synthesis of hybrid transition-metalloproteins via thiol-selective covalent anchoring of Rh-phosphine and Ru-phenanthroline complexes. Dalt Trans 39:8477. doi:10.1039/c0dt00239a
168. Davies RR, Distefano MD (1997) A semisynthetic metalloenzyme based on a protein cavity that catalyzes the enantioselective hydrolysis of ester and amide substrates. J Am Chem Soc 119:11643–11652. doi:10.1021/ja970820k
169. Davies RR, Kuang H, Qi D et al (1999) Artificial metalloenzymes based on protein cavities: exploring the effect of altering the metal ligand attachment position by site directed mutagenesis. Bioorg Med Chem Lett 9:79–84. doi:10.1016/S0960-894X(98)00684-2
170. Sansiaume-Dagousset E, Urvoas A, Chelly K et al (2014) Neocarzinostatin-based hybrid biocatalysts for oxidation reactions. Dalt Trans 43:8344. doi:10.1039/c4dt00151f

Metal Complexes and Imprinted Polymers for Shape-Selective Catalysis

Fosca Mirata and Marina Resmini

Abstract Polymers provide a very interesting matrix for the development of specific functionalities. The incorporation of metal atom complexes in polymer systems using the molecular imprinting approach allows to achieve key specificities that could not otherwise be obtained. This chapter provides an overview of the latest achievements in terms of imprinted polymers containing metal atoms and their applications in the important field of catalysis.

1 Introduction

Biological processes in nature, such as DNA transcriptions, enzyme catalysis, antibody binding, greatly rely on the process of molecular recognition. Researchers have invested considerable time in the study of the mechanisms that are involved in molecular recognition with the aim of understanding the roles played by the different components but also to use such knowledge for the development of new systems. The ability to obtain artificial receptors with specific recognition properties has been an attractive target with great potential for applications in a variety of areas, such as catalysis, sensing, biological assaying, and separation of complex chemical mixtures [1–3].

Among the different approaches that have been investigated for the development of artificial receptors, the molecular imprinting technology has proved very successful, in particular with polymeric materials [4, 5]. A large number of examples are available in literature, covering in particular the use of molecular imprinting for the development of sensors, [6–8] however, in the area of catalysis progress has been much slower, [9, 10] as a result of the challenges linked with obtaining efficient and specific catalysts. In recent years developments in this field have led to important achievements. This chapter will introduce the basic concept of molecular

F. Mirata · M. Resmini (✉)
School of Biological and Chemical Sciences, Queen Mary University of London, Mile End Road, London E1 4NS, UK
e-mail: m.resmini@qmul.ac.uk

R. Poli (ed.), *Effects of Nanoconfinement on Catalysis*,
Fundamental and Applied Catalysis, DOI 10.1007/978-3-319-50207-6_4

imprinting and highlight recent advances in the field of molecularly imprinted polymers (MIPs) for applications in catalysis. It will emphasize how the use of metal complexes embedded into the polymeric cavities has led to matrices with excellent activities. It will also include recent achievements in the important area of surface imprinting to yield surface molecularly imprinted polymers (SMIPs) and highlight some important applications methodologies for the preparation of nano-reactors, and photocatalytic degradations of pollutants.

2 The Molecular Imprinting Technology

The molecular imprinting approach involves a casting process, where functional monomers and cross-linkers are copolymerised in the presence of the target analyte, also known as template. The newly formed polymer is then washed under particular conditions to ensure that the template molecule is removed, leaving cavities or recognition sites that should be complementary in shape and functionality to the template used. Such cavities contain specific functional groups that are covalently linked to the polymeric matrix and are then able to rebind the analyte with affinities and specificities that are dependent on the strength of the interactions. The key step for successful imprinting is the formation of a tight complex between the functional monomer and the template, which can be achieved by either covalent or non-covalent interactions (or combinations of both), thereby allowing considerable flexibility in the choice of monomers and the types of templates that can be imprinted (Fig. 1).

One of the main issues with catalytic MIPs has always been the difficulty in achieving high efficiency. An interesting approach has focused on the use of metals, inserted in the cavities [11, 12]. Metals play important roles in catalysis, they can activate the substrate, stabilize the transition state and any leaving groups or activate the nucleophiles, even at neutral pH [13]. In addition, considerable effort has been put into the application of the imprinting approach with a variety of polymeric formats. Although a large part of the initial literature on catalytic imprinted polymers focused on 'bulk' materials [14, 15], it became clear that such rigid system was not ideal for the mimic for enzymes, which are naturally quite flexible and

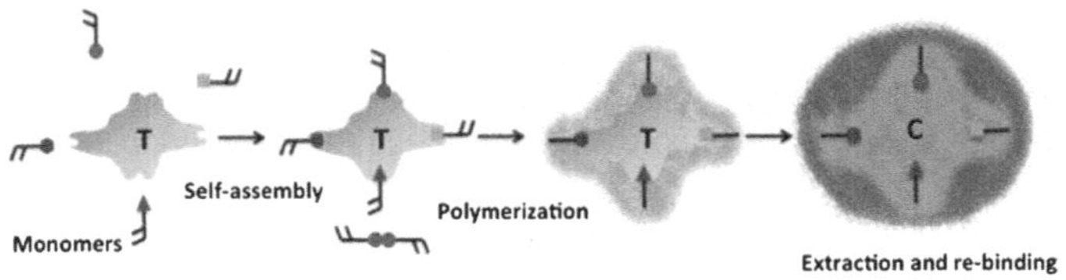

Fig. 1 Schematic representation of the imprinting process where T represent the template and C the complementary cavities *left* behind after removing the template

subject to conformational changes. Some interesting work was developed on nanogels, which showed catalytic activity and selectivity [16–18].

3 Surface Imprinting: An Interesting New Approach

In recent years the surface imprinting approach has been developed in an attempt to overcome the drawbacks of bulk polymers for catalytic applications. These disadvantages include slow rebinding kinetics due to poor accessibility of the polymer matrix, recognition sites to the binding site on the target molecule and difficulty in handling of water-soluble substances. Furthermore, the presence of residual target molecules in the imprinted polymer matrix can be overcome by providing more accessible binding sites on the surface of the matrix. Surface imprinted polymers are prepared by polymerization of water-in-oil emulsions using an emulsion stabilizer in combination with the functional monomers, a cross-linker and the template molecule. The aqueous–organic interface acts as a recognition field for the water-soluble target molecules. These molecules form a complex with the functional monomer and rearrange in a regular pattern by fixing the orientation of the monomer at the interface. After polymerization and template removal, complementary recognition sites for the target molecule can be found on the surfaces of the imprinted polymer. To control the hydrophobicity of the complex and ensure adequate balance, an amphiphilic functional monomer, such as a surfactant is required. In this way, a high imprinting efficiency for the target molecule can be achieved [19–21].

The application of MIPs in catalysis helps to reduce the amount of waste generated during a particularly difficult synthesis, increase the efficiency, and selectivity of the process and obtain a higher catalyst recovery. The synthesis of *p*-chlorotoluene, for instance, is carried out by chlorination of toluene catalyzed by Lewis acid catalysts. Along with *p*-chlorotoluene, which is commercially valuable as an intermediate in the manufacturing of dyes, pharmaceuticals and pesticides, its *ortho-* isomer is also produced. Large volumes of water are required to neutralize the Lewis acid and separate it from the products, generating large amounts of aqueous waste and making recovery of the catalyst impossible. In Meng et al. [22] used the surface imprinting technique to prepare a catalyst with the capacity to overcome the issues in the synthesis of *p*-chlorotoluene. Moreover, the presence of the binding sites on the surface ensures rapid binding kinetics due to a greater number of accessible sites. The authors used silica microspheres as a supporting matrix, where a C=C bond was then introduced onto the surface by silanization of activated silica. Following the formation of the complex between the functional monomer (acrylamide) and the product analo, *p*-hydroxybenzoic acid (*p*-HB) as the template, polymerization of the complex was conducted in the presence of the cross-linker ethylene glycol dimethacrylate (EGDMA), the initiator 2,2′-azobisisobutyronitrile (AIBN) and the silanized silica. The template was then removed to obtain the final surface imprinted polymer catalyst, *p*-HB-SMIP. The catalyst was characterized by SEM, FT-IR, TG and HPLC-UV. UV-Vis spectroscopic studies of interactions and stability demonstrated that the complex with the functional monomer-template ratio of 1:2 predominated in the prepolymerization mixture. Moreover, the absorption kinetic curve showed that the absorption of *p*-HB on the *p*-HB-SMIP increased rapidly with time during the first 50 min and reached equilibrium at approximately 100 min, thus proving the existence of imprinted cavities on the surface of the material and facilitating entry of the template into the cavities to bind the recognition site. The authors carried out additional studies to investigate the adsorption efficiency of the system. The adsorption isotherm of *p*-HB-SMIP showed that the equilibrium adsorption capacity of the catalyst increased gradually with the concentration of *p*-HB in the initial solution until the system reached saturation. For the purposes of comparison, the surface non-imprinted polymer (SNIP) was prepared in the absence of the template and treated according to the same method used for the SMIP. Adsorption capacity analysis showed no significant differences in the adsorption levels between the *p*-HB-SMIP and SNIP although the differences became more marked with increasing *p*-HB concentrations, reaching a maximum adsorption capacity of 41 μmol/g for the SMIP in contrast to 17 μmol/g for the SNIP. Analysis of the catalyst thermodynamic parameters showed that the template adsorption was an endothermic ($\Delta H^\circ > 0$), spontaneous ($\Delta G^\circ < 0$) process with increased randomness at the solid-solution interface ($\Delta S^\circ > 0$). In addition to the studies on catalytic activity, the catalyst selectivity was also evaluated by HPLC. A mixed solution of several substrates was treated with the *p*-HB-SMIP, showing excellent selectivity towards the *p*-HB template relative to the other competitive species in the mixture. Investigations of the catalytic activity showed that the *p*-HB-SMIP exhibited a higher catalytic activity and selectivity for the chlorination of toluene to *p*-chlorotoluene (85.5% conversion of toluene and a *p*/*o* ratio of 1.38) compared to those of the other catalysts used. In conclusion, Meng and co-workers successfully prepared a surface imprinted polymer with high catalytic activity and product selectivity with an increased ratio in favor of the desired isomer.

In the last few years there has been a concerted effort to address the issue of water, soil, or air pollution. Researchers have attempted to develop several efficient physical and chemical approaches to degrade pollutants that are of particular concern given their threat to human health and the environment. These approaches include adsorption [23–25] and chemical precipitation [26, 27]. The most promising degradation technique, however, is based on photocatalysis. This process uses the adsorption of sunlight or artificial light by a suitable substrate to promote molecular degradation. Titanium dioxide (TiO_2), which is the most widely used photocatalyst, is inexpensive, readily available, nontoxic, chemically and mechanically stable, and has a high turnover. TiO_2 is wavelength-selective and its activity is accelerated by UV light, which covers only 4% of the solar light spectrum on earth. However, it is possible to dope the photocatalyst with either metal or nonmetal ions to enhance its utilization efficiency under visible light irradiation, hence shifting its optical response from the UV to the visible range [28, 29]. Furthermore, conductive polymers that have a high absorption coefficient in visible light are also used as stable photosensitizers to modify TiO_2. Polypyrrole (PPy) is the most promising and important conducting polymer due to its electrochemical reversibility, relative environmental stability, and the ease of preparation [30]. In the presence of PPy, TiO_2 can be excited with UV-Vis irradiation leading to capture of the photogenerated electrons by PPy, which are quickly transferred to the counter electrode. Thus, enhanced photoelectrochemical performance is expected from the incorporation of a PPy conducting polymer into a TiO_2 photoelectrode. By 2012, Deng and co-workers had already published a study showing the photocatalytic activity of PPy/TiO_2 nanocomposites prepared by surface molecular imprinting [31]. By using the surface molecular imprinting technique, characteristics, such as high affinity for target contaminants and improved accessibility to binding sites further increase the efficiency with which the catalyst degrades pollutants. The molecularly imprinted PPy was deposited on the surface of TiO_2 nanoparticles to form MIP-PPy/TiO_2, as shown in Fig. 2.

The final material showed higher adsorption capacity and selectivity compared with PPy/TiO_2 nanoparticles, which are hydrophobic and exhibit low affinity toward contaminants, hence confirming the positive effect of the incorporation of molecular imprinting in the catalyst. Moreover, the photocatalytic activity was enhanced due to the improved accessibility of the complementary cavities on the surface of the nanocomposite. Furthermore, the presence of the conducting polymer PPy increased the optical response of TiO_2, allowing a response in visible light. Nevertheless, the catalyst was not easily recovered.

Two years later, inspired by Deng's work, Wei et al. overcame the recovery issue by preparing PPy/TiO_2 magnetic nanoparticles (MNPs) for use as a photocatalyst in the degradation of Congo red (CR). This is a synthetic dye that, when discharged into environmental water, has a severe influence on both the environment and human health [32]. Figure 3 shows how the photocatalyst was prepared. Fe_3O_4 nanoparticles were used as a magnetic core to facilitate the recycling of the catalyst by an external magnetic field. The core was then coated with SiO_2 to reduce the risk of leaching of iron. An outer TiO_2 layer was then applied on the surface

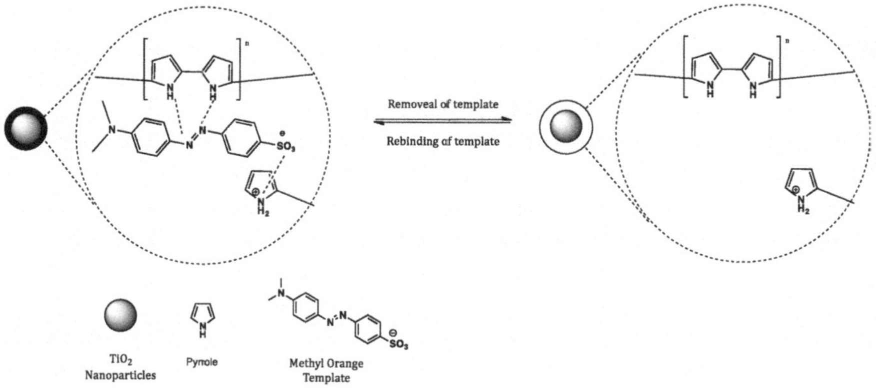

Fig. 2 Schematic preparation of MIP-PPy/TiO_2 photocatalyst. The removal of the template leaves empty cavities on the surface of nanoparticles, which are able to recognize and bind the analyte

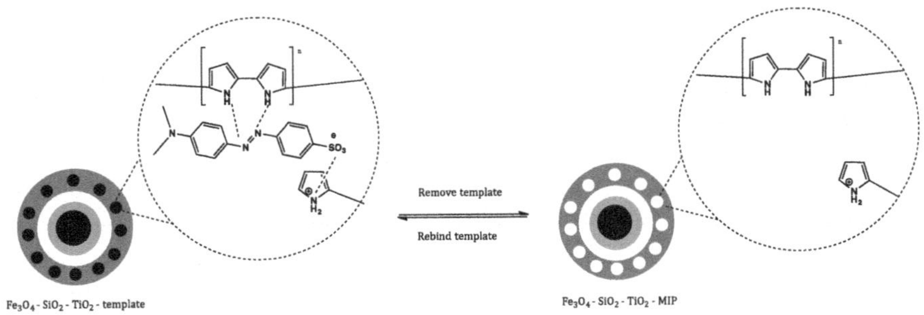

Fig. 3 Schematic preparation of Fe_3O_4-SiO_2-TiO_2-MIP photocatalyst

followed by a coating of a conductive molecularly imprinted PPy obtained by oxidative polymerization with $FeCl_3$. During the imprinting procedure, a precursor was formed between the chosen template, methyl orange (MO), and the pyrrole.

A series of tests was performed on the newly obtained material, after being characterized by FT-IR, TGA, XRD, and TEM. UV-Vis spectrophotometric investigation of the binding properties of the imprinted photocatalyst showed that the amount of MO bound to the MNPs increased rapidly with increases in the initial concentration of MO until it reached saturation (approximately 20 mg/L). The binding capacity of the imprinted MNPs was much higher than that of the non-imprinted MNPs, demonstrating the presence of the cavities. The MNPs were applied for the degradation of CR in water, a process that can be affected by several parameters such as pH, temperature, oxygen level and oscillation. Investigation of the effects of pH showed that at pH < 4.0, the degradation rate decreased from 67.7 to 47%, possibly because the polypyrrole is protonated at these pH values, resulting in the presence of a high positive charge density on the surface of catalyst. This causes increased adsorption of CR via the electrostatic interactions between the protonated polypyrroles and the SO_3^- groups of CR. Within the pH range 6–7, the acidity is close to that of the template solution, consequently allowing the CR molecules to enter the imprinted cavities and bind with the inner polypyrroles through hydrogen bonds. This results in increased adsorption of CR, with degradation rates at these pH values exceeding 90%. However, at pH > 7.0, the degradation rate was reduced to 47%. The negative charge on the surface of the catalyst greatly reduces the adsorption of CR due to the repulsions between negatively charged surface of the catalyst and the SO_3^- groups of CR. Therefore, higher adsorption of CR resulted in higher degradation rates. Investigation of the effect of temperature showed that the degradation rate increased with the temperature in accordance with the theoretical prediction of the influence of temperature on reaction rates. The authors also showed that the oxygen dissolved in solution influenced the rate of CR degradation, with the rate reaching 98% in the presence of oxygen in solution, while the rate was reduced to 67.8% when the solution was saturated with nitrogen. The oscillation rate of the mixture also affected the degradation of CR. With increased oscillation rates of 60, 180, and 300 time/min, the degradation rate increased to 66.8, 79.1, and 92.0%, respectively. The higher oscillation rates were beneficial for the transfer and adsorption of CR molecules to the surface of the catalyst, resulting in higher degradation rates. In conclusion, the authors obtained a new catalyst with several advantages, such as the absence of a requirement for illumination, ease of recycling and high catalytic activity at ambient conditions (in the dark, at room temperature and atmospheric pressure). This catalyst has the potential for application to the rapid, "green" and low-cost degradation of CR produced in industrial printing and as wastewater during dyeing processes.

4 Surface Imprinted Polymers Containing Metals

Led by the aspiration to contribute to the reduction of contaminants in both water and soil, Carboni et al. studied the hydrolysis of the commercially available pesticide, Paraoxon. The organophosphate oxon is one of the most potent acetylcholinesterase-inhibiting insecticides currently available; however, due to its extreme toxicity in humans and animals, it is now rarely used [33]. This group generated molecularly imprinted mesoporous titania film (MIF) characterized by hydrolytic properties toward organophosphates. Lanthanum is known to interact with high affinity with organophosphates; in fact, in 2004 the FDA (Food and Drug Administration) approved lanthanum carbonate as a phosphate binder to absorb excess phosphate in cases of end-stage renal disease. Therefore, Carboni and co-workers doped titania film with lanthanum by dispersing it in the matrix to act as the catalytic center and increase the reaction rate. The mesoporous titania matrix, characterized by high hydrolytic stability, promoted substrate binding and allowed easy removal of the template. The synthesis of MIF was based on the formation of a stable complex between the template, bis-4-nitro-phenyl-phosphate (B4NPP), and lanthanum cations. The addition of titanium tetrachloride to the solution containing the surfactant, pluronic F127, and the imprinting complex, allowed the depositing titania films to incorporate micelles and molecular templates in a "one-pot" system. The removal of B4NPP by buffer treatment at pH 9 left molecular cavities, characterized by lanthanum cations on the inner surface, dispersed in a mesoporous matrix. The authors then investigated the characteristics of the film obtained. FT-IR evaluation of the thermal stability of the template in the mesoporous film, for example, showed that the molecule started decomposing at temperatures above 200 °C. Characterization of the mesoporosity of the titania films using TEM and spectroscopic ellipsometry showed a monodispersed porosity with an average pore size of 5.4 ± 1.0 nm. Spectroscopic ellipsometry was also used to estimate the thickness, surface roughness, and percentage porosity of the MIF samples before and after removal of the template. After buffer treatment, the film thickness was 382 ± 18 nm, while a thickness of 396 ± 15 nm was estimated for the same sample before buffer treatment. Treatment with buffer solution did not affect the film thickness, confirming the hydrolytic stability of the titania matrix in a basic environment. The catalytic activity of the MIF/NIF (non-imprinted film) samples was evaluated by testing their ability to hydrolyze Paraoxon. This was measured by monitoring the concentration of 4-nitrophenolate, which is formed when Paraoxon is hydrolyzed, as a function of time. Kinetic studies indicated that the NIF provided 17% rate acceleration compared to the background hydrolysis, while the MIF showed an increase of 27% of the initial rate. This material was shown to have a very high number of active sites and based on the density of active sites, the authors were also able to determine the catalytic loading corresponding to the area of the film used for the kinetics. These films have shown promising results as catalysts, proving to be versatile tools for the degradation of organophosphate-based pollutants.

This new approach in which Carboni demonstrated that the presence of metal improves the recognition, selectivity and activity of the catalyst was followed by Jakubiak-Marcinkowska et al. [34], who reported a novel MIP with the ability to mimic oxygenases based on lanthanum-doped titania film, in which the metal constituted the catalytic center. Oxygenases, which catalyze insertion of oxygen atoms on substrates, are members of the oxidoreductase family, which includes some of the most important enzymes in living organisms due to their essential role in the respiratory processes. The active centers of these enzymes contain Cu(II) ions coordinated by histidine ligands, among others amino acids residues that mediate electron transfer and reagent-binding. In the study by Jakubiak-Marcinkowska, the surface imprinting technique was employed to synthesize catalytic MIPs, carrying the recognition sites only on the inner surface of the polymer, Fig. 4.

This characteristic solved the diffusion issues and hence, facilitated the transport of reagents in the medium. MIP catalytic activity is indeed strongly related to the kinetics of reagent transport to/from the polymer matrix. Oxidation of hydroquinone (HQ) to quinine (Q) by H_2O_2 was used as the model reaction for this investigation. A complex of Cu(II) and 4-methoxybenzyl alcohol was used as template and 1-vinylimidazole was used as functional monomer due to its similarity to the histidyl ligand present in the natural enzymes. The cross-linker, 1,1,1-tris (hydroxymethyl)propane trimethacrylate provided a material with a porous and

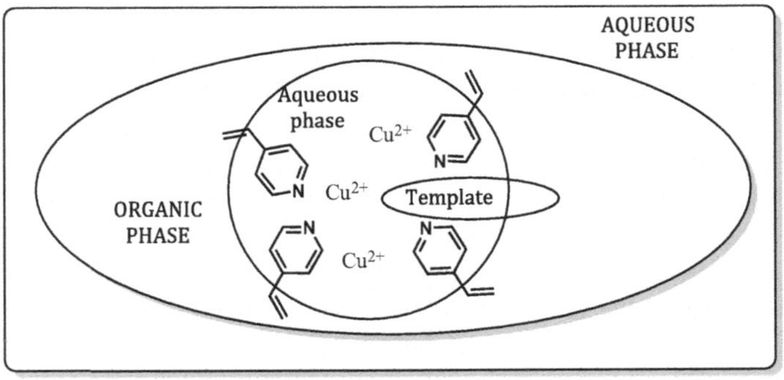

Fig. 4 Schematic representation of the synthesis of surface molecularly imprinted polymer catalyst

rigid structure. Moreover, the monomers, acrylonitrile and 4-vynyloxybutylstearate, were also added to the monomer mixture during polymerization. Acrylonitrile was added to increase the rigidity and porosity without using a high level of cross-linker, while the surface active monomer, 4-vynyloxybutylstearate was included to improve the stability of the water/oil emulsion, allowing easier removal of metal ions. Following removal of the Cu(II)-complex templates, the newly generated surface MIPs were further loaded with Cu(II) ions to prepare the catalysts with active centers. The Cu(II) ions interacted with the imidazole nitrogen and with carboxyl group formed by hydrolysis of the cross-linker during the polymerization. Characterization of the catalyst by EPR highlighted the interactions within the active center. To improve the catalytic efficiency of the materials, the authors modified the active center with ionic liquids (ILs) in a two-step process that involved bromoalkylation followed by ion-exchange of bromine. In the liquid state, ILs are a salt-like material, with excellent thermal and chemical stability, as well as good miscibility with water and organic solvents. Moreover, these substances have high polarity, reactivity toward transition metal ions and function in a two-phase system, which is very useful for catalysis and polymer chemistry. Modified and non-modified series of surface MIPs were synthesized. Comparison of EPR spectra and computer-simulated experimental spectra indicated that protonation of the imidazole nitrogen and the introduction of groups with strong electrostatic properties in the cavities yielded a clear enhancement (>50%) in catalytic activity. This work illustrates how the addition of free active groups in a catalytic site can impact activity by making the metal loading less relevant to high efficiency.

5 Molecular Imprinted Polymers as Nanoreactors

The following section will focus on the new line of work, already mentioned, in which metals are used in combination with MIPs to increase the selectivity and specificity of the catalyst and enhance catalytic activity. Li, in particular, contributed significant time and resources to this specific field, producing several examples of MIP-containing metals. In 2013, for instance, he presented an innovative nanoreactor with the ability to catalyze a cascade of reactions instead of focusing exclusively on a single reaction. This approach results in a significant reduction in the cost of the synthesis and improves streamlining of the whole process [35]. This material provides practical advantages in complicated synthetic processes such as that involved in the production of aminophenol (AP), which involves many separation steps. The proposed reaction is initiated by the hydrolysis of 4-nitrophenyl acetate (NPA) with the formation of nitrophenol (NP) followed by catalytic reduction in the presence of an Ag, Au or Fe catalyst, the steps are shown in Fig. 5.

Since the nanoreactor catalyzes two consecutive reactions, it was imprinted with two molecular templates, NPA and AgNP, which is a complex formed between NP and Ag. After photochemical polymerization, the nanoreactor MIP-NPA-AgNP was isolated by removal of the templates. To compare the activity of the catalysts, the

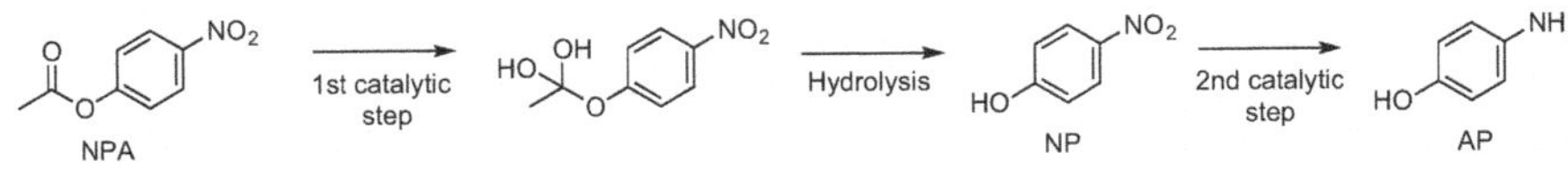

Fig. 5 Outline of the synthesis of aminophenol (AP) starting from 4-nitrophenyl acetate (NPA)

control nanoreactors, MIP-AgNP (without NPA), NIP-Ag (without the NP and NPA templates) and NIP, were prepared using the same experimental conditions. The nanoreactors were characterized by FT-IR, TEM, SEM, and BET and the interactions between the nanoreactors and NP were investigated by temperature programmed desorption (TPD). The template molecule NP, desorbed from MIP-NPA-AgNP, MIP-AgNP, NIP-Ag, and NIP at 262, 263, 192, and 194 °C, respectively, suggesting a stronger interaction between NP and the imprinted nanoreactors (MIP-NPA-AgNP and MIP-AgNP) than between the non-imprinted NIP-Ag and NIP. These results demonstrated that the binding cavities left during the imprinting process have a positive effect in rebinding the substrate. Monitoring of the catalytic activity by UV spectroscopy revealed that MIP-NPA-AgNP significantly accelerated the hydrolysis of NPA, while other nanoreactors failed to do so, showing that the imprinted species in the MIP-NPA-AgNP clearly play a role in promoting the hydrolysis of NPA. Moreover, the authors conducted electrochemical studies to investigate the mechanism of the interaction between the nanoreactors and the substrate. The potential to reduce/oxidize a bound molecule depends upon the binding state, with more energy required to disrupt stronger interactions. Therefore, electrochemical studies provide valuable information on the binding mechanism between nanoreactors and their substrates. Despite the encapsulated Ag nanoparticles, NIP-Ag showed almost the same desorption/potentials as NIP for both NPA and NP. MIP-NPA-AgNP and MIP-AgNP revealed comparable desorption/reduction potentials for the substrate, regardless of whether NPA or NP was used. This result strongly suggests that the cascade reaction catalyzed by MIP-NPA-AgNP is essentially a result of the molecular imprinting of templates, which regulates the access of reactants to the reactive sites, thereby steering the successive-reaction processes. These data demonstrate that successive-reaction catalysis can be realized by using this novel and unique imprinted nanoreactor, which opens up new opportunities for developing functional catalysts for complicated chemical processes.

Two-years later the same group prepared a similar nanoreactor for use in the cascade synthesis of AP by hydrolysis of NPA through the formation of NP and its further catalytic reduction [36]. This new material was composed of a MIP-containing Pt nanoparticles, instead of Ag, and an acidic catalytic site. As in the aforementioned case, NPA and NP, in form of a complex with the metal, were selected as the two templates and 2-acrylamido-2-methylpropanesulfonic acid was used as monomer to obtain the acidic catalytic site during the preparation of MIP-NPA-PtNP. Control nanoreactors, MIP-PtNP, NIP-Pt, NIP, and a nonacidic MIP-NPA-PtNP, were prepared using the same conditions for the purposes of comparison. The materials were characterized by FT-IR, SEM, or TEM and several tests were carried out mimicking those performed on the Ag-based nanoreactors. TPD verification of the interactions between nanoreactors and substrates showed that both NP and NPA have stronger interactions with the imprinted polymers than with the non-imprinted polymers (NIP-Pt and NIP). Moreover, studies on the catalytic activity were performed by monitoring the conversion of NPA as a function of time. The results demonstrated that both the presence of the acidic catalytic site and the binding cavities are necessary for increasing the NPA hydrolysis and improving the performance of the catalyst. MIP-NPA-PtNP yielded the best results, while the nonacidic system showed extremely low conversion compared with that achieved using the other systems. To confirm the cascade nature of the reaction catalyzed by the polymer, the catalytic hydrolysis of NPA was monitored by UV spectroscopy. In the case of the NIP system, a decreasing peak for NPA and an increasing peak of NP were observed. In contrast, the reaction conducted in the MIP-NPA-PtNP system led, as expected, to a decrease in the NPA peak and an increase in the AP peak, thus confirming that the consecutive reaction occurs in the catalyst. The same analysis was carried out using the MIP-PtNP system and in this case, the AP peak was smaller than that achieved with the MIP-NPA-PtNP because of the smaller amount of NP released from the hydrolysis of NPA. Thus, catalytic-cascade reactions become feasible only when the molecular recognition properties have been incorporated into the catalytic sites, which allow the specified reaction, whereas the catalytic sites are responsible for the catalytic processes. Moreover, similar results were obtained in electrochemical studies on the Ag nanoparticle-based catalyst [35]. These results showed that the stronger interactions with NPA and NP offered by MIP-NPA-PtNP are the result of the imprinted polymer carrier, which allows access of the specific substrate. In conjunction with the catalytic study, this result further suggests that the catalytic-cascade reactions occurring in the MIP-NPA-PtNP were due to synergy between the recognition properties and the catalytic sites. Comparison of the results obtained from the two similar studies by Li, suggests that the two materials have similar efficiency. Comparable values have, in fact, been obtained from TPD studies as well as from reduction potential analyses. UV analysis indicated that the two materials promote the cascade reaction equally. However, the analysis of catalytic activity of the two nanoreactors showed a small difference: MIP-NPA-PtNP converted 45% of NPA over a 5-h period, whereas the MIP-NPA-AgNP converted only 36%.

Another example of a reactor produced by Li to provide metal nanoparticle catalysts with selectivity, was based on the nickel nanoparticles equipped with a molecular imprinted polymer [37]. Selectivity and recognition characteristics were incorporated with the metallic catalytic sites into a single entity. The studied reaction involved the reduction of methylene blue, a heterocyclic aromatic chemical compound used in several applications both in biology and chemistry, which can be reduced under mild conditions and is easily monitored by UV-Vis spectroscopy. The substrate, in the form of a complex with nickel ions (MBNi), was chosen as the template molecule in the imprinting process, while acrylamide, EGDMA, and AIBN were used as functional monomer, cross-linker and initiator, respectively. After removing the template from the matrix with acidic ethanol, the reactor

MIP-Ni was obtained and characterized. Three controls, NIP-Ni, MIP, and NIP, were also prepared for comparative study. As with the previously described nanoreactors produced by Li et al., the authors investigated the interactions between the substrate and the reactors, the molecular recognition, the catalytic properties and kinetics, in addition to carrying out electrochemical studies. TPD (temperature programmed desorption) analysis showed that MIP-Ni and MIP had stronger interactions with MB compared with NIP-Ni and NIP, thus demonstrating that the presence of the binding cavities enhances recognition. To confirm that molecular recognition takes place in the reactor, methylene green (MG) was selected as a control substrate. MIP-Ni exhibited higher catalytic activity for MB than for MG. Detailed analysis of catalytic activity by UV-Vis spectroscopy therefore confirmed that the presence of metal nanoparticles is essential for activity. MIP and NIP, in fact, did not show relevant catalysis, whereas MIP-Ni and NIP-Ni revealed significant catalytic activity, where the conversion increased rapidly with time, reaching 90% in 18 min for MIP-Ni and 82% in the case of NIP-Ni. Electrochemical studies of substrate desorption can provide information on the binding behavior between the reactors and substrate. MB attached onto NIP-Ni exhibited a reduction peak at −815 mV. In contrast, the corresponding reduction peak in MIP-Ni was shifted to −853 mV. These results demonstrated a stronger interaction between MB and MIP-Ni compared with its interaction with NIP-Ni. Despite the encapsulated nickel nanoparticles, NIP-Ni (and MIP-Ni) showed almost the same potential as NIP (and MIP). This result further confirms that the catalytic selectivity of MIP-Ni is a result of the molecularly imprinted carrier, which dictates selective catalytic ability. In conclusion, the authors were able to produce a reactor that incorporated both the recognition properties of the polymer carrier and the catalytic sites of the metal nanoparticles.

Sun et al. [38] developed a series of MIP-containing equal amounts of iron ions, taking advantage of the excellent performance of this metal in selective oxidation processes. Nitric acid is usually used to selectively oxidize benzyl alcohol derivatives to the corresponding aldehydes, giving rise to unavoidable issues such as environmental pollution and equipment corrosion. To provide a safer and eco-friendly catalyst, Sun and co-workers prepared molecularly imprinted catalysts, which exhibited high selectivity, substrate recognition and catalytic activity. Three catalysts, *p*-Fe(III)-MIP, *o*-Fe(III)-MIP, and *m*-Fe(III)-MIP, were prepared using acrylamide as a functional monomer, EGDMA as a cross-linker, AIBN as an initiator and *p*-nitrobenzyl alcohol (*p*-NBA), *o*-nitrobenzyl alcohol (*o*-NBA), and *m*-nitrobenzyl alcohol (*m*-NBA), respectively, as templates. For comparison studies, NIP and Fe(III)-NIP controls were synthetized using the same protocol. The amount of iron incorporated into the MIP in each system was calculated to be 0.3 mmol/g by titration with EDTA. Following the removal of template from the polymeric matrix, the systems were characterized by SEM, BET, FT-IR, and UV-Vis spectroscopy. Based on UV spectroscopic investigations of the interactions between the monomer and the metal, the template and metal and the monomer and the template, the acrylamide and template were used at a molar ratio of 4:1 to obtain a matrix with the ability to bind the imprint molecule stoichiometrically; this also produced a highly specific imprint. Studies on the catalytic activity demonstrated that the conversion of nitrobenzyl alcohol to the corresponding benzaldehyde was increased when the MIP catalyst was used. In particular, the conversion of the corresponding substrate was enhanced by a specific catalyst. For example, *p*-Fe(III)-MIP converted 80% of *p*-NBA into the benzaldehyde within 4 h at 80 °C in the presence of 30% H_2O_2 as the oxidizing agent, whereas only 36% of *o*-NBA and 42% of *m*-NBA were converted under the same conditions; the same results were obtained using the other catalysts. Differences in catalytic activity are related to the complementarity of the imprinted cavity for the specific substrate in terms of the shape and position of functional groups, which increase the affinity toward the substrate, therefore enhancing the reactivity of the catalyst. To verify the specificity of substrate recognition, each catalyst was tested using a mixture of *o*-, *p*- and *m*-NBA. The Fe(III)-MIP samples showed highly specific substrate recognition and catalysis toward the imprint species during the oxidation. Again, taking *p*-Fe(III)-MIP as an example, 85% of *p*-NBA was converted, while a mismatch effect was observed for *o*-NBA and *m*-NBA, with only 29 and 40% conversion, respectively. These results confirmed that developing a specific structure and function of the catalytic active center in the imprinted cavities of the MIPs provided the polymer with catalytic activity and specific substrate recognition. The imprinted cavities in the catalysts were responsible for the substrate recognition and the accelerated reaction rate.

An innovative example of a metal used within a MIP to enhance the characteristics of the catalyst was proposed by Yang et al. They reported a molecularly imprinted Ru complex catalyst with an NH_2-binding site on the wall of a molecularly imprinted cavity to achieve shape-selective transfer hydrogenation of *o*-fluorobenzophenone (*o*-F-BP) [39]. Immobilization of metal complexes on support surfaces is a very modern method used in the field between homogeneous and heterogeneous catalysis [40]. The potential of immobilized metal-complex catalysts interplays greatly with the nature of support surfaces, resulting in significant rate enhancements and unique catalytic performances not exhibited by the analogous homogeneous complexes. The alcohol, *o*-hydroxybenzhydrol, which is similar in shape to the product of *o*-fluorobenzophenone hydrogenation, was coordinated to a SiO_2-supported Ru-monomer complex and was utilized as a template in the preparation of the catalyst. A Ru complex, which is known to be active for transfer hydrogenation, was attached chemically to a SiO_2 surface and a template ligand with a carbamate moiety was coordinated to the supported Ru complex. Cleavage of the carbamate moiety, then, leaves an NH_2 group grafted to the wall of the cavity with a short alkyl chain, which can serve as a molecular binding site for *o*-F-BP through hydrogen bonding. To investigate the effects of the template and the NH_2 binding site, three samples were prepared: G-BS, G-noBS, and G-noTemp, with the latter two being controls. G-noBS was prepared by using *o*-methylbenzhydrol as template, as a substitute for the template ligand with the carbamate. This generated an imprinted cavity with a similar shape to the template ligand but without the NH_2 binding site. G-noTemp was prepared without any template and so, lacked the cavity. Similar amounts of SiO_2 matrix overlayers were stacked in a similar way on the three catalysts. Studies on the catalytic activity of the prepared catalyst in comparison with controls showed that G-noTemp was devoid of catalytic activity for the transfer hydrogenation. There was no reaction space on the Ru complex surrounded by the SiO_2 matrix overlayers, hence, it was proven that the catalytic transfer hydrogenation occurred within the molecularly imprinted cavity left behind by the template ligand. In the case of G-noBS, the *o*-F-BP activity was lower than that of G-BS, which contained the NH_2 binding site. Also the enantioselectivity in the transfer hydrogenation increased after the imprinting due to restriction of the mobility of the chiral diamine ligand on the Ru complex by the wall of the SiO_2 matrix overlayers. Therefore, the SiO_2 matrix overlayers serve not only to create a molecular-shaped cavity, but also to regulate the flexibility of the coordinating ligands to enhance enantioselective behavior. The effects of the NH_2 binding site on the molecularly imprinted Ru catalysts were investigated in the transfer hydrogenation of *o*-F-BP and *o*-methylbenzophenone (*o*-Me-BP), which are similar in shape. The former interacts with the NH_2 binding site via hydrogen bonding, while the latter does not. In the case of *o*-Me-BP, the transfer hydrogenation rates catalyzed by G-BS and G-noBS were similar. In contrast, the catalytic activity toward *o*-F-BP was increased by the presence of the NH_2 sites in G-BS, which catalyzed a reaction rate that was two-fold greater than that catalyzed by G-noBS. These results implied the preferential adsorption of this substrate in the molecularly imprinted cavity on G-BS containing the NH_2 binding site. Although, the positive interaction with the NH_2 binding site that provided shape-selective behavior in discriminating the position of the fluoro-substituent was demonstrated, the catalytic activity in the system was not enhanced. Despite the scope for improvement in these results, further changes in the design of molecularly imprinted metal-complex catalysts holds great promise for the creation of artificial enzymatic catalyst surfaces with high catalytic activity and fine molecular recognition properties.

6 Molecular Imprinted Polymers as Photocatalysts

The last field of study attracting particular interest involves the use of MIPs as photocatalysts. The degradation of pollutants in both waters and soils is becoming an increasingly sensitive topic. In fact, due to industrialization, population growth, and urbanization, the amount of pollutants produced and released into the environment has become a serious issue [41]. Most of these contaminants, which are highly toxic, carcinogenic and teratogenic, are extremely persistent. Thus, to limit the risk to the health of humans and animals, it is important to develop a selective and efficient method to remove the toxic pollutants from the environment. As mentioned in the first part of this chapter, photocatalysis is the most efficient method of pollutant degradation and TiO_2 is the most commonly used photocatalyst due to its inexpensive costs, availability, nontoxicity, chemically and mechanically stability, and high turnover. However, it is worth remembering that TiO_2 lacks target selectivity and has low quantum efficiency for sunlight, being sensitive only to the UV radiation [28, 29]. To overcome these drawbacks, TiO_2 materials have been variously modified by doping them with metals or conductive polymers or by coupling with MIP in either inorganic- or organic-framework forms.

6.1 Inorganic MIPs Applied in Photocatalysis

Inorganic-framework MIPs have been used for degradation of 4-nitrophenol (4NP), which is used to manufacture drugs, fungicides, insecticides and dyes and to darken leather. Although the EPA (Environmental Protection Agency) has not classified this compound for potential carcinogenicity, acute inhalation or ingestion of 4-nitrophenol in humans causes headaches, drowsiness, nausea, and cyanosis.

Deng et al. published a simple "one-step sol-gel" method for the production of an inorganic-framework molecularly imprinted TiO_2/WO_3 nanocomposite for photodegradation of 2NP and 4NP [42]. Tetrabutyl orthotitanate was used as a titanium source as well as the precursor of functional monomers, which complex with the template molecules, and ammonium tungstate was used as the precursor of WO_3. In Fig. 6 is shown the framework molecularly imprinted TiO_2/WO_3 nanocomposite with the formation of the recognition sites after the removal of template.

2NP-MIP TiO_2/WO_3, 4NP-TiO_2/WO_3 and NIP-TiO_2/WO_3 were prepared using 2NP and 4NP as templates for the first two and no template for the NIP served as the control. The materials were characterized by SEM, EDS, XRD, and UV-Vis spectroscopy. Analysis of adsorption, selectivity, photocatalytic activity, and reusability for all the materials showed that, compared with NIP-TiO_2/WO_3, MIP-TiO_2/WO_3 not only exhibited higher adsorption capacity and selectivity for the target contaminant, but also showed enhanced photocatalytic activity in degrading the target contaminant. This can be attributed to selective adsorption of target molecules on MIP-TiO_2/WO_3. Moreover, the photocatalyst showed high stability and reusability. However, the materials presented a disadvantage in that some imprinted cavities were destroyed at high-temperature during the sol-gel process. Therefore, it was necessary to develop milder methods for the preparation

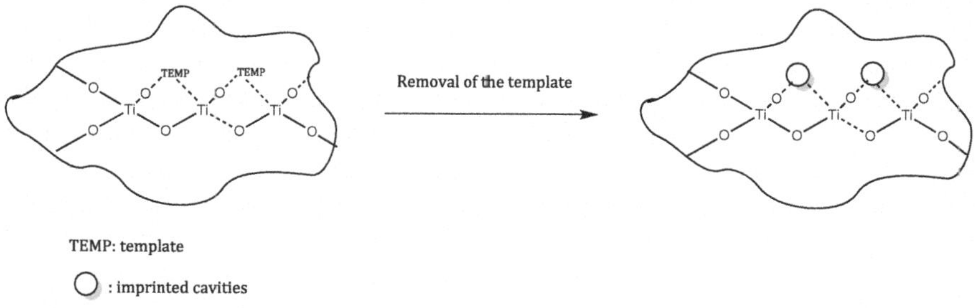

Fig. 6 Schematic preparation of inorganic-framework photocatalyst

of molecularly imprinted photocatalysts with improved performance due a richer content of imprinted cavities.

Two-years later, the same group developed another inorganic-framework molecularly imprinted TiO_2/SiO_2 (MIP-TiO_2/SiO_2) nanocomposite for use in the degradation of 4NP [43]. Instead of using the sol-gel method, the material in this case was prepared by the sol-hydrothermal method, using tetrabutyl orthotitanate as the titanium source and tetraethyl orthosilicate as the precursor of the functional monomer, which complexes with the template molecules. 4NP, as the target contaminant, was also used as the template. In the same way NIP-TiO_2/SiO_2 was prepared for comparison studies. After removing the template the MIP was characterized by TEM, SEM, XRD, UV-Vis and DLS. Studies carried out on the materials showed the superior ability of MIP-TiO_2/SiO_2 to adsorb the target selectively compared to NIP-TiO_2/SiO_2, due to the presence of the complementary cavities left behind during the formation of the MIP. The amount of 4NP adsorbed onto MIP-TiO_2/SiO_2 increased rapidly with time, reaching equilibrium after only 20 min. Moreover, the adsorption capacity of MIP-TiO_2/SiO_2 for 4NP was much higher than that of NIP-TiO_2/SiO_2. To test the selectivity of the catalyst, 2NP (2-nitrophenol) was chosen as a competitor of 4NP. The adsorption capacity of MIP-TiO_2/SiO_2 for 4NP was, as expected, higher than that for 2NP, whereas NIP-TiO_2/SiO_2 showed similar adsorption capacity toward both targets. Investigations were also conducted on the ability of the material to degrade 4NP. MIP-TiO_2/SiO_2 catalyzed complete degradation in 30 min, while NIP-TiO_2/SiO_2 took 110 min to achieve completion. The photocatalytic selectivity was also evaluated using 2NP as a competitor for 4NP. In treatment of a mixture of the two pollutants, MIP-TiO_2/SiO_2 showed higher activity toward 4NP than 2NP, while NIP-TiO_2/SiO_2 did not discriminate between the two targets. Moreover, the authors analyzed the photocatalytic degradation efficiency of 4NP as a function of pH in a range from 3.0 to 9.0. The degradation efficiency increased at pH values ranging from 3.0 to 4.0. The efficiency reached a maximum at pH 4.0 and decreased at higher pH values. When the pH of the system was increased from 3.0 to 4.0, 4NP is in its nonionic form and its solubility in water is minimized, while its adsorption on the catalyst is maximized due to the involvement of surface hydroxyl groups. With a further rise in pH, the surface of MIP-TiO_2/SiO_2 is negatively charged, while 4NP is its anionic form. The electrostatic repulsion between the catalyst surface and 4NP inhibits adsorption, consequently, decreasing the degradation efficiency. To confirm the reusability of the catalyst, the photocatalytic degradation–regeneration cycles were repeated four times. The photocatalytic degradation ability of MIP-TiO_2/SiO_2 was stable over the four cycles, indicating excellent regeneration capacity of the photocatalyst. In conclusion, the authors successfully developed a photocatalyst with molecularly imprinted cavities on the surface of the TiO_2/SiO_2 nanocomposite exhibiting selective affinity and specific molecular recognition ability toward the target contaminant. Because of the inorganic-framework and the simple release of active adsorption sites upon regeneration, MIP-TiO_2/SiO_2 is highly reusable.

6.2 *Organic MIPs Applied in Photocatalysis*

Nitrobenzene (NB), another dangerous nitro-compound pollutant, is used to manufacture aniline, which is a precursor of rubber chemicals, pesticides, dyes, explosives and pharmaceuticals. Acute and chronic inhalation, and oral and dermal exposure of humans to NB results in effects on the blood, such as methemoglobinemia. At higher concentrations, depressed respiration, bluish-gray skin, disturbed vision, and coma may occur. However, the EPA has classified NB as a Group D compound, which means that is not classifiable with regard to human carcinogenicity. To remove NB from wastewater, Shen et al. [44] developed a new method based on organic MIP-coated P25 TiO_2. *o*-phenylenediamine (*o*-PDA) was used as a functional monomer to produce a conducting polymeric skeleton with a polyaniline-like structure, which enhanced the ability of the TiO_2 photocatalyst. Polyaniline provides a nontoxic material with good stability, high instinct redox, and corrosion protection properties and, more importantly, high absorption coefficients in the visible light range as well as high mobility of charge carriers [45]. Given the toxicity of the target molecule, it cannot be used as template molecule; therefore, a transition state analog (TSA) was chosen for this purpose. After investigations on the NB degradation pathway, which led to the identification of mono-nitrophenols as main intermediates, 2NP and 4NP were selected as TSA templates. Two photocatalysts, 2NP-P25, and 4NP-P25 and a control, NIP-P25, were therefore prepared by UV polymerization (Fig. 7).

Photocatalytic activity tests on the prepared systems confirmed NB degradation, with removal reaching 67% over NIP-P25, 86% over neat P25, and more than 95% over TSA-MIP-TiO_2 within 50 min. The non-imprinted NIP-P25 exhibited lower photocatalytic activity than neat P25 TiO_2. The coated layer on the NIP-P25 has no active cavities, and thus hinders the direct contact of NB with TiO_2. In contrast, the coating of TSA-MIP on the surface of TiO_2 played an important role in the photocatalysis, and both 2-NP and 4-NP enhanced the photocatalytic degradation of NB. During the photocatalytic degradation of NB, three mono-nitrophenols, 2NP, 3NP, and 4NP, were generated over both the TSA-MIP-TiO_2 and neat P25

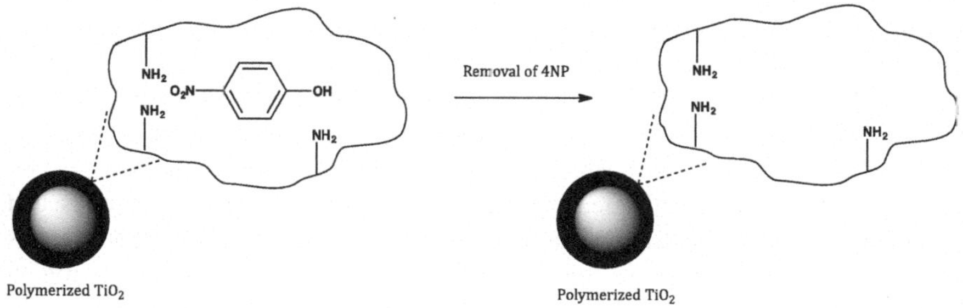

Fig. 7 Schematic representation of the photocatalyst. Here the transition state analog is represented by 4NP

photocatalysts. As these nitrophenols are the disproportionation products of the transition state, they influence the formation of the catalyst-substrate complex, and the consequent degradation of NB and accumulation of intermediates. The concentrations of the three isomers of nitrophenols over each of the photocatalysts increased during the initial reaction stage, and then decreased with prolonged UV irradiation time. This indicated that the TSA-MIP-TiO_2 photocatalysts not only promoted the photodecomposition of the highly toxic parent pollutants, but also inhibited the generation of the toxic intermediates. The photocatalytic selectivity of MIP-coated TiO_2 was evaluated by degrading the NB target pollutant at low concentrations in the presence of nontarget pollutants at high concentrations. BPA was selected as the co-existing pollutant because of its structural similarity to the target NB and its lack of influence on the analysis of NB and its degradation intermediates. The degradation rate of the target NB was much lower than that of BPA over neat P25, showing that BPA is more easily degraded photocatalytically by P25. Using 2NP-TiO_2 and 4NP-TiO_2, the degradation of NB was significantly promoted, while the degradation of the co-existing pollutant BPA was depressed. In conclusion, using an appropriate TSA of the target pollutant as the template molecule, the authors successfully developed a MIP-based photocatalyst, which exhibited enhanced photocatalytic activity and selectivity toward the target. Moreover, the molecular recognition inhibited the accumulation of unwanted intermediates.

Chlorophenols (CPs) are another class of dangerous organic pollutants. CPs are phenol organochloride molecules, which contain a covalently bound chlorine atom at one of three different positions, giving rise to three isomers (2CP, 3CP, 4CP), or up to five atoms, giving rise to pentachlorophenol (PCP). These compounds are widely used in various industries as pesticides, herbicides and insecticides. Both the EPA, the European Parliament and the Council of the European Union listed CPs as priority pollutants. CPs degrade slowly in the environment and can produce more complicated and even more toxic molecules; hence, the need for a system that allows fast purification. Huang et al. developed a similar MIP-coated P25TiO_2 photocatalyst to that reported by Shen and co-workers, for use in the degradation of CPs [46]. The *o*-PDA was used as the functional monomer and a substrate analog (SA), 2,4-dinitrophenol (2,4-DNP), was selected as the template molecule. After removing the template, the photocatalyst (DNP-P25) was obtained and characterized by FT-IR, ATR, TEM, and DLS. The photocatalytic activity and selectivity were investigated by evaluating the degradation of phenol, 4-CP, 2,4-DCP, 2,4,6-TCP, or PCP over the DNP-P25 photocatalyst, neat P25 and NIP-P25. Neat P25 and NIP-P25 had similar activity showing that no positive effect was achieved by the presence of the NIP layer. In contrast, DNP-P25 showed the fastest photocatalytic degradation, with greater selectivity toward the target compounds that contained more chlorine atoms, particularly PCP and 2,4,6-TCP. These results demonstrated that the degree of chlorination of the targets plays an important role in the selectivity. Moreover, by carrying out the photocatalysis in consecutive runs, the stability and reusability of the photocatalyst was confirmed. The MIP layer on the P25 photocatalyst enhanced its ability to mineralize the CPs. Indeed, in 45 min, dechlorination of CPs over DNP-P25 exceeded 90%, while that catalyzed by P25 was less than 60%, demonstrating that DNP-P25 has the ability to degrade CPs completely by total dechlorination. In conclusion, the authors developed a green method for mineralization of harmful chlorophenols based on a substrate analog imprinted photocatalyst. Moreover, tests carried out on water samples confirmed that the common inorganic ions showed no interference in the detection of chlorine ions.

7 Conclusions

Work in the last decade has led to very significant achievements in the area of imprinted polymers for catalytic applications. The drive to obtain new catalysts with novel and/or improved functionalities that can complement the activity of enzymes is still very strong. The results shown in this chapter are some of the most interesting examples involving metal atoms. There is a wealth of catalytic units that can be easily incorporated into polymer matrices, taking advantage of the imprinting approach, to generate new and more sophisticated catalysts with improve activity and specificity. The development of these systems, together with the advantages brought on by the new nanosized polymers described in this chapter are expected to bring very interesting applications in the coming years.

References

1. Danielsson B (2008) Artificial receptors. In: Scheper T (ed) Advances in biochemical engineering/biotechnology, vol 109. Springer, Berlin, pp 97–122
2. Kataev EA, Muller C (2014) Recent advances in molecular recognition in water: artificial receptors and supramolecular catalysis. Tetrahedron 70:137–167. doi:10.1016/j.tet.2013.11.010
3. Mirsky VM, Yatsimirsky A (eds) (2010) Artificial receptors for chemical sensors. Wiley-VCH, Germany
4. Uzun L, Turner AP (2016) Molecularly-imprinted polymer sensors: realising their potential. Biosens Bioelectron 76:131–144. doi:10.1016/j.bios.2015.07.013
5. Ramstrom O, Ansell RJ (1998) Molecular imprinting technology: challenges and prospects for the future. Chirality 10:195–209. doi:10.1002/(SICI)1520-636X(1998)10:3<195:AID-CHIR1>3.0.CO;2-9
6. Kriz D, Mosbach K (1995) Competitive amperometric morphine sensor based on an agarose immobilised molecularly imprinted polymer. Anal Chim Acta 300:71–75. doi:10.1016/0003-2670(94)00368-V
7. Bui BTS, Haupt K (2010) Molecularly imprinted polymers: synthetic receptors in bioanalysis. Anal Bioanal Chem 398:2481–2492. doi:10.1007/s00216-010-4158-x
8. Gu Y, Yan X, Li C, Zheng B, Li Y, Liu W, Zhang Z, Yang M (2016) Biomimetic sensor based on molecularly imprinted polymer with nitroreductase-like activity for metronidazole detection. Biosens Bioelectron 77:393–399. doi:10.1016/j.bios.2015.09.060

9. Matsui J, Nicholls IA, Mosbach K (1996) Carbon–Carbon bond formation using substrate selective catalytic polymers prepared by molecular imprinting: an artificial class II aldolase. J Org Chem 61:5414–5417. doi:10.1021/jo9516805
10. Wulff G (2002) Enzyme-like catalysis by molecularly imprinted polymers. Chem Rev 102:1. doi:10.1021/cr980039a
11. Liu JQ, Wulff G (2004) Molecularly imprinted polymers with strong carboxypeptidase A-like activity: combination of an amidinium function with a zinc-ion binding site in transition-state imprinted cavities. Angew Chem Int Ed Engl 43(10):1287–1290. doi:10.1002/anie.200352770
12. Czulak J, Jakubiak-Marcinkowska A, Trochimczuk A (2013) Polymer catalysts imprinted with metal ions as biomimics of metalloenzymes. Adv Mater Sci Eng. 2013, Article ID 464265. doi:10.1155/2013/464265 (Hindawi Publishing Corporation)
13. Beller M, Bolm C (eds) (2004) Transition metals for organic synthesis: building blocks and fine chemicals, 2nd edn. Wiley-VCH. doi:10.1002/9783527619405
14. Strikovsky AG, Kasper D, Grun M, Green BS, Hradil J, Wulff G (2000) Catalytic molecularly imprinted polymers using conventional bulk polymerization or suspension polymerization: selective hydrolysis of diphenyl carbonate and diphenyl carbamate. J Am Chem Soc 122 (26):6295–6296. doi:10.1021/ja994269y
15. Alexander C, Davidson L, Hayes W (2003) Imprinted polymers: artificial molecular recognition materials with applications in synthesis and catalysis. Tetrahedron 59:2025–2057. doi:10.1016/S0040-4020(03)00152-2
16. Bonomi P, Servant A, Resmini M (2012) Modulation of imprinting efficiency in nanogels with catalytic activity in the Kemp elimination. J Mol Recogn 25(6):352–360. doi:10.1002/jmr.2180
17. Resmini M (2012) Molecularly imprinted polymers as biomimetic catalysts. Anal Bioanal Chem 402(10):3021–3026. doi:10.1007/s00216-011-5671-2
18. Chen Z, Hua Z, Wang J, Guan Y, Zhao M, Li Y (2007) Molecularly imprinted soluble nanogels as a peroxidase-like catalyst in the oxidation reaction of homovanillic acid under aqueous conditions. Appl Catal A 328(2):252–258. doi:10.1016/j.apcata.2007.05.040
19. Toorisaka E, Uezu K, Goto M, Furusaki S (2003) A molecularly imprinted polymer that shows enzymatic activity. Biochem Eng J 14:85–91. doi:10.1016/S1369-703X(02)00155-9
20. Yoshida M, Hatate Y, Uezu K, Goto M, Furusaki S (2000) Chiral-recognition polymer prepared by surface molecular imprinting technique. Colloids Surf A 169:259–269. doi:10.1016/S0927-7757(00)00468-4
21. Yoshida M, Hatate Y, Uezu K, Goto M, Furusaki S (1999) Metal ion imprinted microsphere prepared by surface molecular imprinting technique using water-in-oil-in-water emulsions. J Appl Polym Sci 73(7):1223–1230. doi:10.1002/(SICI)1097-4628(19990815)73:7<1223:AID-APP16>3.0.CO;2-Y
22. Meng M, Bao L, He M, Sun K, Li W, Zhao D, Feng Y, Yan Y (2014) Preparation, characterization, and adsorption performance of p-hydroxybenzoic acid imprinted polymer and selective catalysis of toluene to para-chlorotoluene. J Appl Polym Sci 131(8):40118. doi:10.1002/app.40118
23. Jia ZG, Liu JH, Wang QZ, Li SB, Qi Q, Zhu RS (2015) Synthesis of 3D hierarchical porous iron oxides for adsorption of Congo red from dye wastewater. J Alloy Compd 622:587–595. doi:10.1016/j.jallcom.2014.10.125
24. Ouasif H, Yousfi S, Bouamrani ML, Kouali M, Benmokhtar SM, Talbi M (2013) Removal of a cationic dye from wastewater by adsorption onto natural adsorbents. J Mater Environ Sci 4 (1):1–10
25. Santhy K, Selvapathy P (2006) Removal of reactive dyes from wastewater by adsorption on coir pith activated carbon. Bioresour Technol 97(11):1329–1336. doi:10.1016/j.biortech.2005.05.016
26. Zhu M, Lee L, Wang H, Wang Z (2007) Removal of an anionic dye by adsorption/precipitation processes using alkaline white mud. J Hazard Mater 149(3):735–741. doi:10.1016/j.jhazmat.2007.04.037

27. Oladoja NA, Aliu YD, Ofomaja AE (2011) Evaluation of snail shell as a coagulant aid in the alum precipitation of aniline blue from aqueous solution. Environ Technol 32(6):639–652. doi:10.1080/09593330.2010.509868
28. Kumar SG, Devi LG (2011) Review on modified TiO_2 photocatalysis under UV/visible light: selected results and related mechanisms on interfacial charge carrier transfer dynamics. J Phys Chem A 115:13211–13241. doi:10.1021/jp204364a
29. Wang YS, Shen JH, Horng JJ (2014) Chromate enhanced visible light driven TiO_2 photocatalytic mechanism on acid orange 7 photodegradation. J Hazard Mater 274:420–427. doi:10.1016/j.jhazmat.2014.04.042
30. Gao J, Heeger AJ, Lee JY, Kim CY (1996) Soluble polypyrrole as the transparent anode in polymer light-emitting diodes. Synth Met 82(3):221. doi:10.1016/S0379-6779(96)03794-0
31. Deng F, Li YX, Luo XB, Yang LX, Tu XM (2012) Preparation of conductive polypyrrole/TiO_2 nanocomposite via surface molecular imprinting technique and its photocatalytic activity under simulated solar light irradiation. Colloids Surf A 395:183–189. doi:10.1016/j.colsurfa.2011.12.029
32. Wei S, Hu X, Liu H, Wang Q, He C (2015) Rapid degradation of Congo red by molecularly imprinted polypyrrole-coated magnetic TiO_2 nanoparticles in dark at ambient conditions. J Hazard Mater 294:168–176. doi:10.1016/j.jhazmat.2015.03.067
33. Carboni D, Malfatti L, Pinna A, Lasio B, Tokudome Y, Takahashi M, Innocenzi P (2013) Molecularly imprinted La-doped mesoporous titania films with hydrolytic properties toward organophosphate pesticides. New J Chem 37:2995–3002. doi:10.1039/C3NJ00291H
34. Jakubiak-Marcinkowska A, Legan M, Jezierska J (2013) Molecularly imprinted polymeric Cu (II) catalysts with modified active centres mimicking oxidation enzymes. J Polym Res 20:317. doi:10.1007/s10965-013-0317-z
35. Li S, Luo Y, Whitcombeb MJ, Piletsky SA (2013) A successive-reaction nanoreactor made of active molecularly imprinted polymer containing Ag nanoparticles. J Mater Chem A 1:15102–15109. doi:10.1039/C3TA13454G
36. Wang J, Zhu M, Shen X, Li S (2015) A cascade-reaction nanoreactor composed of a bifunctional molecularly imprinted polymer that contains pt nanoparticles. Chem Eur J 21 (20):7532–7539. doi:10.1002/chem.201406285
37. Zhang X, Zhu M, Li S (2014) "Key-vs.-Lock"-like polymer reactor made of molecularly imprinted polymer containing metal nanoparticles. J Inorg Organomet Polym 24:890–897. doi:10.1007/s10904-014-0061-9
38. Sun W, Tan R, Zheng Yin D (2013) Molecularly imprinted polymer containing Fe(III) catalysts for specific substrate recognition. Chin J Catal 34(8):1589–1598. doi:10.1016/S1872-2067(12)60624-X
39. Yang Y, Weng Z, Muratsugu S, Ishiguro N, Ohkoshi S, Tada M (2012) Inside cover: the excess electron in a boron nitride nanotube: pyramidal NBO charge distribution and remarkable first hyperpolarizability. Chem Eur J 18(36):1142–1153. doi:10.1002/chem.201290154
40. Thomas JM, Raja R (2008) Exploiting nanospace for asymmetric catalysis: confinement of immobilized, single-site chiral catalysts enhances enantioselectivity. Acc Chem Res 41 (6):708–720. doi:10.1021/ar700217y
41. Capra A, Scicolone B (2004) Emitter and filter tests for wastewater reuse by drip irrigation. Agric Water Manage 68(2):135–149. doi:10.1016/j.agwat.2004.03.005
42. Luo X, Deng F, Min L, Luo S, Guo B, Zeng G, Au C (2013) Facile one-step synthesis of inorganic-framework molecularly imprinted TiO_2/WO_3 nanocomposite and its molecular recognitive photocatalytic degradation of target contaminant. Environ Sci Technol 47:7404–7412. doi:10.1021/es4013596
43. Denga F, Liua Y, Luoa X, Wua S, Luoa S, Auc C, Qia R (2014) Sol-hydrothermal synthesis of inorganic-framework molecularly imprinted TiO_2/SiO_2 nanocomposite and its preferential photocatalytic degradation towards target contaminant. J Hazard Mater 278:108–115. doi:10.1016/j.jhazmat.2014.05.088

44. Shen X, Zhu L, Wang N, Zhang T, Tang H (2014) Selective photocatalytic degradation of nitrobenzene facilitated by molecular imprinting with a transition state analog. Catal Today 225:164–170. doi:10.1016/j.cattod.2013.07.011
45. Lin Y, Li D, Hu J, Xiao G, Wang J, Li W, Fu X (2012) Highly efficient photocatalytic degradation of organic pollutants by PANI-modified TiO_2 composite. J Phys Chem C 116 (9):5764–5772. doi:10.1021/jp211222w
46. Huanga C, Tu Z, Shen X (2013) Molecularly imprinted photocatalyst with a structural analogue of template and its application. J Hazard Mater 248–249:379–386. doi:10.1016/j.jhazmat.2013.01.037

Catalysis Inside Folded Single Macromolecules in Water

Müge Artar and Anja R.A. Palmans

Abstract Enzymes are dynamic, folded macromolecules with a perfectly defined three-dimensional structure that is highly active and selective in converting substrates into products. In contrast, synthetic polymers typically adopt random coil conformations in dilute solution or, in poor solvents, dense globular structures. Yet, the capability to control the global conformation of a synthetic polymer opened up many new applications that result from a three-dimensional structure. In recent years, a novel class of polymers, dynamic single-chain polymeric nanoparticles (SCPNs), were developed to control global conformations of synthetic polymers. SCPNs result from the thermodynamically controlled folding of synthetic polymers via directional interactions encoded in pendant supramolecular motifs. This marriage between polymer chemistry and supramolecular chemistry afforded a series of compartmentalised, nanometre-sized polymeric particles, which were studied in detail. In this review, we address the folding of an amphiphilic polymer in water around a catalytic centre. We highlight recent results obtained in oxidations, reductions and C–C bond forming reactions with these SCPNs. In addition, we will discuss the benefits of the ordered hydrophobic interior in the SCPNs to achieve *selective* catalysis in water.

1 Introduction

Synthetic polymers adopt random coil conformations in dilute solution or, in poor solvents, dense globular structures [1]. If a polymer backbone is capable of adopting a preferential conformation as a result of intramolecular interactions or steric hindrance, helical superstructures can be obtained [2]. Helical superstructures, such as helical polymers and foldamers, are highly interesting for applications in,

M. Artar · A.R.A. Palmans (✉)
Laboratory for Macromolecular and Organic Chemistry, Institute for Complex Molecular Systems, TU Eindhoven, PO Box 513, 5600 MB Eindhoven, The Netherlands
e-mail: a.palmans@tue.nl

R. Poli (ed.), *Effects of Nanoconfinement on Catalysis*,
Fundamental and Applied Catalysis, DOI 10.1007/978-3-319-50207-6_5

amongst others, asymmetric catalysis and sensing [3]. Alternatively, intramolecular covalent [4] and dynamic covalent bond [5] formation as well as fully reversible, non-covalent bond [6] formation between pendant groups attached to a polymer backbone have been investigated as an alternative way to "lock" the global conformation of a synthetic polymer and herewith enhance a function.

We here concentrate on the application of reversible, non-covalent bonds to fold one single polymer chain into a single chain polymeric nanoparticle (SCPN), with an emphasis on hydrogen bonds because of their adaptable and dynamic nature. The process to "lock" a flexible polymer into a specific conformation by directional hydrogen-bond formation is often referred to as a "folding" process because of the reminiscence in which a polypeptide folds into α-helical and β-sheet structures, providing a tertiary, well-defined superstructure. This makes the folding–unfolding of the polymers reversible and we are convinced that this potential for adaptability is beneficial for enhancing activity and selectivity in catalytic applications. It is important to achieve the folding of one single polymer chain into a SCPN because this permits access to well-defined, unimolecular nanoreactors of nanometre size. These afford homogeneous catalysis solutions in which the hydrophobic pockets formed by the SCPNs provide a reaction space in which hydrophobic substrates accumulate and catalysts that normally only function in organic media can still operate.

Several reviews have been recently written on the topic of SCPNs [7]. In addition, the application of SCPNs in catalysis, [8] sensing, [9] and imaging [10] has been reported. We here mostly focus on polymers that form reversible, dynamic SCPNs in organic media and in water. The current understanding of how and when the polymers fold into SCPNs and which size and shape they adopt will be discussed. Then, we elaborate on the application of SCPNs as nanoreactors for efficient and selective catalysis in organic media but especially in water. Although this is still a very young field, it is becoming evident that dynamically folded polymers are efficient and selective catalysts. In addition, their compartmentalised structure and enzyme-like selectivity are only at the beginning of being exploited.

2 Folding of a Single Polymer Chain into a Dynamic Nanoparticle

Single polymer chains that have "sticky" grafts attached as pendants to the polymer backbone can form single SCPNs. When these sticky grafts aggregate via directional interactions into predetermined supramolecular complexes, a change in the conformation of the polymer backbone occurs (Fig. 1), which we refer to as a folding process. Temperature, pH, light or the addition of solvents are all triggers that can induce folding, and also unfolding, of these polymers. The nature and the directionality of the interactions in the supramolecular complex formation in combination with conflicts in solubility of the sticky pendants versus the backbone

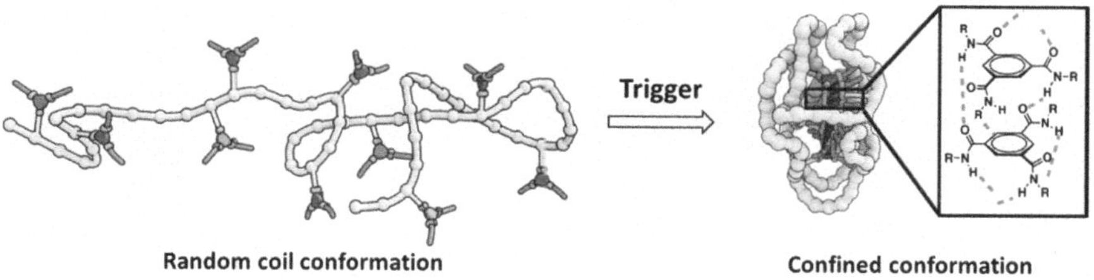

Fig. 1 Schematic representation of a polymer with pendant supramolecular motifs that, upon a trigger, self-assemble via triple hydrogen bonding into a helical aggregate. Hereby the backbone mobility of the random coil becomes restricted and a confined, compact conformation results

polymer in the selected solvent govern the global conformation—shape, size and compactness—of the polymer.

The first polymers with pendant supramolecular motifs capable of intramolecular dimerization via hydrogen-bond formation were reported by Hawker and co-workers [6b]. Their results showed that attaching benzamide-based groups, capable of dimerizing via twofold hydrogen bonding, to a polymethacrylate backbone resulted in the formation of well-defined, stable nanoparticles. At higher loadings of the benzamide groups, intramolecular hydrogen bonding was favoured over intermolecular interactions. By shifting to the ureidopyrimidinone (UPy) group, a self-dimerising unit, by Meijer and co-workers (see Fig. 2 for chemical structures), the dimerization strength of the motif was significantly increased [6b]. The change in polymer size before (when the UPy was protected with a UV-cleavable *o*-nitrobenzyl group, preventing dimerization) and after deprotection using UV light was pronounced, as evidenced by size exclusion chromatography (SEC) measurements. In addition, the dynamic nature of the SCPN formation was highlighted by the addition of acid, capable of breaking hydrogen bonds and increasing the hydrodynamic volume of the polymer. Stals et al. performed a systematic study on the balance between intramolecular UPy dimerization, polymer solubility and backbone rigidity in the formation of SCPNs (Fig. 2) [6g]. A combination of SEC, dynamic light scattering (DLS), proton nuclear magnetic resonance (^{1}H-NMR) and atom force microscopy (AFM) was used to elucidate all aspects of the folding processes. While ^{1}H-NMR indicated the formation of UPy dimers after UV deprotection, SEC and DLS revealed the reduction of the hydrodynamic diameters after hydrogen-bond formation and the single-chain character of the formed particles, respectively. Finally, AFM visualised the formed particles on a surface [11]. Interestingly, a wide variety of polymer backbones (polyacrylates, polymethacrylates, polystyrenes and polynorbornenes) were similarly suitable for SCPN formation via UPy dimerization. Crucially, the results revealed that the solvent choice was the determining factor in the intramolecular folding process; solvents that enhanced intramolecular hydrogen bonding, at the same time suppressed interparticle interactions. Other strongly dimerising self-complementary motifs [6l] but also hetero-complementary supramolecular motifs such as the cyanuric acid—Hamilton wedge, [12] thymine—diamidopyridine [13] and

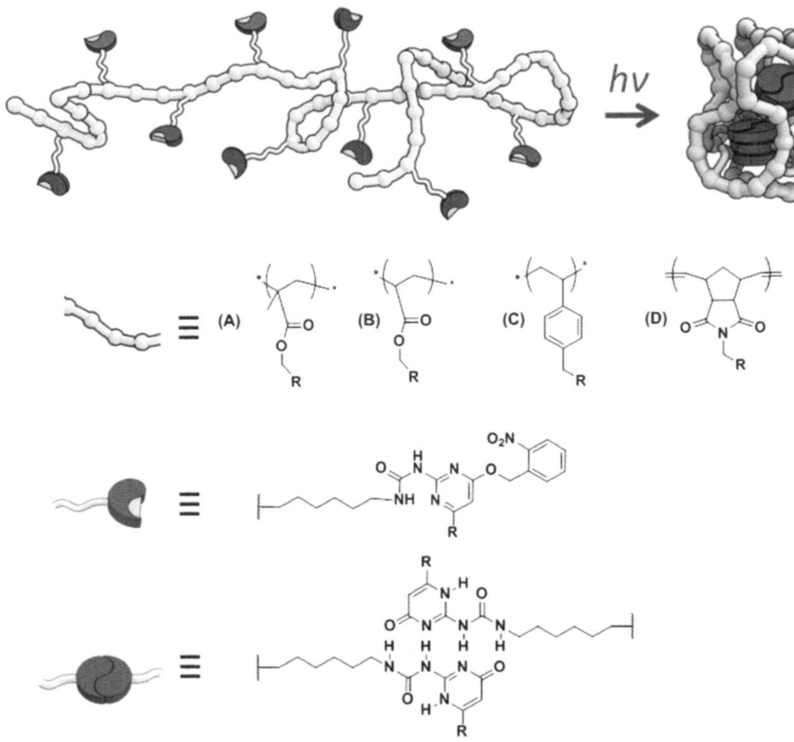

Fig. 2 Schematic representation of a polymer backbone (polyacrylate-, polymethacrylate-, polystyrene- or polynorbornene-based) with pendant *o*-nitrobenzyl protected (*yellow dots*) UPy motifs (*blue half moon*) that upon UV-induced cleavage of the *o*-nitrobenzyl group forms UPy dimers (*blue circles*) that induce the formation of a compact conformation (Reproduced from Ref. [6g] with permission from the Royal Society of Chemistry.) (Color figure online)

ureidoguanosine—diaminonaphthyridine [14] have been investigated and all units promoted single-chain folding.

Apart from supramolecular motifs that form dimeric complexes, motifs capable of cooperative supramolecular polymerizations are also highly interesting. Notably, the benzene-1,3,5-tricarboxamide (BTA) motif was introduced, which cooperatively self-assembles via directional hydrogen-bond formation into helical, cylindrically shaped aggregates (Fig. 1) [15]. Circular dichroism (CD) spectroscopy is a particularly useful tool to follow the self-assembly of BTAs because their helicity can be biased by introducing a stereogenic centre, which results in a bias for either *P* or *M* helical structures. By attaching *o*-nitrobenzyl protected BTA units, which are incapable of helical aggregation, to a poly(methacrylate) backbone, Mes et al. showed that the stepwise deprotection of the BTA units using UV light afforded particles with an increasingly structured inner compartment [6d]. Whereas CD

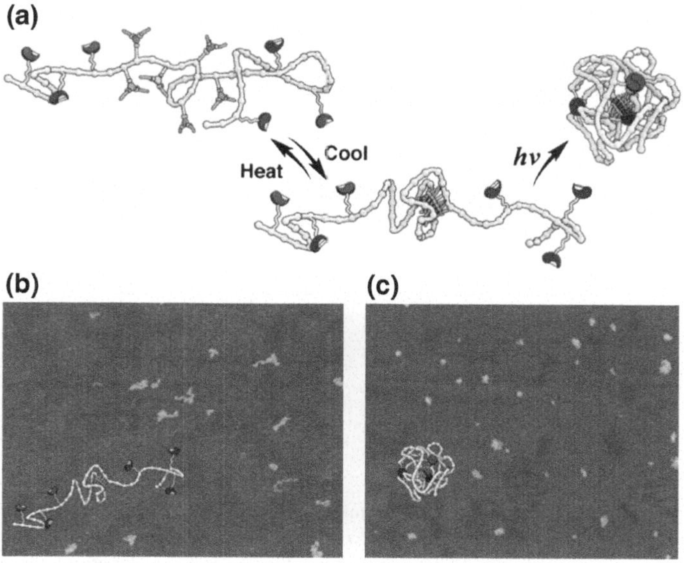

Fig. 3 **a** pUPy-BTA-pUPy triblock copolymers that undergo a two step folding process upon first a heating/cooling step (resulting in aggregation of the BTA moieties) followed by UV deprotection of the pUPy group affording UPy dimers; **b** AFM image of pUPy-BTA-pUPy showing more open structures on the mica surface; **c** AFM image of fully deprotected UPy-BTA-UPy showing compact structures on the mica surface

measurements clearly revealed this helical inner structure formation, X-ray scattering studies to confirm the single-chain character of the formed particles proved difficult because the system required the use of halogenated solvent mixtures. Later, results from Hosono et al. indicated that not all BTAs self-assembled in the same aggregate within the SCPN but rather that small domains of aggregated BTAs separated by disordered regions were formed [16]. This affords polymer conformations with more elongated structures. The combination of UPys and BTAs as pendants in triblock copolymers (either BTA-UPy-BTA [17] or UPy-BTA-UPy [6j]) showed that orthogonally self-assembling units could exist on one polymer chain and turning on both types of interactions resulted in a more compact conformation (Fig. 3a). This compaction was visualised by AFM measurements (Fig. 3b, c) and corroborated by small angle X-ray (SAXS) measurements.

In water, the hydrophobic interaction often dominates the aggregation behaviour of molecules and polymers. Also, water can interfere with hydrogen-bond formation, making it a challenging solvent for hydrogen-bond driven self-assembly processes. Terashima et al. [18] prepared the first series of amphiphilic BTA-based polymers that showed self-folding in water. The polymers consisted of a random poly(*o*EGMA-*co*-BTAMA) copolymer (*o*EGMA = oligoethylene glycol

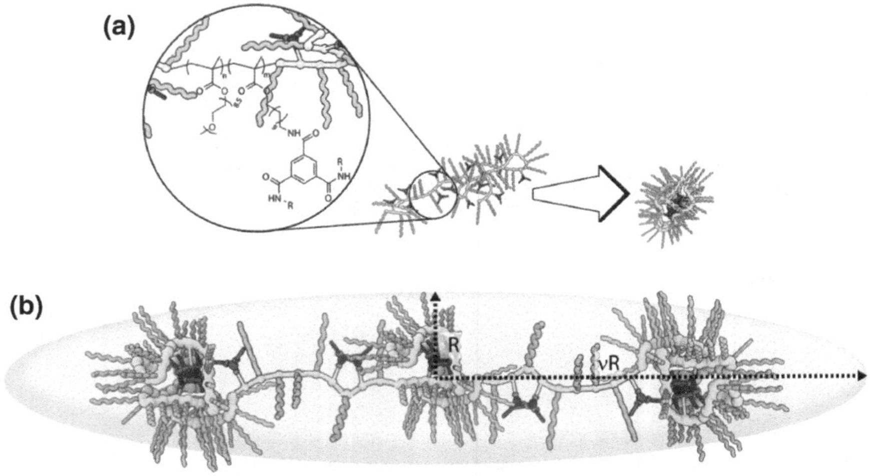

Fig. 4 **a** Schematic representation of the structure of amphiphilic poly(*o*EGMA-co-BTA) random copolymers that fold in water into compact conformations; **b** Representation of the global, ellipsoidal conformation of folded SCPNs in water; when the polymer length increases, only the major axis vR increases in length with the minor axis *R* remains constant (Reprinted with permission from Ref. [19]. Copyright 2014 American Chemical Society.)

methacrylate, BTAMA = BTA methacrylate), in which the *o*EG part compatibilized the polymer in water while the BTA pendant afforded the driving force for self-assembly via directional hydrogen bonding (Fig. 4a). The conformational behaviour of this series of polymers was studied in detail using a combination of spectroscopy (NMR, fluorescence, UV, CD and IR), SEC, microscopy and scattering studies (dynamic and static light scattering (DLS, SLS) small angle neutron scattering (SANS) and SAXS) [19, 20]. The results showed that the amphiphilic polymers adopted elongated conformations. In addition, SANS studies revealed that increasing the chain length of the polymers while keeping the BTAMA/*o*EGMA ratio constant at 1/10 resulted in an increase of the aspect ratio, while the diameter of the particle remained constant (Fig. 4b). Remarkably, these systems kept a predominantly single-chain character up to concentrations of 100 mg/mL. Importantly, fluorescence studies with the hydrophobic, solvatochromic dye Nile Red—widely used to probe the formation of hydrophobic pockets within self-assembled structures—confirmed the formation of a hydrophobic pocket [21].

The SCPNs unfolded upon the addition of isopropanol, a solvent that competes with the hydrogen-bond formation, or by increasing the temperature [19]. The latter proved to be a fully reversible process, indicating that the SCPNs adopt thermodynamically stable conformations in water. Remarkably, the temperature-dependent cooling curves revealed that not the total BTA concentration but the local BTA concentration determined the unfolding temperature, another indication that these are intramolecular processes. Evaluating the cooling curves of polymers with increasing chain lengths but constant ratio between *o*EGMA/BTAMA suggested that the folding/unfolding of the SCPNs is a non-cooperative process. This is highly remarkable in view of the highly cooperative nature of BTA aggregation [22]. The elongated shape of the SCPN was proposed to account for the non-cooperative folding observed as such a shape results in a constant local BTA concentration as the copolymer increases in length (Fig. 4b).

Apart from SCPNs with BTAs attached, a series of amphiphilic polymers was prepared by ter Huurne et al. which comprised bipyridine-based supramolecular motifs attached to the polymer backbone [6k]. These motifs are highly hydrophobic and show green fluorescence when self-assembled. Although they predominantly interact via π-stacking interactions, some additional structuring hydrogen bonding is also present. The combination of SAXS and CD spectroscopy revealed that the folding of the polymer was dependent on the solvent quality and temperature. The folding process in water was fine-tuned via the addition of a good co-solvent (tetrahydrofuran—THF), resulting in an optimal balance between the conformational freedom of the polymer's backbone and the stability of the π-stacked units (Fig. 5). SAXS experiments showed that the shape of the SCPNs was controlled by the formation of a chiral, internal secondary structure. Furthermore, the use of the electron-rich BiPy-BTA groups allowed, for the first time, the visualisation of the synthetic nanoparticle's internal structure using SAXS.

Finally, the BTAs that drive the folding process can also function as a recognition motif to capture "free" BTAs into the hydrophobic pocket. As a proof of concept, Huerta et al. incorporated a dye-labelled BTA into the SCPN, rendering them fluorescent [23]. This mix-and-match approach is an easy way to introduce functionalities into the SCPNs. In addition, Artar et al. showed that the folding driven by hydrogen bonding was fully orthogonal with intrachain metal–ligand complexation (Ru complexed to pendant triphenylphosphine moieties) [21]. The latter is highly important in view of rendering these systems catalytically active and will be discussed in the next paragraph.

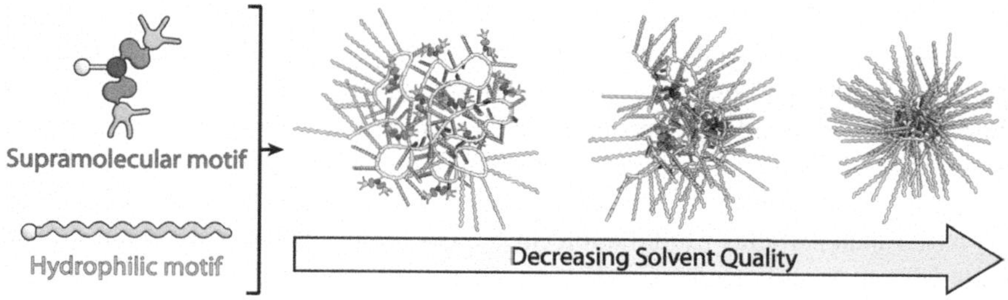

Fig. 5 Schematic representation of the folding of a hydrophilic BiPy-BTA containing copolymer into SCPNs as a function of solvent quality (Reproduced with permission from Ref. [6k]. Copyright 2015 American Chemical Society.)

3 Single-Chain Polymeric Nanoparticles as Compartmentalised Catalysts

3.1 Catalysis in Organic Media

Polymeric nanoparticles have been evaluated as catalysts for a variety of organic reactions [8]. The first example focussed on the use of star polymers—chemically cross-linked polymers usually consisting of more than one polymer chain—with embedded catalytic centres. Hawker and co-workers reported the formation of Pd loaded nanoparticles for hydrogenations and Heck reactions, [8a] whereas Sawamoto and co-workers focussed on Ru-loaded nanoparticles for oxidations and reductions [24]. The effectiveness of shielding catalyst sites within these nanoparticles was highlighted by the work of Fréchet and co-workers [25]. Either acid or basic sites were embedded in the interior of the star polymers. Efficient cascade catalytic reactions, in which an acid catalysed deprotection was followed by a base-catalysed condensation, were conducted. The site-isolation in the interior of the particles prevented the acid and base to negatively affect each other.

Later, single polymer chains were folded into single-chain nanoparticles, mostly using the metal itself to fold the particles via interaction with pendant ligands or reactive groups in the polymer main chain (Fig. 6). Elegant examples have been presented in which SCPNs showed catalytic activity in organic media, some even with high selectivity for specific substrates. For example, Pomposo and co-workers showed SCPNs active in diketone reductions and THF ring-opening polymerizations by entrapping $B(C_6F_5)_3$ units to oxygen-containing functional groups (ether, carbonyl) of the intramolecularly cross-linked nanoparticles via B···O interactions [8c]. Later, SCPNs were developed based on a poly(methylmethacrylate) backbone in which the interaction of Cu(II) with pendant diketo-units induced self-folding of the chains (Fig. 6a). These metallo-folded nanoparticles were active at low concentrations of Cu(II) ions in oxidative coupling reactions of terminal acetylenes in bulk conditions [8d]. Interestingly, a high substrate selectivity was found in the reactivity of the terminal acetylenes: some were reactive whereas others were not, despite the fact that all tested acetylenes were active when a "free" Cu(II) catalysts was applied. The substrate specificity exhibited by the SCPNs was tentatively attributed to the presence of multiple, compartmentalised local catalytic sites composed of Cu-complexes surrounded by an environment of methyl methacrylate repeat units allowing an optimum transition state stabilisation for the preferred substrates. Apart from these examples, Lemcoff and co-workers created Rh(I)- and Ir(I)-based nanoparticles by intramolecular crosslinking of the main chain double bonds in poly(cyclo-octadiene) (Fig. 6b) [8e]. The obtained particles were catalytically active in cross-coupling reactions and benzamide reductions for Rh(I)- and Ir(I)-based systems, respectively, albeit less active than the "free" organometallic complexes. In addition, the proximity of the metal centres promoted side reactions in case of the Rh(I) catalysed cross-coupling reactions. Intriguingly, it was also possible to prepare nanoparticles containing both Ir(I) and Rh(I) metal

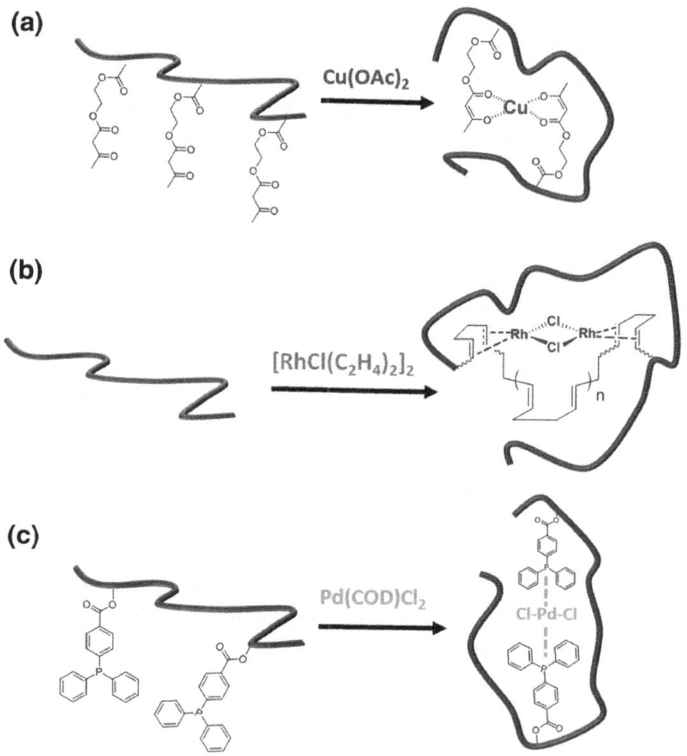

Fig. 6 **a** SCPNs based on a poly(methylmethacrylate) backbone in which folding is induced by Cu(II) complexation to pendant diketo-units [8d]; **b** SCPNs based on poly(cyclo-octadiene) in which folding is induced by complexation of the main chain double bonds to Rh(I) [8e]; **c** SCPNs based on polystyrene with pendant triphenylphosphine ligands in which folding is induced by complexation of the triphenylphosphines to Pd(II) [8f]

centres. Such bimetallic nanoparticles have the potential to achieve cooperative or tandem catalysis, in which a single particle combines the catalytic efficiency of both embedded metals. The group of Barner-Kowollik presented polystyrene-based copolymers with pendant triphenylphosphine ligands. These were reported to fold intramolecularly with Pd(II). The Pd(II) loaded nanoparticles were successfully applied in the Sonogashira coupling reaction between 2-bromopyridine and phenylacetylene, although the conversion was lower than the free catalyst in the same conditions (Fig. 6c) [8f].

3.2 *Catalysis in Water*

A major challenge is to prepare SCPNs that show high activity and selectivity in complex *aqueous* media. This is especially relevant in view of future biomedical applications and cascade catalytic conversions in water. Achieving catalytic conversions in water with rates and selectivities approaching those of enzymes has attracted considerable attention and many approaches have been followed to increase activity and selectivity of synthetic catalysts [26]. Promising are the utilisation of shell cross-linked nanoparticles [27], micellar structures [28], hydrogels, [29] star polymers, [24c] self-assembled systems [30] and dendrimers [31] which not only show high activity in water due to the so-called "concentrator effect" but also can give high selectivities. Among these, *homogeneous* water-soluble systems offer a number advantages over heterogeneous systems such as providing a relatively more uniform micro-environment around the catalytic sites, facilitating efficient diffusion of reagents and products and allowing easy recyclability [32].

The water-soluble, BTA-based SCPNs discussed above offer a compact yet dynamic and responsive inner structure; in contrast to the rather dense and "frozen" interior reaction spaces that are present in chemically cross-linked nanoparticles. These dynamic SCPNs could therefore be very interesting for efficient and selective catalysis in water. In addition, the "concentrator effect" can localise hydrophobic substrates in the interior of the particle, enhancing reactions rates. It is likely that both the enhanced conformational mobility as well as the concentration of substrates within the particles enhance the turnover frequencies of the catalysts applied compared to "free" catalytic complexes in water. In addition, the presence of the catalyst within the hydrophobic compartments of the SCPN may shield it from undesired interactions with other catalysts or reagents.

As discussed above, supramolecular folding of BTA-based amphiphilic polymer chains into SCPNs is an attractive way to prepare compartmentalised, water-soluble, nanometre-sized particles with a hydrophobic interior (Fig. 7). There are two main approaches that have been evaluated to access water-soluble, catalytically active SCPNs. In the first, the catalysts or ligands that bind a metal-based catalyst contain a polymerisable group and are copolymerised with water-soluble monomers and hydrophobic/structuring monomers. Frequently, (meth)acrylates and/or styrenes have been employed in combination with controlled radical polymerisation techniques. When there are no reactivity differences in the polymerisable monomers, random copolymers are obtained. Alternatively, segmented copolymers can be prepared by taking advantage of the reactivity differences between styrene- and methacrylate-based monomers [18, 21]. However, polymerizing highly functional monomers can be rather troublesome and needs careful optimisation since functional units (i.e. ligands, organocatalysts, charged units, radicals, etc.) can hamper controlled radical polymerization processes resulting in poorly defined products and broad molecular weight distributions of the prepared polymers [33]. As an alternative, the post-polymerization modification of poly(pentafluorophenyl acrylate) (pPFPA) [34] with amine derivatives carrying functional groups has been investigated [35]. This post-functionalization approach is a convenient way to prepare polymers with tuneable amounts of functional side chains. Using these approaches, a library of catalytically active SCPNs capable of reductions, oxidations, C–C bond forming reactions, deprotection reactions and click reactions have been evaluated (Fig. 8).

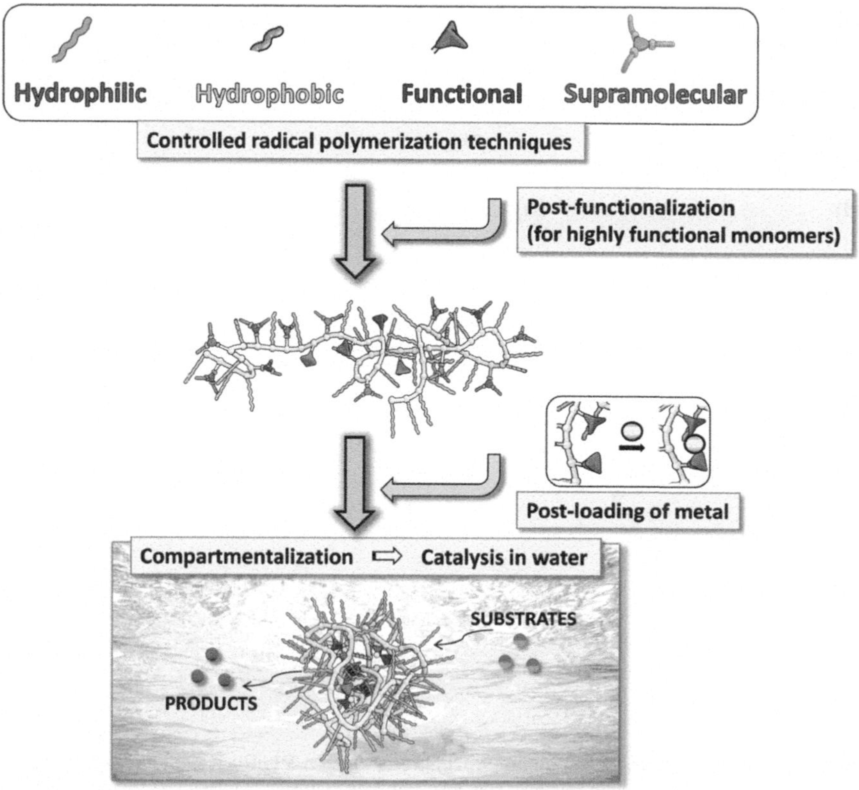

Fig. 7 General approach for preparing catalytically active SCPNs in water

Terashima et al. prepared a water-soluble segmented copolymer using the benzene-1,3,5-tricarboxamide motif as the supramolecular unit and triphenylphosphine (SDP) pendants as ligands for complexing Ru(II) (**P1**, Fig. 8). This system efficiently catalysed the transfer hydrogenation of ketones in the presence of sodium formate as the hydrogen source with a turnover frequency (TOF) of 20 h^{-1}, which was high compared to previous examples of similar Ru-based complexes in water [18]. Next, Artar et al. performed transfer hydrogenation experiments with SCPN-based catalysts in which the BTAMA was partially or complete replaced by lauryl methacrylate (LMA), which is hydrophobic but not capable of hydrogen-bond interactions [21]. The almost identical TOFs for **P1** and **P2** indicated that a tightly structured micro-environment around Ru sites via coordination of polymer pendant SDP ligands in combination with hydrophobic pendants were sufficient for the efficient shielding of the Ru catalyst within the SCPNs. While the additional structuring by BTA-induced stack formation was not essential for catalytic activity in water, the presence of hydrophobic groups was: poly(*o*EGMA-*co*-SDP) complexed to Ru(II) but lacking hydrophobic units did not catalyse transfer

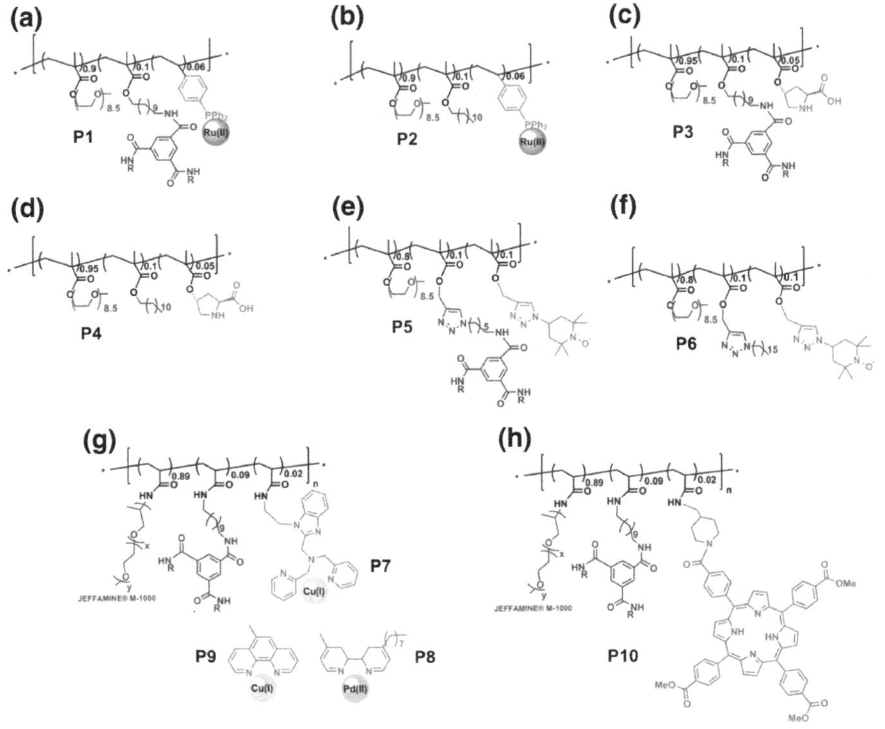

Fig. 8 Chemical structures of amphiphilic SCPNs employed **a P1**: Ru(II)@SCPN$_{BTA}$, **b P2**: Ru (II)@SCPN$_{LMA}$, **c P3**: Proline@SCPN$_{BTA}$, **d P4**: Proline@SCPN$_{LMA}$, **e P5**: TEMPO@SCPN$_{BTA}$, **f P6**: TEMPO@SCPN$_{LMA}$, **g P7**-**P8/P9**: Cu(I)/Pd(II)@SCPN, **h P10**: Porphyrin@SCPN (*R* = (S)-3,7-dimethyloctyl)

hydrogenations in water. Moreover, the **P1** and **P2** systems were uniquely active in the oxidation of secondary alcohols in the presence of *t*-BuOOH as the oxidant [36]. The locally increased concentration of substrates and active sites afforded a faster and more efficient catalyst at much lower catalyst and substrate concentrations with a TOF of 171 h^{-1} for oxidation of cyclohexanol to cyclohexanone compared to the previously reported non-immobilised catalyst (TOF = 6 h^{-1}) [37].

In another example, Huerta et al. reported L-proline functionalized SCPNs as highly active catalysts for the aldol reaction of cyclohexanone and *p*-nitrobenzaldehyde in water [33]. The BTA-based L-proline functionalized SCPN (**P3**) showed a diastereomeric excess up to 94% and an enantiomeric excess up to 70%. The activity and stereoselectivity of the reaction were intricately linked to the L-proline loading on the polymer chain and the amount of BTA as a structuring motif: higher L-proline loading gave higher activity but lower selectivity while replacing BTAMA with LMA (**P4**) resulted in an inactive organocatalyst. In a later study, in which a BTA functionalized L-proline was incorporated to a BTA-based SCPN via molecular recognition, a higher stereoselectivity was observed [23].

For both Ru@SCPN and proline@SCPN examples, the origin of the efficient catalysis was attributed to the high effective molarity of substrates and catalytic sites and shielding of the active sites via compartmentalization in water. Yet, the loss of catalytic activity in the absence of structuring motif in L-proline functionalized SCPNs indicated the need for a structured, conformationally adaptable pocket via BTA-induced helical stack formation and required further investigation. Recent studies on enzyme catalysis using 2,2,6,6-tetramethyl-1-piperidinyloxyl (TEMPO) radical spin labels have helped to elucidate the role of structure and dynamics of hydration water in relation to their activity [38]. Therefore, polymers **P5** and **P6** were prepared as models for **P3** and **P4**, respectively. Analysis of the polymers was performed using electron paramagnetic resonance (EPR) line-shape analysis and the local water translational diffusion dynamics within 0.5–1 nm of tethered TEMPO spin labels by solution-state Overhauser dynamic nuclear polarisation (ODNP) NMR relaxometry under ambient solution conditions by Stals and Cheng et al. [39]. A stronger retardation of surface water was reported for **P5** compared to **P6** as a result of a more ordered packing of the folded core by BTA self-assembly which consequently displays a more ordered polymer surface and solvation structure in water. The surface water diffusivity serves as an important parameter that differs for a structured and non-structured SCPN, however further research is needed to reveal the exact reason why a more retarded hydration shell would be so critical for the activity.

Recently, a new family of BTA-based amphiphilic SCPNs was prepared using the post-polymerization modification of poly(pentafluorophenyl acrylate) (pPFPA). Ligand-containing BTA-based amphiphilic SCPNs capable of coordinating to Cu(I) (**P7**, **P8**) or Pd(II) (**P9**) were prepared for bio-orthogonal organometallic catalysts while porphyrin-containing SCPNs were prepared as photosensitizers (**P10**) [35]. **P7** and **P8** significantly accelerated azide-alkyne cycloaddition reactions while **P9** efficiently catalysed depropargylation reactions. Most importantly, these catalytic reactions proceeded efficiently in phosphate buffer at physiological pH and at low substrate concentrations, demonstrating that SCPNs are promising systems to function in complex media, e.g. cellular environments.

Finally, Artar et al. showed that the SCPN folding around an intrinsically non-selective Ru(II) centre successfully created a selective hydrophobic reaction space in water [35]. A competition experiment was performed in which a mixture of alcohol substrates that varied in hydrophobicity (tetrahydropyranol, cyclohexanol and 4-*tert*-butylcyclohexanol) was oxidised to the corresponding ketones. The SCPN structure allowed for efficient diffusion and conversion of even the most water-soluble substrate (4-tetrahydropyranol). In addition, a significant selectivity was observed: both the rate as well as the end conversion differed between the three substrates (Fig. 9a). The results correlated nicely with the logP values of the alcohols: higher logP values resulted in higher rates and end conversions. In contrast, no selectivity was observed when the competition experiment was performed in the presence of $RuCl_2(PPh_3)_3$ (without polymer) at RT in acetone (Fig. 9b). A similar selectivity was observed in the reverse reaction, namely the transfer hydrogenation of the corresponding ketones. These results demonstrated that SCPN

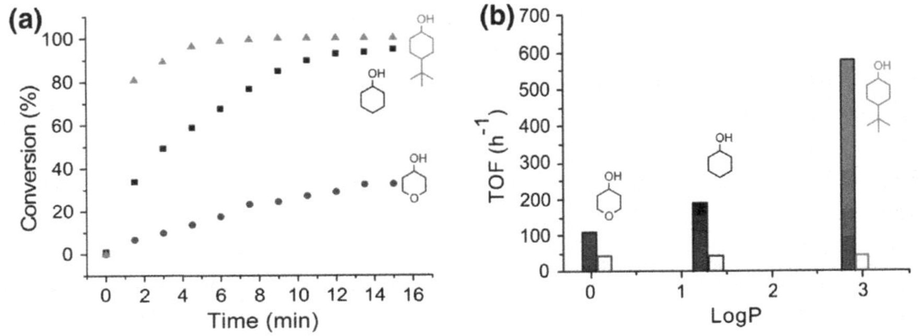

Fig. 9 **a** Conversion as a function of time of alcohols in the oxidation using $SCPN_{LMA}$@Ru(II) as the catalyst in a competition experiment in water. **b** TOF versus logP comparison in competition experiment using **P2** in water (*filled blocks*) and $Ru(II)Cl_2(PPh_3)_3$ without **P2** in acetone (*empty blocks*)

folding around the active site provided very efficient reaction spaces to achieve selectivity, based on hydrophobic effects.

4 Conclusions and Outlook

Folding a single amphiphilic polymer chain around a catalyst to create compartmentalised structures is an attractive approach to prepare highly active reaction spaces for catalysis in water. Advances in controlled radical polymerization techniques offer a huge potential to access polymeric backbones of controlled length, composition and, most importantly, sequence. This will permit to create highly defined polymer chains and, combined with post-functionalisation techniques, enable the preparation SCPNs with almost every functional unit of interest. Whereas many supramolecular motifs have been developed and studied in detail in the supramolecular chemistry community, also in water, only few have been evaluated for inducing single-chain folding of amphiphilic polymers in water. It is clear that the nature of the self-assembly process in combination with the selected polymer backbone will determine the global conformation of the formed nanoparticle, but also the hydrophobicity and compactness of the formed inner compartment. In addition, although the mechanisms of supramolecular self-assembly processes have been studied in detail, the nature of the folding process of single-chain polymers (cooperative or non-cooperative) has barely been explored.

There is much to be gained by a further integration of supramolecular chemistry, polymer chemistry and homogeneous catalysis. At the interface of these different fields of research, crucial next steps can be taken to develop active and selective enzyme mimics. Evidently, our ability to fold synthetic macromolecules into defined, compartmentalised structures is still far from the perfection achieved in the folding of polypeptides into functional conformations. The current state of the art in forming compartmentalised structures with an ordered interior is not enough. Precise location of self-assembling motifs within the polymer chain and enhanced control over molar mass dispersities are of crucial importance to improve the scope of synthetic analogues and achieve activities and selectivities in water that can match those of natural enzymes.

References

1. (a) Wu C, Wang X (1998) Globule-to-coil transition of a single homopolymer chain in solution. Phys Rev Lett 80:4092–4094; (b) Fetters LJ, Hadjichristis N et al (1994) Molecular weight dependence of hydrodynamic and thermodynamic properties for well-defined linear polymers in solution. J Phys Chem Rev Data 23:619–638
2. (a) Nakano T, Okamoto Y (2001) Synthetic helical polymers: conformation and function. Chem Rev 101:4013–4038; (b) Hill DJ, Mio MJ et al (2001) A field guide to foldamers. Chem Rev 101:3893–4011
3. (a) Yashima E, Maeda K et al (2009) Helical polymers: synthesis, structures, and functions. Chem Rev 109:6102–6211; (b) Yashima E, Maeda K (2008) Chirality-responsive helical polymers. Macromolecules 41:3–12; (c) Yamamoto T, Yamada T et al (2010) High-molecular weight polyquinoxaline-based helically chiral phosphine (PQXphos) as chirality-switchable, reusable, and highly enantioselective monodentate ligand in catalytic asymmetric hydrosilylation of styrenes. J Am Chem Soc 132:7899–7900; (d) Wang H, Li N et al (2015) Synthesis and properties of a novel Cu(II)-pyridineoxazoline containing polymeric catalyst for asymmetric Diels-Alder reaction. RSC Adv 5:2882–2890
4. (a) Mecerreyes D, Lee V, Hawker CJ et al (2001) A novel approach to functionalized nanoparticles: self-crosslinking of macromolecules in ultradilute solution. Adv Mater 13:204–208; (b) Harth E, van Horn B et al (2002) A facile approach to architecturally defined nanoparticles via intramolecular chain collapse. J Am Chem Soc 124:8653–8660; (c) Croce TA, Hamilton SK et al (2007) Alternative *o*-quinodimethane cross-linking precursors for intramolecular chain collapse nanoparticles. Macromolecules 40:6028–6031; (d) Cherian AE, Sun FC et al (2007) Formation of nanoparticles by intramolecular cross-linking: following the reaction progress of single polymer chains by atomic force microscopy. J Am Chem Soc 129:11350–11351; (e) de Luzuriaga AR, Ormategui N et al (2008) Intramolecular click cycloaddition: an efficient room-temperature route towards bioconjugable polymeric nanoparticles. Macromol Rapid Commun 29:1156–1160; (f) Beck JB, Killops KL et al (2009) Facile preparation of nanoparticles by intramolecular cross-linking of isocyanate functionalized copolymers. Macromolecules 42:5629–5635; (g) Zhu B, Ma J et al (2001) Formation of polymeric nanoparticles via Bergman cyclization mediated intramolecular chain collapse. J Mater Chem 21:2679–2683; (h) He J, Tremblay L et al (2011) Preparation of polymer single chain nanoparticles using intramolecular photodimerization of coumarin. Soft Matter 7:2380–2386; (k) Ormategui N, García I et al (2012) Synthesis of single chain thermoresponsive polymer nanoparticles. Soft Matter 8:734–740; (i) Dirlam PT, Kim HJ et al (2013) Single chain polymer nanoparticles via sequential ATRP and oxidative polymerization. Polym Chem 4:3765–3773; (k) Altintas O, Willenbacher J et al (2013) A mild and efficient approach to functional single-chain polymeric nanoparticles via photoinduced Diels-Alder ligation. Macromolecules 46:8092–8101; (l) Perez-Baena I, Asenjo-Sanz I et al (2014) Efficient route to compact single-chain nanoparticles: photoactivated synthesis via thiol-yne coupling reaction. Macromolecules 47:8270–8280; (m) Roy RK, Lutz J-F (2014) Compartmentalization of single polymer chains by stepwise intramolecular cross-linking of sequence-controlled macromolecules. J Am Chem Soc 136:12888–12891; (n) Frank PG, Tuten BT et al (2014) Intra-chain photodimerization of

pendant anthracene units as an efficient route to single-chain nanoparticle fabrication. Macromol Rapid Commun 35:249–253; (o) Wong EHH, Qiao GC (2015) Factors influencing the formation of single-chain polymeric nanoparticles prepared via ring-opening polymerization. Macromolecules 48:1371–1379; (p) Jeong J, Lee Y-J et al. (2015) Colored single-chain polymeric nanoparticles via intramolecular copper phthalocyanine formation. Polym Chem 6:3392–3397

5. (a) Murray BS, Fulton DA (2011) Dynamic covalent single-chain polymer nanoparticles. Macromolecules 44:7242–7252; (b) Tuten BT, Chao D et al (2012) Single-chain polymer nanoparticles via reversible disulfide bridges. Polym Chem 3:3068–3071; (c) Whitaker DE, Mahon CS et al (2013) Thermoresponsive dynamic covalent single-chain polymer nanoparticles reversibly transform into a hydrogel. Angew Chem Int Ed 52:956–959; (d) Sanchez-Sanchez A, Fulton DA et al (2014) pH-responsive single-chain polymer nanoparticles utilising dynamic covalent enamine bonds. Chem Commun 50:1871–1873
6. (a) Ghosh S, Ramakrishnan S (2005) Small-molecule-induced folding of a synthetic polymer. Angew Chem 117:5577–5583; (b) Seo M, Beck JB et al (2008) Polymeric nanoparticles via noncovalent cross-linking of linear chains. Macromolecules 41:6413–6418; (c) Foster EJ, Berda EB et al (2009) Metastable supramolecular polymer nanoparticles via intramolecular collapse of single polymer chains. J Am Chem Soc 131:6964–6966. (d) Mes T, van der Weegen R et al (2011) Single-chain polymeric nanoparticles by stepwise folding. Angew Chem Int Ed 50:5085–5089; (e) Appel EA, del Barrio J et al (2012) Metastable single-chain polymer nanoparticles prepared by dynamic cross-linking with nor-seco-cucurbit 10 uril. Chem Sci 3:2278–2281. (f) Altintas O, Lejeune E et al (2012) Bioinspired dual self-folding of single polymer chains via reversible hydrogen bonding. Polym Chem 3:640–651; (g) Stals PJM, Gillissen MAJ et al (2013) The balance between intramolecular hydrogen bonding, polymer solubility and rigidity in single-chain polymeric nanoparticles. Polym Chem 4:2584–2597; (h) Stals PJM, Li Y et al (2013) How far can we push polymer architectures? J Am Chem Soc 135:11421–11424; (i) Gillissen MAJ, Voets IK et al (2012) Single chain polymeric nanoparticles as compartmentalised sensors for metal ions. Polym Chem 3:3166–3174; (j) Hosono N, Gillissen MAJ et al (2013) Orthogonal self-assembly in folding block copolymers J Am Chem Soc 135:501–510; (k) ter Huurne GM, Gillissen MAJ et al (2015) The coil-to-globule transition of single-chain polymeric nanoparticles with a chiral internal secondary structure. Macromolecules 48:3949–3956; (l) Wang F, Pu H et al (2015) From single-chain folding to polymer nanoparticles via intramolecular quadruple hydrogen-bonding interaction. J Polym Sci A Polym Chem 53:1832–1840
7. (a) Altintas O, Barner-Kowollik C (2012) Single chain folding of synthetic polymers by covalent and non-covalent interactions: current status and future perspectives. Macromol Rapid Commun 33:958–971. (b) Sanchez-Sanchez A, Perez-Baena I et al (2013) Advances in click chemistry for single-chain nanoparticle construction. Molecules 18:3339–3355; (c) Lyon CK, Prasher AAM et al (2015) A brief user's guide to single-chain nanoparticles. Polym Chem 6:181–197; (d) Gonzalez-Burgos M, Latorre-Sanchez A, Pomposo JA (2015) Advances in single chain technology. Chem Soc Rev 44:6122–6142; (e) Artar M, Huerta E, Meijer EW, Palmans ARA (2014) In: Lutz JF, Meyer TY, Ouchi M, Sawamoto M (eds) Sequence-controlled polymers: synthesis, self-assembly, and properties. American Chemical Society, Washington, DC, pp 313–325; (f) Altintas O, Barner-Kowollik C (2015) Single-chain folding of synthetic polymers: a critical update. Macromol Rapid Commun 37:29–46; (g) Mavila S, Eivgi O, Berkovich I, Lemcoff NG (2016) Intramolecular cross-linking methodologies for the synthesis of polymer nanoparticles. Chem Rev 116:878–961
8. (a) Bosman AW, Vestberg R et al (2003) A modular approach toward functionalized three-dimensional macromolecules: from synthetic concepts to practical applications. J Am Chem Soc 125:715–728. (b) Wulff G, Chong BO (2006) Soluble single-molecule nanogels of controlled structure as a matrix for efficient artificial enzymes. Angew Chem Int Ed 45:2955–2958; (c) Perez-Baena I, Barroso-Bujans F et al (2013) Endowing single-chain polymer nanoparticles with enzyme-mimetic activity. ACS Macro Lett 2:775–779;

(d) Sanchez-Sanchez A, Arbe A et al (2015) Metallo-folded single-chain nanoparticles with catalytic selectivity. ACS Macro Lett 3:439–443; (e) Mavila S, Rozenberg I et al (2014) A general approach to mono- and bimetallic organometallic nanoparticles. Chem Sci 5:4196–4203; (f) Willenbacher J, Altintas O et al (2015) Pd-complex driven formation of single-chain nanoparticles. Polym Chem 6:4358–4365; (g) Sanchez-Sanchez A, Arbe A, Kohlbrecher J, Colmenero J, Pomposo JA (2015) Efficient synthesis of single-chain globules mimicking the morphology and polymerase activity of metalloenzymes. Macromol Rapid Commun 6:1592–1597
9. Gillissen MAJ, Meijer EW, Voets IK, Palmans ARA (2012) Single chain polymeric nanoparticles as compartmentalised sensors for metal ions. Polym Chem 3:3166–3174
10. Perez-Baena I, Loinaz I, Padro D et al (2010) Single-chain polyacrylic nanoparticles with multiple Gd(III) centres as potential MRI contrast agents. J Mater Chem 20:6916–6922
11. van Roekel HWH, Stals PJM et al (2013) Evaporative self-assembly of single-chain, polymeric nanoparticles. Chem Commun 49:3122–3124
12. Altintas O, Gerstel P et al (2010) Single chain self-assembly: preparation of α, ω-donor-acceptor chains via living radical polymerization and orthogonal conjugation. Chem Commun 46:6291–6293
13. Altintas O, Rudolph T et al (2011) Single chain self-assembly of well-defined heterotelechelic polymers generated by ATRP and click chemistry revisited. J Polym Sci A Polym Chem 49:2566–2576
14. Romulus J, Weck M (2013) Single-chain polymer self-assembly using complementary hydrogen bonding units. Macromol Rapid Commun 34:1518–1523
15. Smulders MMJ, Schenning APHJ et al (2008) Insight into the mechanisms of cooperative self-assembly: the "sergeants-and-soldiers" principle of chiral and achiral C_3-symmetrical discotic triamides. J Am Chem Soc 130:606–611
16. Hosono N, Palmans ARA et al (2014) "Soldier-sergeant-soldier" triblock copolymers: revealing the folded structure of single-chain polymeric nanoparticles. Chem Commun 50:7990–7993
17. Hosono N, Stals PJM et al (2014) Consequences of block sequence on the orthogonal folding of triblock copolymers. Chem Asian J 9:1099–1107
18. Terashima T, Mes T et al (2011) Single-chain folding of polymers for catalytic systems in water. J Am Chem Soc 133:4742–4746
19. Gillissen MAJ, Terashima T et al (2013) Sticky supramolecular grafts stretch single polymer chains. Macromolecules 46:4120–4125
20. Stals PJM, Gillissen MAJ et al (2014) Folding polymers with pendant hydrogen bonding motifs in water: the effect of polymer length and concentration on the shape and size of single-chain polymeric nanoparticles. Macromolecules 47:2947–2954
21. Artar M, Terashima T et al (2014) Understanding the catalytic activity of single-chain polymeric nanoparticles in water. J Polym Sci A Polym Chem 52:12–20
22. Nakano Y, Hirose T et al (2012) Conformational analysis of supramolecular polymerization processes of disc-like molecules. Chem Sci 3:148–155
23. Huerta E, van Genabeek B et al (2014) A modular approach to introduce function into single-chain polymeric nanoparticles. Macromol Rapid Commun 35:1320–1325
24. (a) Terashima T, Ouchi M, Ando T et al (2011) Transfer hydrogenation of ketones catalyzed by PEG-armed ruthenium-microgel star polymers: microgel-core reaction space for active, versatile and recyclable catalysis. Polym J 43:770–777; (b) Terashima T, Ouichi M, AndoT et al (2010) Thermoregulated phase-transfer catalysis via PEG-armed Ru(II)-bearing microgel core star polymers: efficient and reusable Ru(II) catalysts for aqueous transfer hydrogenation of ketones. J Polym Sci A Polym Chem 48:373–379; (c)Terashima T, Ouchi M et al (2011) Oxidation of sec-alcohols with Ru(II)-bearing microgel star polymer catalysts via hydrogen transfer reaction: unique microgel-core catalysis. J Polym Sci A Polym Chem 49:1061–1069

25. (a) Helms B, Guillaudeu SJ, Xie Y et al (2005) One-pot reaction cascades using star polymers with core-confined catalysts. Angew Chem Int Ed 44:6384–6387; (b) Chi Y, Scroggins ST, Frechet JMJ (2008) One-pot multi-component asymmetric cascade reactions catalyzed by soluble star polymers with highly branched non-interpenetrating catalytic cores. J Am Chem Soc 130:6322–6323
26. (a) Liang C, Fréchet JMJ (2005) Applying key concepts from nature: transition state stabilization, pre-concentration and cooperativity effects in dendritic biomimetics. Prog Polym Sci 30:385–402; (b) Kofoed J, Reymond J-L (2005) Dendrimers as artificial enzymes. Curr Opin Chem Biol 9:656–664; (c) Dwars T, Paetzold E et al (2005) Reactions in micellar systems. Angew Chem Int Ed 44:7174–7199; (d) Kirkorian K, Ellis A et al (2012) Catalytic hyperbranched polymers as enzyme mimics: exploiting the principles of encapsulation and supramolecular chemistry. Chem Soc Rev 41:6138–6159
27. Liu Y, Wang Y et al (2011) Shell cross-linked micelle-based nanoreactors for the substrate-selective hydrolytic kinetic resolution of epoxides. J Am Chem Soc 133:14260–14263
28. (a) Ge Z, Xie D et al (2007) Stimuli-responsive double hydrophilic block copolymer micelles with switchable catalytic activity. Macromolecules 40:3538–3546; (b) Lu A, Cotanda P et al (2012) Aldol reactions catalyzed by L-proline functionalized polymeric nanoreactors in water. Chem Commun 48:9699–9701; (c) van Oers MCM, Abdelmohsen LKEA et al (2014) Aqueous asymmetric cyclopropanation reactions in polymersome membranes. Chem Commun 50:4040–4042; (d) Cardozo AF, Julcour C et al (2015) Aqueous phase homogeneous catalysis using core-shell nanoreactors: application to rhodium catalyzed hydroformylation of 1-octene. J Catal 324:1–8
29. (a) Berdugo C, Miravet JF et al (2013) Substrate selective catalytic molecular hydrogels: the role of the hydrophobic effect. Chem Commun 49:10608–10610; (b) Haimov A, Neumann R (2006) An example of lipophiloselectivity: the preferred oxidation, in water, of hydrophobic 2-alkanols catalyzed by a cross-linked polyethyleneimine–polyoxometalate catalyst assembly. J Am Chem Soc 128:15697–15700; (c) Moore BL, Moatsou D et al (2014) Studying the activity of the MacMillan catalyst embedded within hydrophobic cross-linked polymeric nanostructures. Polym Chem 5:3487–3494; (d) Lu A, Moatsou D (2014) Recyclable L-proline functional nanoreactors with temperature-tuned activity based on core-shell nanogels. ACS Macro Lett 3:1235–1239
30. (a) Huerta E, van Genabeek B et al (2015) Triggering activity of catalytic rod-like supramolecular polymers. Chem Eur J 21:3682–3690; (b) Neumann LN, Baker MB et al (2015) Supramolecular polymers for organocatalysis in water. Org Biomol Chem 13:7711–7719
31. (a) Bosman AW, Janssen HM et al (1999) About dendrimers: structure, physical properties and applications. Chem Rev 99:1665–1688; (b) Grayson SM, Fréchet JMJ et al (2001) Convergent dendrons and dendrimers: from synthesis to applications. Chem Rev 101:3819–3868; (c) Helms B, Meijer EW et al (2006) Dendrimers at work. Science 313:929–930; (d) Yu J, RajanBabu TV et al (2008) Conformationally driven asymmetric induction in a catalytic dendrimer. J Am Chem Soc 130:7845–7847
32. (a) Frenzel T, Solodenko W, Kirschning A (2003) In: Buchmeiser MR (ed) Polymeric materials in organic synthesis and catalysis. Wiley-VCH, Weinheim, pp 201–240; (b) Dickerson TJ, Reed NN et al (2002) Soluble polymers as scaffolds for recoverable catalysts and reagents. Chem Rev 102:3325–3344; (c) Helms B, Fréchet JMJ (2006) The dendrimer effect in homogeneous catalysis. Adv Synth Catal 348:1125–1148
33. Huerta E, Stals PJM et al (2013) Consequences of folding a water-soluble polymer around an organocatalyst. Angew Chem Int Ed 52:2906–2910
34. Eberhardt M, Mruk R et al (2005) Postpolymerization modification of poly(pentafluorophenyl methacrylate): synthesis of a diverse water-soluble polymer library. Eur Polym J 41:1569–1575

35. Liu Y, Pauloehrl T et al (2015) A modular synthetic platform for the construction of functional single-chain polymeric nanoparticles: from aqueous catalysis to photo-sensitization. J Am Chem Soc 137:13096–13105
36. Artar M, Souren ERJ et al (2015) Single chain polymeric nanoparticles as selective hydrophobic reaction spaces in water. ACS Macro Lett 4:1099–1103
37. Murahashi SI, Naota T et al (1993) Ruthenium-catalyzed oxidations for selective syntheses of ketones and acyl cyanides—selective acylation of amino-compounds with acyl cyanides. Synthesis 4:433–440
38. (a) Armstrong BD, Han S et al (2009) Overhauser dynamic nuclear polarization to study local water dynamics. J Am Chem Soc 131:4641–4647; (b) Franck JM, Pavlova A et al (2013) Quantitative Overhauser effect dynamic nuclear polarization for the analysis of local water dynamics. Prog Nucl Magn Reson Spectrosc 74:33–65
39. Stals PJM, Cheng C-Y et al (2016) Surface water retardation around single-chain polymeric nanoparticles: critical for catalytic function? Chem Sci 7:2011–2015

Microgel Star Polymer Catalysts as Active and Functional Nanoreactors for Organic Reactions and Polymerizations

Takaya Terashima and Mitsuo Sawamoto

Abstract In this chapter, we present recent advances on microgel star polymer catalysts as active and functional nanoreactors for organic reactions and polymerizations. Microgel star polymers are soluble core-shell gels consisting of a covalently crosslinked microgel core that is covered by multiple linear arm chains. The central cores thus serve as nano-confined spaces for various applications. Actually, microgel star polymers are effective as polymeric scaffolds to immobilize metal catalysts and organocatalysts since the catalysts are not only dissolved in a homogeneous phase but also effectively protected within the crosslinked network and isolated with arm chains from outer environments. A series of functional microgel star polymer catalysts are efficiently synthesized by the linking reaction of linear polymers with functional linking agents and monomers in living radical polymerization. Metal-bearing microgel star polymer catalysts have high activity, functionality tolerance, recyclability, and provide easy product recovery in *sec*-alcohol oxidation, ketone hydrogenation, and living radical polymerization; the compatibility of catalytic activity and practicability is much better than those of conventional heterogeneous supported catalysts and homogeneous nonsupported counterparts. Because of their inherent structure, star polymer catalysts further afford unique thermoregulated phase-transfer catalysis and one-pot cascade reactions. Thus, microgel star polymers are one of the most advanced polymer supports with nano-confined designer spaces for catalysis.

1 Introduction

Ideal catalysis is an active, efficient, and selective transformation of substrates into desired products, compatible with high practicability and environmental adaptability. Various metal or organocatalysts have been developed so far to selectively

---

T. Terashima (✉) · M. Sawamoto
Department of Polymer Chemistry, Graduate School of Engineering, Kyoto University, Katsura, Nishikyo-ku, Kyoto 615-8510, Japan
e-mail: terashima@living.polym.kyoto-u.ac.jp

R. Poli (ed.), *Effects of Nanoconfinement on Catalysis*,
Fundamental and Applied Catalysis, DOI 10.1007/978-3-319-50207-6_6

obtain products in high yield, with special focus on the design of catalytic sites including metal species and ligands. However, the practical use and industrial applications of catalytic systems require high efficiency, reproducibility, sustainability, and cost-effectiveness; easy product recovery, catalyst separation, catalyst recycle, and high catalyst stability are thus crucial.

To improve these factors, various polymer-supported catalysts and related nanoreactor systems have been created [1–10]. They generally include two categories, dependent on the scaffold solubility and the system homogeneity: (1) heterogeneous or dispersed systems with insoluble polymeric scaffolds such as macrogels and microgels, or with block copolymer micelles, and (2) homogeneous systems with soluble polymeric or macromolecular scaffolds such as linear random copolymers, hyperbranched polymers, microgel star polymers, and dendrimers (Fig. 1). Importantly, polymer-supported catalysts afford easy product recovery and catalyst recycle by using insolubility, precipitation, amphiphilic, or thermoresponsive solubility, while they can further provide inherent reaction spaces around the catalytic centers to modulate the reactivity to unique catalysis.

Insoluble gel-supported catalysts, typically based on polystyrene gels and silica gels, can be easily separated from products after reactions by simple filtration to provide virtually pure products without any further purification and afford easy catalyst recycle. However, such insoluble catalysts often have low catalytic activity and selectivity owing to reduced accessibility of substrates to the catalytic sites. Soluble polymer-supported catalysts, typically using linear random copolymers, in turn induce active and selective catalysis comparable to the original non-supported catalysts but often suffer from catalyst leaching from the polymer support owing to insufficient catalyst binding.

Microgel-core star polymers [7–13] are one of the most promising soluble polymers to enclose catalysts. They have core-shell structure comprising a crosslinked globular core (<50 nm) and multiple linear arms on the polymer shell

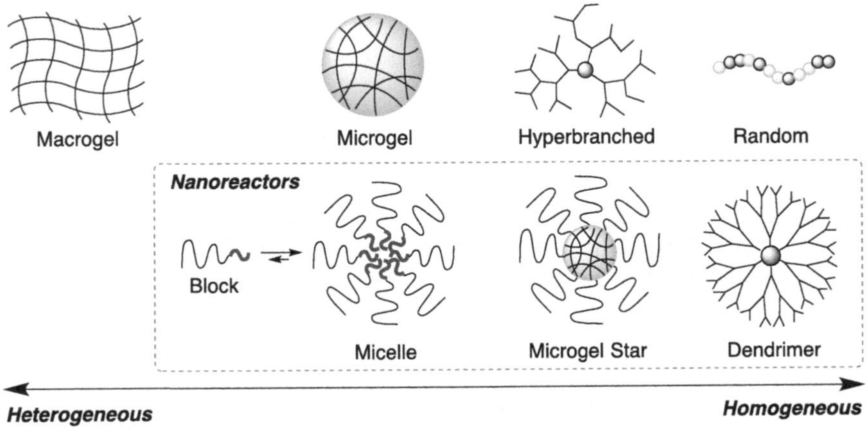

Fig. 1 Macromolecular scaffolds to immobilize catalysts

(10–1000 arms) [14, 15]. Thus, encapsulation of catalysts into the microgel cores affords active, selective, and robust catalysis compatible with high practicability: (1) high activity and selectivity by homogeneous solubility thanks to the polymer shell arms, (2) high stability and functionality tolerance by catalyst protection within the microgel network, (3) easy product recovery and catalyst recycle by precipitation or thermoresponsive solubility, and (4) efficient and unique catalysis within nano-confined designer spaces. Importantly, microgel-core star polymers give totally homogeneous but locally crosslinked spaces around the catalytic centers; they thus work as active and stable nanoreactors, [7–10] different from the others typically based on block copolymer micelles [16, 17] and dendrimers [18–20].

Given these features, we have recently developed catalyst-embedded microgel star polymer catalysts as active, efficient, and recyclable nanoreactors for organic reactions and for polymerization [7–10, 21–28]. This chapter reviews recent advances on microgel star polymer catalysts, with special focus on the design of star polymer catalysts via living radical polymerization and on their catalytic applications [21–31].

2 Microgel Star Polymers as Functional Polymeric Nanospaces

Microgels include several types with different three-dimensional and/or crosslinked structure: totally crosslinked microgels and nanogels, shell-crosslinked microgels (shell-crosslinked micelles), and core-crosslinked microgels (microgel-core star polymers, core-crosslinked micelles) [6, 7, 32, 33]. Microgel-core star polymers are one class of core-crosslinked microgels solubilized by arm polymer chains. The solubility of the star polymers is thus highly dependent on that of the arms and thereby can be easily controlled by the arm monomer species.

In general, microgel star polymers are obtained from the crosslinking reaction of linear polymers (living polymers, macroinitiators, and macromonomers) with multifunctional linking agents such as divinyl compounds (e.g., divinylbenzene: DVB, ethylene glycol dimethacrylate:EGDMA) via living polymerization systems (Fig. 2a) [11, 12, 34]. This smart strategy was first developed in the 1960s by living anionic polymerization [34]. For example, living polystyryl lithium ($PSt^{-}Li^{+}$) was directly crosslinked with DVB to form star-shaped polystyrene with a DVB microgel core [35]. The arm number and total molecular weight can be typically controlled by the feed ratio of the living polymers and DVB. Such arm-linking method is now widely applied to several living polymerization systems to afford various microgel star polymers with different types of arms, microgels, and functional groups [36, 37].

The emergence of "living" radical polymerization systems in the 1990s and the subsequent development of the related initiating/catalytic systems have opened, through versatile design, access to functional polymeric materials with

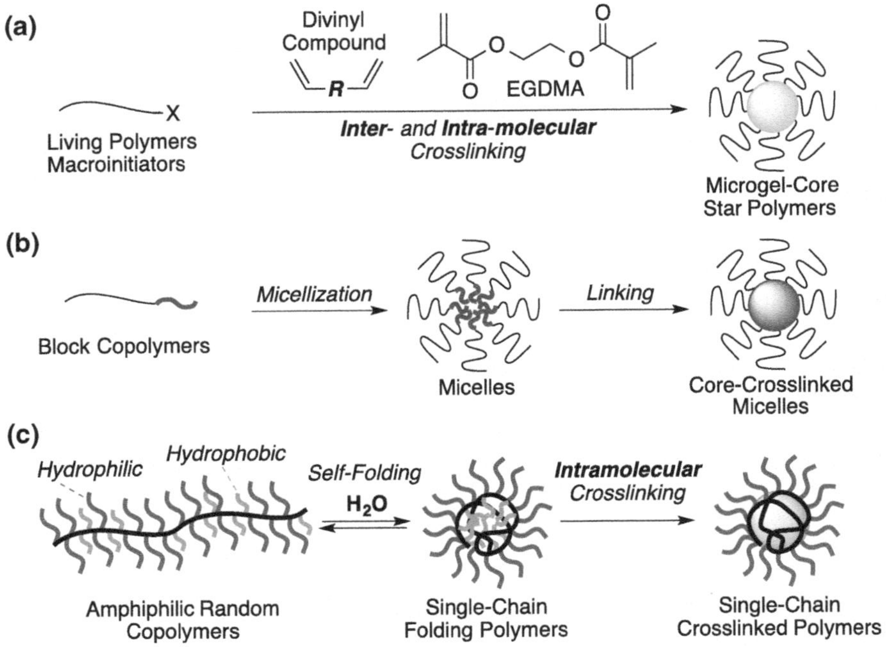

Fig. 2 Various core-crosslinked microgels: **a** microgel-core star polymers via the crosslinking of arm polymers with linking agents, **b** core-crosslinked micelles via the crosslinking of block copolymer micelles, and **c** single-chain crosslinked polymers via the intramolecular crosslinking of single-chain folding amphiphilic random copolymers in water

well-controlled primary structure and three-dimensional architecture [38–41]. In particular, the high functionality tolerance originating from the neutral radical active species allows us to directly functionalize polymers with polar functional monomers and employ polar solvents (e.g., alcohols and water) [42–45]. Thus, functional star polymers can be also tailor-made by living radical polymerizations more efficiently and conveniently than by the other living polymerization systems. The position-selective functionalization of star polymers is available via living radical polymerization as follows.

1. Surface-functionalized star: Crosslinking of end-functional linear polymers obtained from living radical polymerization with functional initiators.
2. Arm-functionalized star: Crosslinking of functional linear polymers obtained from living radical polymerization of functional monomers.
3. Core-functionalized star: Crosslinking of linear arm polymers with functional linking agents and/or monomers, in which functional groups are selectively accumulated within the crosslinked cores.

In particular, core functionalization is attractive to produce unique "functional nanospaces" (Fig. 3) [8, 9, 13]. The key is to accumulate functional groups (**1**–**8**) into the solubilized microgel cores; such star polymers serve as nanocapsules for selective recognition, encapsulation, and stimuli-responsive release of molecules, [46–48], while they also work as nanoreactors for catalysis [21–31] as described in this chapter.

Microgel-core star polymers have a core-shell structure almost identical to that of core-crosslinked micelles, while the microscopic structure and crosslinking density would be slightly different because of the distinct synthetic approaches. Core-crosslinked micelles are obtained by the crosslinking of block copolymer micelles (Fig. 2b), as the related examples applicable to catalytic systems described in other chapters [49, 50]. One advantage of this method would be high yield synthesis of resulting core-crosslinked microgels; additionally, the size and molecular weight are in principle controlled with the pre-organized micelle template. But such micelle-mediated process may involve the limitation of design versatility and monomer selection. In contrast, recent advance of the catalytic systems in living radical polymerization affords the efficient and easy synthesis and versatile design of functional microgel star polymers with on-targeted structure and narrow size distribution in high yield (>90%).

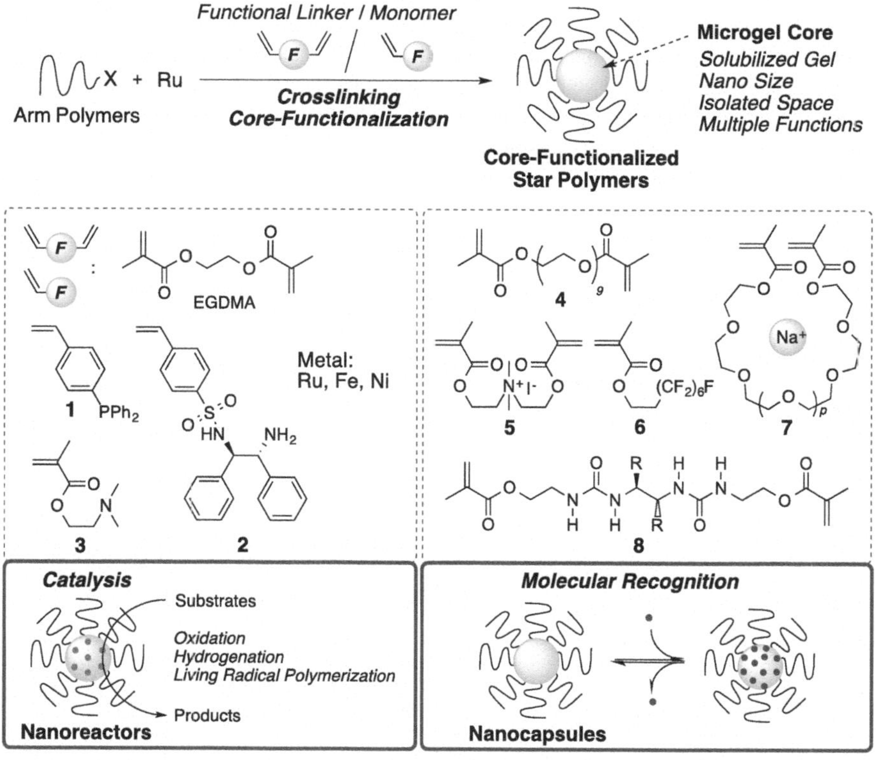

Fig. 3 Core-functionalized star polymers via arm-linking with functional linkers and monomers (EGDMA, (**1**–**8**): nanoreactors for catalysis and nanocapsules for molecular recognition

Quite recently, single-chain crosslinked polymers have also been created via the selective intramolecular crosslinking of self-folding amphiphilic random copolymers in water (Fig. 2c); they are regarded as a new crosslinked star polymers consisting of a single main chain [51]. Related single-chain nanoparticles are also employed as polymeric scaffolds to enclose catalysts for unique catalysis [42], as described in Chap. 5.

3 One-Pot Synthesis of Metal-Embedded Microgel Star Polymer Catalysts

Given these features, we have produced various metal-bearing microgel star polymer catalysts by ruthenium-catalyzed living radical polymerization with a phosphine ligand monomer (**1**) to create unique nanoreactors for active, selective, and recyclable catalytic systems (Fig. 4) [21–28]. Using living radical polymerization, catalysts can be loaded onto the desired sites of the star polymers by adjusting the timing to introduce functional monomers with ligands, catalysts, and their precursors or using such functional initiators during the synthetic process [21, 42, 49, 52]. In our works, we selectively introduced metal catalysts into "microgel cores" of star polymers because of the following prospective advantages: catalyst-enclosed microgel star polymers would effectively stabilize and protect the catalysts with the microgel cores, while the catalysts could have high activity and selectivity owing to the high solubility.

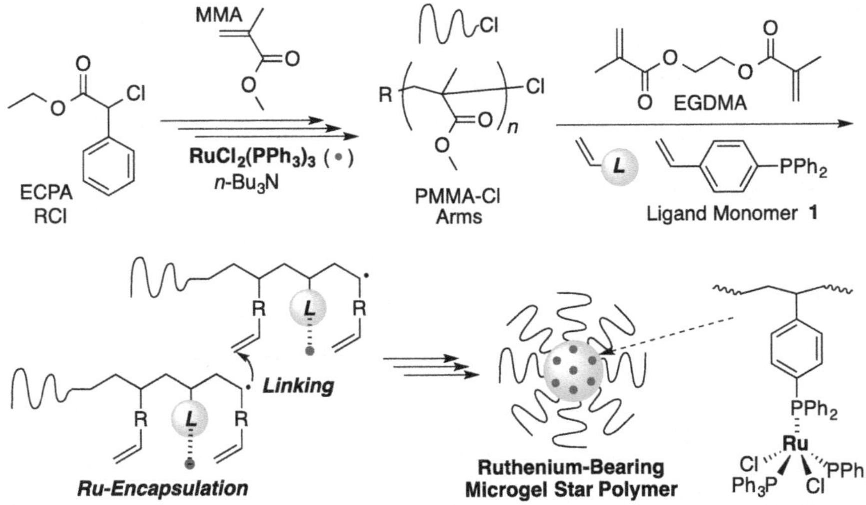

Fig. 4 One-pot synthesis of ruthenium-bearing microgel star polymers via the $RuCl_2(PPh_3)_3$-catalyzed linking reaction of chlorine-capped poly(methyl methacrylate)s (PMMA-Cl) with EGDMA and a phosphine ligand monomer (**1**)

For this, we developed a one-pot synthesis leading to the encapsulation of metal catalysts into the microgel star polymer core [21–25]. If the desired in-core metal catalysts are identical to those used for the star polymer synthesis (polymerization), this one-pot method is more efficient and convenient than the conventional synthetic strategies of polymer-supported metal catalysts: (1) post immobilization of metal catalysts (metal complex precursors or salts) into ligand-bearing polymers that are obtained from (co)polymerization of ligand monomers or (2) polymerization of metal-bearing monomers that are synthesized with ligand monomers and metal precursors. The former has the advantage of versatility to incorporate metal complexes, while it sometimes suffers from the low efficiency to load the metals owing to the steric hindrance and/or solubility of the polymeric ligands. The latter requires high compatibility of the metal-containing monomers and the polymerization systems.

To directly synthesize ruthenium-bearing star polymer catalysts via one-pot metal encapsulation, we employed a phosphine-bearing styrene derivative (**1**) as a ligand monomer during the arm-linking process in $RuCl_2(PPh_3)_3$/n-Bu_3N-catalyzed living radical polymerization [21, 22, 24]. Typically, chlorine-capped poly (methyl methacrylate)s with narrow molecular weight distribution (PMMA-Cl: $M_n = \sim 10{,}000$, $M_w/M_n < 1.2$) are first prepared by $RuCl_2(PPh_3)_3$/n-Bu_3N-mediated polymerization of MMA with a chlorine-based initiator [chlorine-capped MMA dimer: H-$(MMA)_2$-Cl or ethyl 2-chloro-2-phenylacetate: ECPA] in toluene at 80 °C. Into the polymerization mixture ($\sim$90% MMA conversion), the toluene solution of EGDMA, **1**, and $RuCl_2(PPh_3)_3$ was directly added under argon to subsequently induce the crosslinking reaction of the PMMA-Cl arms at 80 °C (a typical condition: $[\text{PMMA-Cl}]_0/[\text{EGDMA}]_0/[\mathbf{1}]_0/[RuCl_2(PPh_3)_3]_{all} = 1/10/5/1$, $[RuCl_2(PPh_3)_3]_{all}$: all ruthenium concentration during arm linking). During the microgelation process, the phosphine ligand monomer (**1**) simultaneously undergoes ligand exchange with the ruthenium polymerization catalyst $[RuCl_2(PPh_3)_3]$ to form ruthenium-bearing monomers in situ [proposed monomer complexes: $RuCl_2(PPh_3)_{3-n}\mathbf{1}_n$]. As a result, the ruthenium complexes are directly embedded into microgel cores of star polymers with PMMA arms. The purified star polymers (after the removal of free $RuCl_2(PPh_3)_3$) still exhibited red-brown and ultraviolet-visible (UV-vis) absorption derived from in-core ruthenium complexes ($\lambda_{max} = 475$ nm). By this method, ruthenium-bearing microgel star polymers were obtained in high yields (over 80%), confirmed by size-exclusion chromatography (SEC). The additional feed of $RuCl_2(PPh_3)_3$ in the microgelation step is required to enhance the arm-crosslinking reaction and ruthenium content loaded into the resulting cores. The star polymer was characterized by SEC-coupled with multi-angle laser light scattering detector (SEC-MALLS) as follows: absolute weight average molecular weight (M_w) = 650,000, average arm numbers (N_{arm}) = 35, and radius of gyration (R_g) = 10 nm.

This one-pot synthetic strategy further allows us to efficiently control arm length, M_w, N_{arm}, and the amount of in-core ligand and ruthenium atoms [22]. The key is to change the feed ratio of EGDMA and/or **1** to PMMA-Cl ($r_{linking} = [\text{EGDMA}]_0/[\text{PMMA-Cl}]_0 = 5, 10, 20$, $r_{ligand} = [\mathbf{1}]_0/[\text{PMMA-Cl}]_0 = 1.25, 2.5, 5$) and the degree of polymerization of MMA (arm length: DP = $[MMA]_0/[initiator]_0$ = 50, 100, 200). To keep an identical concentration of active radical intermediates, we employed a constant concentration of PMMA-Cl chains (Cl terminals) and $RuCl_2(PPh_3)_3$ catalysts ([PMMA-Cl] = 13 mM, $[RuCl_2(PPh_3)_3]$ = 13 mM). Under these conditions, various ruthenium-bearing microgel star polymer catalysts are systematically synthesized in high yield (mostly in 80–90% yield) (Fig. 5a) and characterized by SEC-MALLS, UV-vis, and inductively coupled plasma-atomic emission spectroscopy (ICP-AES). The scope, structure control, and features of the star polymer catalysts are summarized as follows.

1. **Arm length**. This is tunable by the feed ratio of MMA to chlorine initiator (DP = 50, 100, and 200).
2. $\boldsymbol{M_w}$ **and** $\boldsymbol{N_{arm}}$. M_w increases with increasing $r_{linking}$ from 5 to 10 or 20 under constant r_{ligand} (2.5) owing to the efficient intermolecular crosslinking reaction of PMMA arms: M_w = 170, 592, 1080 K. As a result, N_{arm} also increased from 11 to 31 or 56. M_w and N_{arm} also increase upon increasing r_{ligand} from 1.25 to 2.5 or 5 under constant $r_{linking}$ (10). This importantly means that the ruthenium complexes within the microgel cores also enhance the intermolecular crosslinking of arm polymers via the coordination of ruthenium metals to phosphine ligands; the ruthenium coordination also works as physical crosslinking sites for the arm polymer chains. In contrast, keeping $r_{linking}$ (10) and r_{ligand} (2.5) constant, the arm length (DP = 50, 100, 200) does not significantly affect N_{arm}, which remained almost constant ($\sim$30).
3. $\boldsymbol{N_{ligand}}$ **and** $\boldsymbol{N_{Ru}}$. The Ru concentration in the polymer product increased with r_{ligand} under constant $r_{linking}$ (10) and arm DP (100): e.g., [Ru] = 31, 45,

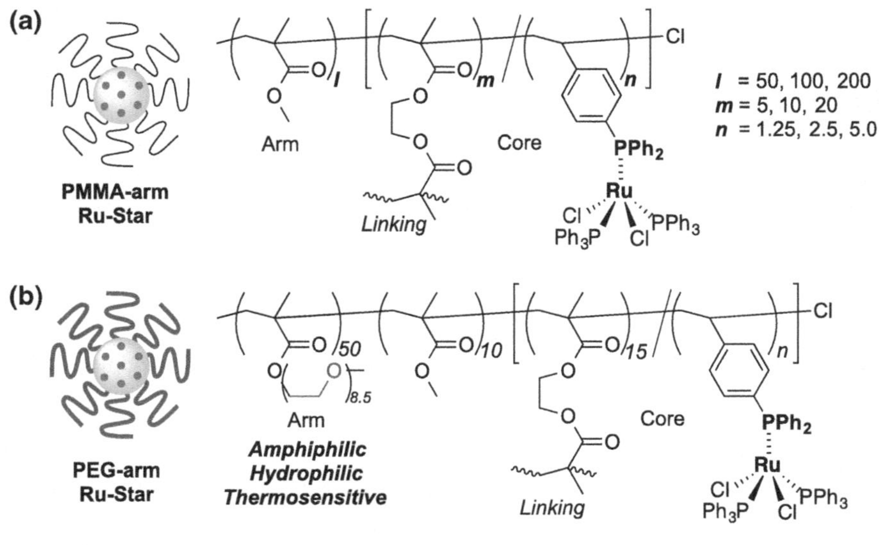

Fig. 5 Design of arm polymers [**a** PMMA arms, **b** poly(PEOMA) arms] and primary structure (*l* arm length, *m* crosslinking density, *n* ligand number) for ruthenium-bearing star polymers

74 μmol/g-polymer for r_{ligand} = 1.25–2.5 or 5. Thus, the number of ligands and ruthenium atoms per a single star polymer (N_{ligand} and N_{Ru}) similarly increased with r_{ligand}. The Ru content loaded in star polymer microgels just depended on r_{ligand}, i.e., it is independent of arm DP and EGDMA feed ratio ($r_{linking}$). This supports the idea that the Ru encapsulation originates from the coordination of **1** to the ruthenium metals, clearly distinct from the physical anchoring of free ruthenium complexes within the microgel network. It should also be noted that the use of a relatively large amount of **1** (r_{ligand} = 5; $[RuCl_2(PPh_3)_3]_0$/ $[\mathbf{1}]_0$ = 1/5) affords quantitative one-pot encapsulation of $RuCl_2(PPh_3)_3$. The star polymer carries approximately 150 phosphine ligands (N_{ligand}) per macromolecule.

Thus, the one-pot ruthenium encapsulation methodology is quite effective to design and synthesize star polymer catalysts with multiple microgel-enclosed ruthenium atoms. In this system, free (non-encapsulated) ruthenium can be easily removed via silica-gel column chromatography of the crude products of star polymer catalysts: just passing the toluene mixture through the column under nitrogen or argon (the ruthenium-bearing star polymer is eluted while the free ruthenium complex is adsorbed at the top of the column). Compared with the micelle-crosslinking methods described in Chap. 7 [49, 50], the star polymer catalysts obtained by our one-pot method have the drawback of the incorporation of residual small amounts of unreacted arms (~10%). However, because the arm residue contains a very small number of phosphine ligands and ruthenium atoms, the effect of the arm residue on catalysis could be virtually ignored [21, 26]. If necessary, the arm residue can be removed by fractional precipitation with toluene and methanol [24, 25].

The globular structure of the star polymer catalysts and the core-bound ruthenium can be further directly observed by atomic force microscopy (AFM) and transmission electron microscopy (TEM). Interestingly, the star polymer catalysts showed black dots of 2–3 nm even without any external staining agents, because the core-bound ruthenium complexes effectively stain the core [21, 22].

4 Versatile Design and Functionalization of Metal-Bearing Microgel Star Polymer Catalysts

Microgel star polymers allow versatile design and functionalization by living radical polymerization to afford unique nanoreactors with on-demand solubility and functions. Ruthenium-bearing star polymer catalysts with PMMA arms are fully soluble in various organic solvents including toluene, acetone, tetrahydrofuran (THF), chloroform ($CHCl_3$), dichloromethane (CH_2Cl_2), and *N*,*N*-dimethylformamide (DMF). Thus, they can work as homogenous catalysts in such organic solvents. However, organic reactions often employ water and polar organic solvents such as alcohols. For example, hydrogenation often requires alcohols (e.g., 2-propanol) as hydrogen source [53, 54].

We thus further synthesized water-soluble, hydrophilic, and thermoresponsive PEG-armed star polymer catalysts (Fig. 5b) [23]. For this, poly(ethylene glycol) methyl ether methacrylate (PEGMA: M_n = 475; average 8.5 ethylene oxide units) was employed to prepare hydrophilic poly(PEGMA) arms by $RuCl_2(PPh_3)_3$/*n*-Bu_3N-mediated polymerization. Like the hydrophobic PMMA counterparts, the chlorine-capped poly(PEGMA)-Cl arms were efficiently crosslinked with EGDMA and **1** in the presence of $RuCl_2(PPh_3)_3$, giving PEG-armed star polymer catalysts with Ru-bearing microgels. It should be noted that ruthenium encapsulation was successfully achieved with **1** even in the presence of polar PEG arms. The resulting star polymers are fully soluble in water and various alcohols (e.g., methanol, ethanol). Uniquely, the star polymers exhibited upper-critical solution temperature (UCST)-type solubility in 2-propanol because of the thermosensitive poly (PEGMA) arms (UCST = ~31 °C).

In addition to the solubility tuning with arm polymers, the core-bound metal can also be changed (e.g., to iron, nickel) via ligand exchange reaction to produce various metal-bearing microgel star polymer catalysts (Fig. 6a) [24]. Typically, a ruthenium-bearing star polymer with PMMA arms was treated with a large excess of a hydrophilic and basic phosphine [$P(CH_2OH)_3$] in toluene/alcohol mixtures at 80 °C ([in-core Ru]/[$P(CH_2OH)_3$] = 1/60 mol/mol), followed by the precipitation into methanol, to give a phosphine ligand-bearing and metal-free microgel star polymer. The star polymer turned into a white powder after precipitation; the solution exhibited no UV-vis absorption of in-core Ru, indicating the efficient removal of ruthenium from the core. Confirmed by ^1H and ^{31}P NMR, the star polymer ligand clearly showed phenyl proton and phosphine signals originating from the in-core ligand (**1**), whereas a ruthenium-bearing star polymer exhibited broad proton and no phosphine signals because of low mobility. This demonstrates that ruthenium coordination works as physical crosslinking. Additionally, NMR measurements have recently allowed a detailed investigation of the mobility and structure of in-core functional groups [48].

Such empty-core star polymer ligands allow the encapsulation of various metal complexes and salts [FeX_2, NiX_2, $FeX_2(PPh_3)_2$, $NiX_2(PPh_3)_2$, $NiBr_2(PBu_3)_2$, X = Br or Cl] by simply mixing the star polymer ligands and the metals. However, star polymer ligands and catalysts obtained via living radical polymerization still carry a small amount of unreacted olefins and living chlorine termini within the microgel cores, which may induce side reactions and/or macroscopic gelation during catalytic applications. Thus, hydrogenation of microgel cores was also investigated to prepare star polymer ligands and catalysts with "inert" microgel cores [24, 25]. Taking advantage of the hydrogenation activity of $RuCl_2(PPh_3)_3$ in the presence of K_2CO_3 and 2-propanol [54], we developed an intriguing "tandem catalyst interchange" method via $RuCl_2(PPh_3)_3$-mediated polymerization to sequentially synthesize star polymer ligands with hydrogenated microgels. For this, the crude solution of $RuCl_2(PPh_3)_3$-bearing star polymer catalysts after polymerization was directly added into a 2-propanol/toluene solution of K_2CO_3 under

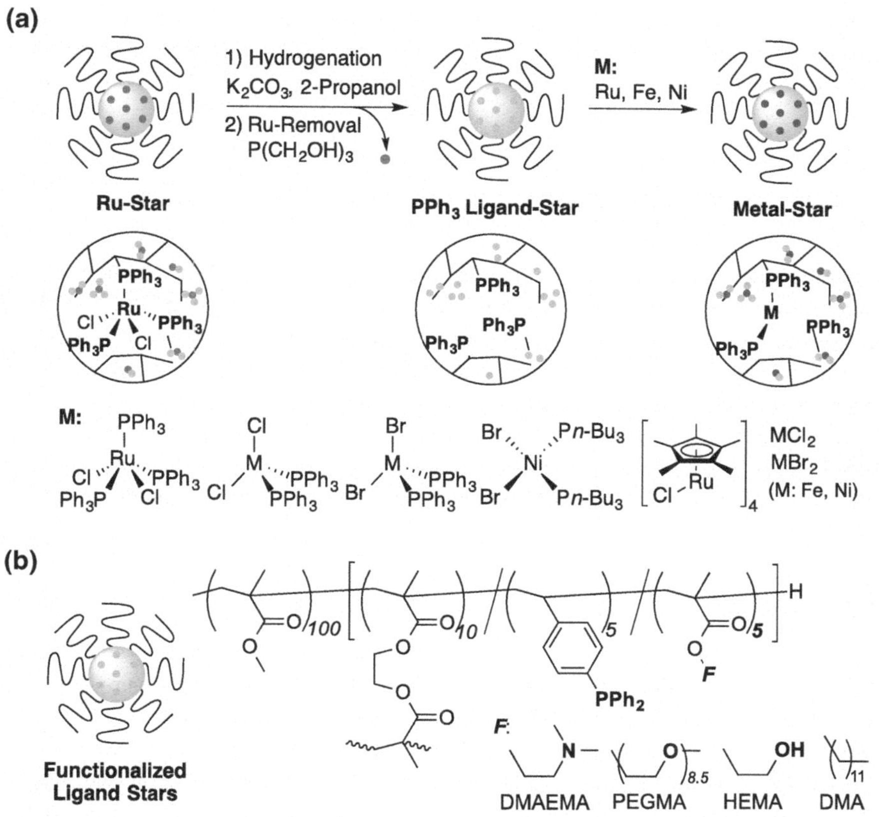

Fig. 6 **a** Various metal-bearing star polymer catalysts via in-core hydrogenation and metal interchange and **b** functionalized ligand star polymers

argon; the mixture was then stirred at 80 °C. The in-core ruthenium chloride was immediately transformed into ruthenium hydride to efficiently induce hydrogenation of the in-core olefins and chlorine termini. Then, the in-core ruthenium was removed with $P(CH_2OH)_3$ to give hydrogenated star polymer ligands.

The design and functionalization of microgel networks is also quite important to make unique reaction spaces around catalytic sites. The polarity and functions of phosphine ligand-bearing microgel spaces are tunable by the copolymerization of EGDMA and **1** with other functional monomers [2-(dimethylamino)ethyl methacrylate (DMAEMA), 2-hydroxyethyl methacrylate (HEMA), PEGMA: hydrophilic; dodecyl methacrylate (DMA): hydrophobic] (Fig. 6b) [25]. The introduction of amino groups into the microgel spaces actually enhanced the activity and stability of ruthenium catalysts for living radical polymerization. The one-pot synthesis is further applicable to the generation of chiral ruthenium-bearing star polymer catalysts. A chiral diamine-bearing styrene (**2**: Fig. 3) efficiently afforded chiral Ru complexes within the microgel cores, confirmed by circular dichroism (CD) and UV-vis [9]. Thus, ruthenium-catalyzed living radical polymerization is one of the most powerful tools to design microgel star polymer catalysts and ligands.

5 Organic Reactions with Star Polymer Metal Catalysts

To examine potential and features as catalyst-embedded nanoreactors, ruthenium-bearing microgel star polymers obtained via one-pot $RuCl_2(PPh_3)_3$ encapsulation were employed for oxidation of *sec*-alcohols in acetone [21, 26] and for the hydrogenation of ketones in 2-propanol [27, 28] via the hydrogen transfer mechanism. $RuCl_2(PPh_3)_3$ is well known to be an effective catalyst (precursor) for these reactions [53]. Here, we investigated the activity, stability, versatility, and recyclability of the star polymer catalysts, compared with conventional polymer-supported catalysts and homogeneous (non-supported) counterparts.

5.1 Oxidation

$RuCl_2(PPh_3)_3$-embedded microgel star polymers with PMMA arms were effective as catalysts for the oxidation of various *sec*-alcohols (**A1**–**A7**) (Fig. 7) [21, 26]. Typically, a ruthenium-bearing star polymer catalyst [Ru-Star: M_w = 388,000, 24 arms (M_n = 10,000, DP = ~100), **1**-Ligand (N_{ligand}): 60/star, core-Ru (N_{Ru}): 15/star] was applied to the oxidation of 1-phenylethanol (**A1**) in acetone in the presence of K_2CO_3 at 65 °C (under reflux), using a very small catalyst charge: $[substrate]_0/[Ru]_0$ = 1000/1 (mol/mol). The star polymer homogenenously induced the oxidation of **A1** into acetophenone in high yield (90%, 8 h); the turnover frequency reached 110 h^{-1}. The catalytic activity was superior to that of insoluble polystyrene gel-supported Ru (64%, 8 h) or linear MMA/**1** random copolymer-supported Ru (52%, 8 h) under the same conditions of temperature and substrate/catalyst ratio. This indicates the following important aspects: (1) in-core ruthenium complexes show high activity even within the crosslinked network, (2) the substrate efficiently diffuses within the core to encounter the in-core catalysts, and (3) the microgel core is fully soaked with acetone (solvent) as hydrogen acceptor for oxidation.

We further investigated the effects of arm length (DP = 50, 100, 200), crosslinking ratio (EGDMA: $r_{linking}$ = 5, 10, 20), and in-core ligand (**1**)/Ru ratio (N_{ligand}/N_{Ru} = 2.5, 4.0) on the catalytic activity for the oxidation of **A1**. The catalytic activity was highly dependent on the in-core ligand (**1**)/Ru ratio: the activity (TOF) increased from 110 to 300 h^{-1} with decreasing the **1**/Ru ratio from 4.0 to 2.5 [21, 26], indicating that a ruthenium catalyst with empty coordination site would be an active intermediate for the oxidation. Surprisingly, the activity is independent of other factors such as arm length (DP) and crosslinking ratio of EGDMA ($r_{linking}$).

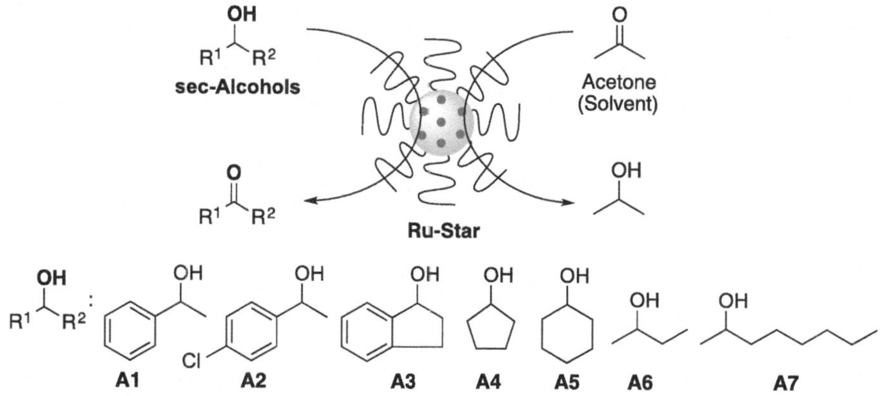

Fig. 7 Oxidation of sec-alcohols (**A1–A7**) with ruthenium-bearing microgel PMMA-arm star polymer catalysts (Ru-Star) and K_2CO_3 in acetone at 65 °C: $[substrate]_0/[Ru]_0 = 1000/1$

This demonstrates that ruthenium-embedded microgel cores are fully swollen by acetone and have enough void spaces in the network structure. Quite recently, we have revealed that EGDMA-based microgel cores of star polymers, isolated from arm-cleavable microgel star polymers, are soluble in various organic solvents and have swollen and globular structure [15]. This finding is fully consistent with the proposed reasons for such high catalytic activity.

The star polymer catalysts further exhibited high stability and recyclability. The star polymer catalysts can be reused three times for the oxidation of **A1** without any retardation of activity even after catalyst recovery in air. Easy product recovery was also achieved: pure products virtually without ruthenium residue were obtained just by precipitation of the star polymer catalysts in hexane. Additionally, the star polymer structure and core-bound ruthenium were still maintained after oxidation, respectively confirmed by SEC and UV-vis spectroscopy.

5.2 *Hydrogenation*

Given these results, we further examined transfer hydrogenation of ketones (**K1–K12**) with star polymer catalysts and K_2CO_3 in 2-propanol at 100 °C (Fig. 8) [27]. For homogeneous catalysis, hydrophilic poly(PEGMA) arm star polymers with $RuCl_2(PPh_3)_3$-embedded microgels (PEG Ru-Star) were employed. This hydrogenation is the reverse reaction of the oxidation examined above. In fact, PEG Ru-Star [$M_w = 772{,}000$, 16 $PEOMA_{50}$-MMA_{10} block arms ($M_n = 17{,}000$), **1**-Ligand (N_{ligand}): 20/star, core-Ru (N_{Ru}): 20/star] homogeneously catalyzed the hydrogenation of acetophenone (**K1**, 86%, 4 h) with a low catalyst feed ratio ($[substrate]_0/[Ru]_0 = 1000/1$). The reaction was faster than with either a hydrophobic PMMA-arm star polymer catalyst, a polystyrene gel-supported

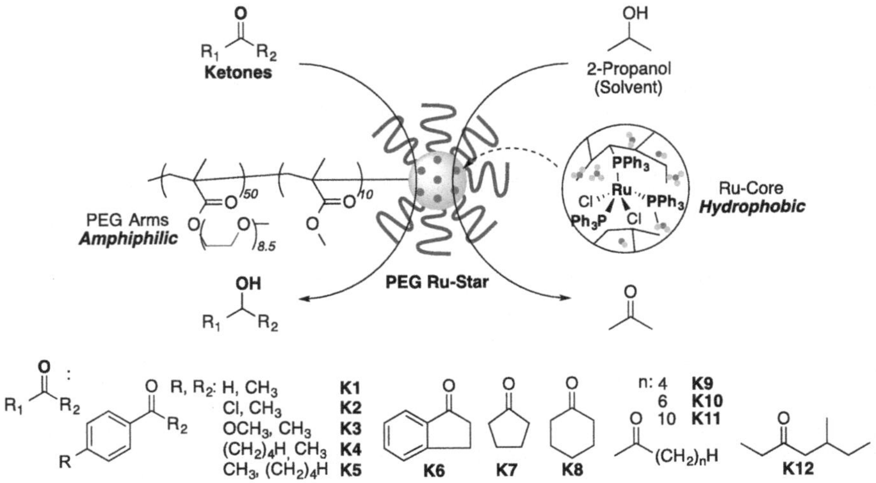

Fig. 8 Transfer hydrogenation of ketones (**K1–K12**) with PEG-arm Ru-bearing microgel star polymer catalysts (PEG Ru-Star) and K_2CO_3 in 2-propanol at 100 °C: $[\text{substrate}]_0/[\text{Ru}]_0 = 1000/1$

ruthenium, or the original $RuCl_2(PPh_3)_3$. This would be attributed to the following two reasons: (1) active ruthenium catalysts with empty coordination sites are formed, and (2) **K1** (substrate) efficiently enters the microgel spaces and the resulting product 1-phenylethanol goes out to the outer environments. Owing to the high solubility, the PEG-arm star polymer catalyst afforded homogenous hydrogenation for all of the substrates, though $RuCl_2(PPh_3)_3$ was sometimes precipitated at the latter stages of hydrogenation for aliphatic ketones. High substrate versatility is one of the attractive features of star polymer catalysts. It should also be noted that the TOF for *p*-chloroacetophenone (**K2**) and cyclohexanone (**K8**) reached approximately 1000 h^{-1}.

Additionally, the asymmetric transfer hydrogenation of ketones was also achieved using a chiral diamine (**2**)/ruthenium-bearing microgel star polymer catalyst that was obtained via one-pot ruthenium encapsulation in $RuCl_2(PPh_3)_3$-mediated polymerization with **2** [9]. Toward active, selective, and unique asymmetric catalysis, various star polymer catalysts and ligands are now developed by tuning the microgel environments (functionality) and arm polymers.

5.3 Thermoregulated Phase-Transfer Catalysis

Modulation and functionalization of arm polymers allow us to give thermoresponsive solubility to microgel star polymers. To achieve efficient catalyst recycle and product recovery, thermoregulated phase-transfer catalysis is one promising approach [1]. We thus further employed PEG Ru-Star for hydrogenation of ketones

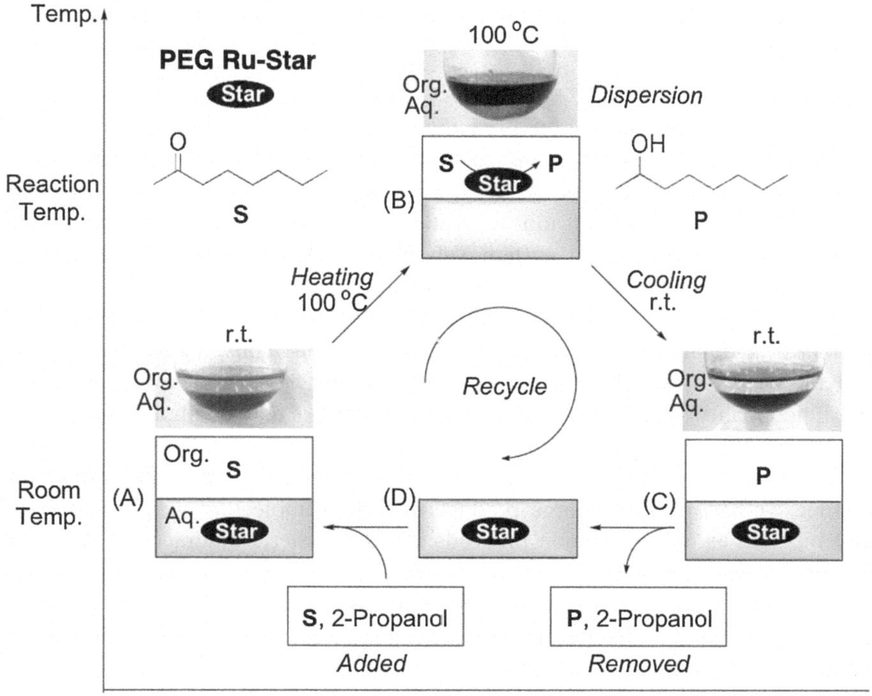

Fig. 9 Thermoregulated phase-transfer catalysis: transfer hydrogenation of 2-octanone with a PEG-arm Ru-bearing microgel star polymer catalyst (PEG Ru-Star) and K_2CO_3 in 2-propanol/H_2O (1/1, v/v) at 100 °C; $[\text{substrate}]_0/[\text{Ru}]_0 = 1000/1$. Adapted with permission from Ref. [28]. Copyright 2009 Wiley

in water and 2-propanol mixture; this system led to thermoregulated phase-transfer catalysis for hydrogenation of hydrophobic 2-octanone (Fig. 9) [28].

First, PEG Ru-Star [M_w = 772,000, 16 PEOMA_{50}-MMA_{10} block arms (M_n = 17,000), **1**-Ligand: N_{ligand} = 20/star, N_{Ru} = 20/star] was solubilized in a mixture of 2-propanol and K_2CO_3 water solution, giving a homogeneous phase. Into this, hydrophobic 2-octanone was added to result in a biphasic mixture of an organic layer (upper: 2-octanone and 2-propanol) and an aqueous layer (lower: water and 2-propanol). The star polymer was placed in the lower layer at room temperature. Upon heating to 100 °C, the star polymer was immediately transferred to the upper organic layer to efficiently induce the hydrogenation of 2-octanone into 2-octanol. After cooling to room temperature, the star polymer again returned to the lower aqueous layer. Thus, the simple recovery and evaporation of the upper organic phase provided a product without contamination of ruthenium catalysts. Efficient catalyst recycle was also achieved by replacing the organic phase via fresh feed of the substrate and 2-propanol.

6 Living Radical Polymerization with Star Polymer Metal Catalysts

Given the high performance in organic reactions, metal-bearing microgel star polymer catalysts were further applied to living radical polymerization of various monomers, i.e., polymer synthesis (Fig. 10). Metal-catalyzed living radical polymerization is now a powerful tool to synthesize various functional polymers, [38–41] but the polymerization system yet includes the following disadvantages: (1) expensive transition metal catalysts and/or ligands are required, (2) catalyst removal from polymeric products is difficult, and (3) metal catalysts are deactivated by polar functional monomers such as methacrylic acid. Thus, development of the catalytic systems with efficient catalyst recyclability, easy product recovery, and high-functionality tolerance is required. Such drawbacks may be improved with star polymer catalysts owing to the unique microgel reaction spaces. However, in sharp contrast to organic reactions with small molecules (substrates and products), polymer synthesis involves large products with high molecular weight that gradually increase as polymerization proceeds. Thus, the steric hindrance of microgel networks around the catalytic sites might decrease the polymerization activity and controllability. However, in spite of such concerns, the star polymer catalysts were actually quite effective for the living radical polymerization of various monomers including methacrylic acid [24, 25].

To suppress side reactions such as macroscopic gelation, we utilized hydrogenated, "inert", microgel star polymer ligands as nanoreactors. A hydrogenated microgel star polymer phosphine ligand with PMMA arms (M_w = 698,000, 38 PMMA arms

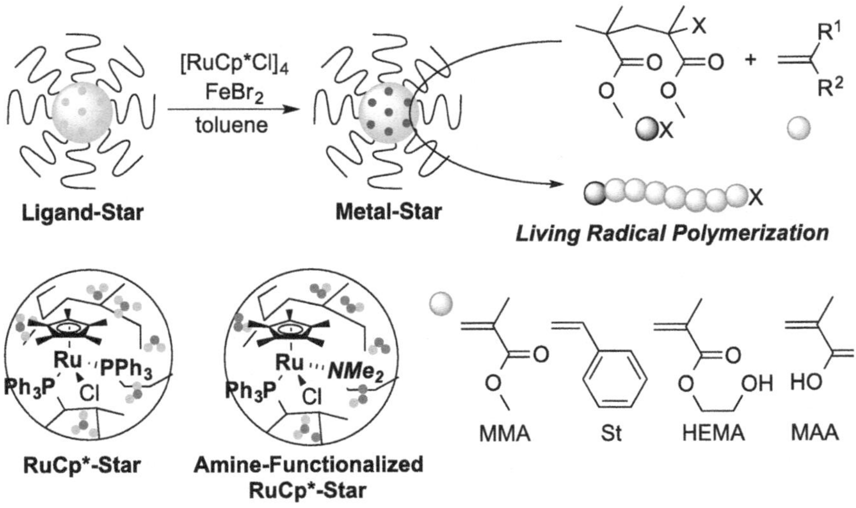

Fig. 10 Star polymer-catalyzed living radical (co)polymerization of various monomers (MMA, St, HEMA, and MAA)

(M_n = 10,000), **1**-Ligand: N_{ligand} = 180/star) was mixed with a ruthenium tetramer ($[RuCp^*Cl]_4$, Cp*: pentamethylcyclopentadienyl) in toluene at 80 °C to give a RuCp*-bearing star polymer catalyst (RuCp*-Star: [in-core **1**]$_0$/[Ru]$_0$ = 2/1). Efficient complexation was confirmed by UV-vis and ^{1}H and ^{31}P NMR spectroscopy, and by ICP-AES.

RuCp*-Star efficiently induced polymerization of methyl methacrylate (MMA) with a chloride initiator in toluene to provide better-controlled PMMA with higher Cl end-functionality than the molecular analog $[RuCp^*Cl(PPh_3)_2]$. The star polymer catalyst is further effective for 2-hydroxyethyl methacrylate (HEMA), methacrylic acid (MAA), and styrene. It should be noted that the direct copolymerization of MMA and MAA (MAA ~ 5 mol%) was achieved without any protection of the acidic functional group. The high tolerance to functional groups probably results from the tight coordination of in-core multiple phosphine ligands onto the ruthenium metals. In contrast to the methacrylate polymerization, styrene polymerization underwent phase separation of the star polymer catalysts in toluene after a certain monomer conversion. But, rather fortunately, the facile recovery of the homogeneous liquid layer gave virtually pure well-controlled polystyrene (M_w/M_n < 1.1, Ru residue: <56 ppm). The precipitated star polymer catalyst can be reused for styrene polymerization without any loss of activity. These results demonstrate that microgel cores have enough space not only for monomers but also for resulting polymers; i.e., growing polymer chains can easily diffuse within the microgel cores and access to the catalytic sites.

The catalytic activity can be further tuned by the design and functionalization of microgel reactions spaces [25]. Typically, a phosphine (**1**) and amine (**3**, DMAEMA)-functionalized microgel star polymer catalyst with RuCp* was more effective for direct copolymerization of MMA and MAA than a phosphine-functionalized microgel counterpart: the former gave well-controlled MMA/MAA copolymers with up to 25 mol% MAA content (M_w/M_n < 1.2) [25]. In addition to ruthenium, iron catalysts are also applicable to star polymer ligands; the resulting star polymer iron catalysts ($FeBr_2/PPh_3$ Ligand Star) had better controllability, stability, and functionality tolerance than an original iron catalyst. In polymerization with star polymer catalysts, compatibility between the arm polymer chains and generating polymers is also quite important to maintain the homogenous and controllable polymerization. Thus, the tailor-made design of arms and microgel spaces would create unique and selective polymerization systems beyond conventional counterparts.

7 Cascade Reactions with Star Polymer Organocatalysts

Another important function of microgel star polymers as polymeric scaffolds is "compartmentalization" of catalysts, in addition to the homogenous solubilization, stable immobilization, and functionalization of metal catalysts for active, versatile, and recyclable systems that we developed. In nature, enzyme and natural polymers often employ compartmentalization of catalytic sites to realize multiple cascade reactions for direct and/or sequential production of complex molecules; the steric and/or special segregation of the catalytically active species and intermediates can effectively prevent the deactivation [18]. This "site isolation" concept has been applied to microgel star polymer organocatalysts for one-pot multicomponent cascade reactions by Fréchet groups (Fig. 11) [29–31].

Typically, they designed and synthesized two types of microgel star polymer catalysts via nitroxide-mediated living radical polymerization: an acid-core star polymer with *para*-toluenesulfonic acid (TSA) catalyst and a base-core star polymer with 4-(dimethylamino)pyridine (DMAP) catalyst [29]. The two star polymer organocatalysts were employed as compartmentalized catalysts for the one-pot cascade acid-catalyzed deprotection of 4-nitrobenzaldehyde dimethyl acetal (**A**) to 4-nitorobenzaldehyde (**B**) followed by nucleophilic amine-catalyzed Baylis–Hillman reaction of the resulting **B** with methyl vinyl ketone (**C**) to the corresponding adduct (**D**). The two star polymers actually gave the final product (**D**) in 65% yield.

Generally, TSA and DMAP immediately form the salt complex, which is inactive for both reactions. However, owing to multiple arm polymers, the star polymer microgels effectively isolate the acidic TSA and basic DMAP without deactivation to successfully induce the cascade reaction. Such cascade reaction proceeded only with the combination of TSA-core star and DMAP-core star, while it did not take place when using a combination of small molecular catalysts (TSA, DMAP) and block copolymer-based catalysts (TSA-block, DMAP-block): e.g., TSA + DMAP, TSA + DMAP-core star, TSA-block + DMAP-block, TSA-block + DMAP-core star.

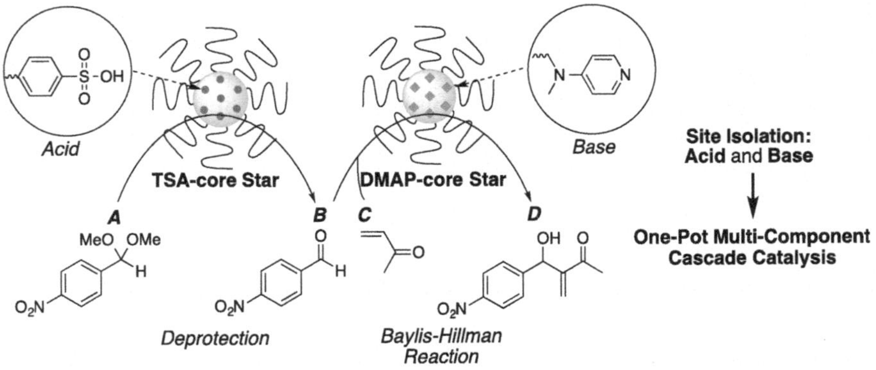

Fig. 11 One-pot cascade reaction of deprotection and Baylis-Hillman reaction with an acid-bearing microgel star polymer catalyst (TSA-core Star) and a base-bearing microgel star polymer catalyst (DMAP-core Star)

8 Concluding Remarks and Perspectives

In this chapter, we have summarized recent advances on the synthesis and catalytic functions of catalyst-embedded microgel star polymers. By introducing metal catalysts and organocatalysts into the cores, star polymers serve as unique nanoreactors for organic reactions (oxidation, hydrogenation), multicomponent cascade reactions, and polymerization. Such star polymer catalysts can be efficiently synthesized by living radical polymerization. With the versatile and on-demand design and selective functionalization, star polymer nanoreactors provided efficient, practical, and innovative catalytic systems far better and more attractive than those with conventional polymer-supported catalysts.

For example, an intriguing one-pot synthetic strategy of ruthenium-bearing microgel star polymer catalysts is developed via arm-linking reaction with a phosphine ligand monomer (**1**) in ruthenium-catalyzed living radical polymerization. The star polymer catalysts showed high activity, stability, substrate versatility, and catalyst recyclability for the oxidation of *sec*-alcohols and for the hydrogenation of ketones. With PEG-based arms, star polymer catalysts further induced thermoregulated phase-transfer catalysis in the aqueous hydrogenation of aliphatic ketones. Owing to the relatively low density and fully swollen microgel cores, star polymer catalysts were further effective for polymerization of various monomers: e.g., direct copolymerization of methacrylic acid was first achieved with metal-catalyzed system. Catalyst-enclosed microgel star polymers thus afford the following advantages to catalysis:

1. **High activity and versatility**: homogenous solubility by arms, microgel functionalization.
2. **High stability and functionality tolerance**: protection of catalysts within microgel networks.
3. **Easy product recovery and catalyst recycle**: precipitation or thermoresponsive properties.
4. **Phase-transfer catalysis**: functionalization with thermoresponsive arm polymers.
5. **One-pot cascade catalysis**: compartmentalization of catalytic sites.

Given these attractive features, star polymer catalysts are one of the most promising nanoreactors widely applicable to chemical transformations. For future perspective, precision control of microgel networks and environments around the catalytic sites would be a key to further advance star polymer catalysts. The possible strategies are molecular imprinting [47] and supramolecular self-assembly [9, 42] of linking agents or functional monomers within microgel cores, resulting in star polymer catalysts with in-core precision cavities or well-defined networks, respectively. Such star polymers may open the highly selective catalysis like enzymes and proteins and/or cause innovative catalysis beyond our understanding. Star polymer catalysts yet keep on evolving forever.

References

1. Bergbreiter DE, Tian J, Hongfa C (2009) Using soluble polymer supports to facilitate homogeneous catalysis. Chem Rev 109:530–582
2. Akiyama R, Kobayashi S (2009) "Microencapsulated" and related catalysts for organic chemistry and organic synthesis. Chem Rev 109:594–642
3. Leadbeater NE, Marco M (2002) Preparation of polymer-supported ligands and metal complexes for use in catalysis. Chem Rev 102:3217–3274
4. Yoshida J, Itami K (2002) Tag strategy for separation and recovery. Chem Rev 102:3693–3716
5. Lu A, O'Reilly RK (2013) Advances in nanoreactor technology using polymeric nanostructures. Curr Opin Biotech 24:639–645
6. Welsch N, Ballauff M, Lu Y (2010) Microgels as nanoreactors: applications in catalysis. Adv Polym Sci 234:129–163
7. Terashima T (2013) Polymer microgels for catalysis. In: Seidel A (ed) Encyclopedia of polymer science and technology, 4th edn. Wiley, New York, pp 1–31
8. Terashima T, Sawamoto M (2012) Microgel-core star polymers as functional compartments for catalysis and molecular recognition. ACS Symp Ser 1101:65–80
9. Terashima T (2014) Functional spaces in star and single-chain polymers via living radical polymerization. Polym J 46:664–673
10. Terashima T (2013) Functional star polymers via living radical polymerization: designer functional nanospaces. Kobunshi Ronbunsyu 70:432–448
11. Gao H, Matyjaszewski K (2009) Synthesis of functional polymers with controlled architecture by CRP of monomers in the presence of cross-linkers: from stars to gels. Prog Polym Sci 34:317–350
12. Blencowe A, Tan JF, Goh TK, Qiao GG (2009) Core cross-linked star polymers via controlled radical polymerization. Polymer 50:5–32
13. Gao H (2012) Development of star polymers as unimolecular containers for nanomaterials. Macromol Rapid Commun 33:722–734
14. Terashima T, Motokawa R, Koizumi S, Sawamoto M, Kamigaito M, Ando T, Hashimoto T (2010) In situ and time-resolved small-angle neutron scattering observation of star polymer formation via arm-linking reaction in ruthenium-catalyzed living radical polymerization. Macromolecules 43:8218–8232
15. Terashima T, Nishioka S, Koda Y, Takenaka M, Sawamoto M (2014) Arm-cleavable microgel star polymers: a versatile strategy for direct core analysis and functionalization. J Am Chem Soc 136:10254–10257
16. Dwars T, Paetzold E, Oehme G (2005) Reactions in micellar systems. Angew Chem Int Ed 44:7174–7199
17. Manabe K, Iimura S, Sun XM, Kobayashi S (2002) Dehydration reactions in water. Brønsted acid-surfactant-combined catalyst for ester, ether, thioether, and dithioacetal formation in water. J Am Che Soc 124:11971–11978
18. Hecht S, Fréchet JMJ (2001) Dendritic encapsulation of function: applying nature's site isolation principle from biomimetics to materials science. Angew Chem Int Ed 40:74–91
19. Oosterom GE, Reek JNH, Kamer PCJ, van Leeuwen PWNM (2001) Transition metal catalysis using functionalized dendrimers. Angew Chem Int Ed 40:1828–1849
20. Helms B, Fréchet JMJ (2006) The dendrimer effect in homogeneous catalysis. Adv Synth Catal 348:1125–1148
21. Terashima T, Kamigaito M, Baek KY, Ando T, Sawamoto M (2003) Polymer catalysts from polymerization catalysts: direct encapsulation of metal catalyst into star polymers core during metal-catalyzed living radical polymerization. J Am Chem Soc 125:5288–5289
22. Terashima T, Ouchi M, Ando T, Kamigaito M, Sawamoto M (2006) Metal complex-bearing star polymers by metal-catalyzed living radical polymerization: synthesis and characterization

of PMMA star polymers with Ru(II)-embedded microgel cores. J Polym Sci, Part A: Polym Chem 44:4966–4980
23. Terashima T, Ouchi M, Ando T, Kamigaito M, Sawamoto M (2007) Amphiphilic, thermosensitive ruthenium(II)-bearing star polymer catalysts: one-pot synthesis of PEG armed star polymers with ruthenium(II)-enclosed microgel cores via metal-catalyzed living radical polymerization. Macromolecules 40:3581–3588
24. Terashima T, Nomura A, Ouchi M, Sawamoto M (2011) Star polymer-catalyzed living radical polymerization: microgel-core reaction vessel via tandem catalyst interchange. Angew Chem Int Ed 50:7892–7895
25. Terashima T, Nomura A, Ouchi M, Sawamoto M (2012) Efficient and robust star polymer catalysts for living radical polymerization: cooperative activation in microgel-core reactors. Macromol Rapid Commun 33:833–841
26. Terashima T, Ouchi M, Ando T, Sawamoto M (2011) Oxidation of sec-alcohols with Ru(II)-bearing microgel star polymer catalysts via hydrogen transfer reaction: unique microgel-core catalysis. J Polym Sci Part A Polym Chem 49:1061–1069
27. Terashima T, Ouchi M, Ando T, Sawamoto M (2011) Transfer hydrogenation of ketones catalyzed by PEG-armed ruthenium-microgel star polymers: microgel-core reaction space for active, versatile and recyclable catalysis. Polym J 43:770–777
28. Terashima T, Ouchi M, Ando T, Sawamoto M (2010) Thermoregulated phase-transfer catalysis via PEG-armed Ru(II)-bearing microgel core star polymers: efficient and reusable Ru (II) catalysts for aqueous transfer hydrogenation of ketones. J Polym Sci, Part A: Polym Chem 48:373–379
29. Helms B, Guillaudeu SJ, Xie Y, McMurdo M, Hawker CJ, Fréchet JMJ (2005) One-pot reaction cascades using star polymers with core-confined catalysts. Angew Chem Int Ed 44:6384–6387
30. Chi Y, Scroggins ST, Fréchet JMJ (2008) One-pot multi-component asymmetric cascade reactions catalyzed by soluble star polymers with highly branched non-interpenetrating catalytic cores. J Am Chem Soc 130:6322–6323
31. Rodionov V, Gao H, Scroggins S, Unruh DA, Avestro AJ, Fréchet JMJ (2010) Easy access to a family of polymer catalysts from modular star polymers. J Am Chem Soc 132:2570–2572
32. Funke W, Okay O, Joos-Müller B (1998) Microgels—intermolecularly crosslinked macromolecules with a globular structure. Adv Polym Sci 136:139–234
33. Sanson N, Rieger J (2010) Synthesis of nanogels/microgels by conventional and controlled radical crosslinking copolymerization. Polym Chem 1:965–977
34. Hsieh HL, Quirk RP (eds) (1996) Anionic polymerization: principles and practical applications. Marcel Dekker, Inc., New York
35. Eschwey H, Burchard W (1975) Star polymers from styrene and divinylbenzene. Polymer 16:180–184
36. Bosman AW, Vestberg R, Heumann A, Fréchet JMJ, Hawker CJ (2003) A modular approach toward functionalized three-dimensional macromolecules: from synthetic concepts to practical applications. J Am Chem Soc 125:715–728
37. Shibata T, Kanaoka S, Aoshima S (2006) Quantitative synthesis of star shaped poly(vinyl ether)s with a narrow molecular weight distribution by living cationic polymerization. J Am Chem Soc 128:7497–7504
38. Ouchi M, Terashima T, Sawamoto M (2008) Precision control of radical polymerization via transition metal catalysis: from dormant species to designed catalysts for precision functional polymers. Acc Chem Res 41:1120–1132
39. Ouchi M, Terashima T, Sawamoto M (2009) Transition metal-catalyzed living radical polymerization: toward perfection in catalysis and precision polymer synthesis. Chem Rev 109:4963–5050
40. Matyjaszewski K, Tsarevsky NV (2009) Nanostructured functional materials prepared by atom transfer radical polymerization. Nat Chem 1:276–288
41. Matyjaszewski K (2012) Atom transfer radical polymerization (ATRP): current status and future perspectives. Macromolecules 45:4015–4039

42. Terashima T, Mes T, De Greef TFA, Gillissen MAJ, Besenius P, Palmans ARA, Meijer EW (2011) Single-chain folding of polymers for catalytic systems in water. J Am Chem Soc 133:4742–4745
43. Nakatani K, Ogura Y, Koda Y, Terashima T, Sawamoto M (2012) Sequence-regulated copolymers via tandem catalysis of living radical polymerization and in situ transesterification. J Am Chem Soc 134:4373–4383
44. Terashima T, Kawabe M, Miyabara Y, Yoda H, Sawamoto M (2013) Polymeric pseudo-crown ether for cation recognition via cation template-assisted cyclopolymerization. Nat Commun 4:2321
45. Koda Y, Terashima T, Sawamoto M, Maynard HD (2015) Amphiphilic/fluorous random copolymers as a new class of non-cytotoxic polymeric materials for protein conjugation. Polym Chem 6:240–247
46. Fukae K, Terashima T, Sawamoto M (2012) Cation-condensed microgel-core star polymers as polymeric nanocapsules for molecular capture and release in water. Macromolecules 45:3377–3386
47. Terashima T, Kojima H, Sawamoto M (2014) Core-imprinted star polymers via living radical polymerization: precision cavity microgels for selective molecular recognition. Chem Lett 43:1690–1692
48. Koda Y, Terashima T, Sawamoto M (2014) Fluorous microgel star polymers: selective recognition and separation of polyfluorinated surfactants and compounds in water. J Am Chem Soc 136:15742–15748
49. Zhang X, Cardozo AF, Chen S, Zhang W, Julcour C, Lansalot M, Blanco J-F, Gayet F, Delmas H, Charleux B, Manoury E, D'Agosto F, Poli R (2014) Core–shell nanoreactors for efficient aqueous biphasic catalysis. Chem Eur J 20:15505–15517
50. Chen S, Cardozo AF, Julcour C, Blanco J-F, Barthe L, Gayet F, Lansalot M, D'Agosto F, Delmas H, Manoury E, Poli R (2015) Amphiphilic core-cross-linked micelles functionalized with bis(4-methoxyphenyl)phenylphosphine as catalytic nanoreactors for biphasic hydroformylation. Polymer 72:327–335
51. Terashima T, Sugita T, Sawamoto M (2015) Single-chain crosslinked star polymers via intramolecular crosslinking of self-folding amphiphilic copolymers in water. Polym J 47:667–677
52. Cardozo AF, Manoury E, Julcour C, Blanco JF, Delmas H, Gayeta F, Poli R (2013) Preparation of phosphine-functionalized polystyrene stars by metal catalyzed controlled radical copolymerization and their application to hydroformylation catalysis. Dalton Trans 42:9148–9156
53. Naota T, Takaya H, Murahashi S-I (1998) Ruthenium-catalyzed reactions for organic synthesis. Chem Rev 98:2599–2660
54. Terashima T, Ouchi M, Ando T, Sawamoto M (2006) In-situ hydrogenation of terminal halogen in poly(methyl methacrylate) by ruthenium-catalyzed living radical polymerization: direct transformation of "polymerization catalyst" into "hydrogenation catalyst". J Am Chem Soc 128:11014–11015

Core-Cross-Linked Micelles and Amphiphilic Nanogels as Unimolecular Nanoreactors for Micellar-Type, Metal-Based Aqueous Biphasic Catalysis

Eric Manoury, Florence Gayet, Franck D'Agosto, Muriel Lansalot, Henri Delmas, Carine Julcour, Jean-François Blanco, Laurie Barthe and Rinaldo Poli

Abstract Biphasic homogeneous protocols are attractive for catalyzed transformations in industry, especially when conducted with water as the catalyst phase as exemplified by the large-scale Rhône-Poulenc/Ruhrchemie hydroformylation process, but can only be applied when the substrate is sufficiently soluble in the aqueous phase to sustain sufficiently fast mass transport. Different solutions to reduce mass transport limitations include the use of additives to increase the substrate solubility in water or increase the water/organic interface, anchoring the catalyst onto a lower critical solution temperature (LCST) polymer to implement thermomorphic behavior, and anchoring the catalyst to the hydrophobic part of surfactants or amphiphilic block copolymers that self-assemble in the form of micelles in water. The use of catalytic micelles appears as the most attractive approach but is limited by the potential formation of stable emulsions and by loss of free macromolecules during separation. These limitations are removed by cross-linking the macromolecules into a unimolecular nanoreactor. This chapter covers the emerging area of unimolecular catalytic nanoreactors, focusing on

E. Manoury · F. Gayet · R. Poli (✉)
Laboratoire de Chimie de Coordination (LCC), CNRS, Université de Toulouse, UPS, INPT, 205 route de Narbonne, BP 44099, 31077 Toulouse Cedex 4, France
e-mail: rinaldo.poli@lcc-toulouse.fr

F. D'Agosto · M. Lansalot
Team LCPP, Bat 308F, Univ Lyon, Université Claude Bernard Lyon 1, CPE Lyon, CNRS, UMR 5265, C2P2 (Chemistry, Catalysis, Polymers and Processes), 43 Bd du 11 Novembre 1918, 69616 Villeurbanne, France
e-mail: franck.dagosto@univ-lyon1.fr

H. Delmas · C. Julcour · J.-F. Blanco · L. Barthe
Laboratoire de Génie Chimique (LGC), Université de Toulouse, CNRS, INPT, UPS, 4 Allée Emile Monso, CS 84234, 31432 Toulouse Cedex 4, France

R. Poli
Institut Universitaire de France, 1 rue Descartes, 75231 Paris Cedex 05, France

R. Poli (ed.), *Effects of Nanoconfinement on Catalysis*,
Fundamental and Applied Catalysis, DOI 10.1007/978-3-319-50207-6_7

transition metal-based catalytic applications. It will also present the synthesis of new types of catalytic unimolecular nanoreactors developed in our laboratories, conceived to function on the basis of the micellar catalysis principle. These nanoreactors consist of either core-cross-linked micelle (CCM) or amphiphilic functionalized nanogels (NG). The proof of principle of their catalytic performance in the aqueous biphasic hydroformylation of 1-octene will also be presented. The catalyst confinement objective which is highlighted in this chapter is process optimization in terms of the catalyst phase recovery and recycling.

1 Introduction

A major concern of industrial catalysis is catalyst recovery and recycling, particularly when using expensive metals (Rh, Ir, Pt, Ru, etc.) and elaborate ligands needed to attain required turnover frequencies and product selectivities [1, 2]. Heterogeneous catalysis, a term typically restricted to the catalytic action of a solid-phase surface in combination with one or more fluid reactants–products phase (liquid and/or gas), offers obvious advantages because the catalyst can be recovered by filtration or sedimentation in batch processes. Continuous-flow protocols are also easily implemented with minimal or no physical catalyst losses. Homogeneous catalysis offers superior mass transport kinetics and precise control of the active site through the monodispersed (single-site) molecular nature and facile tuning through molecular engineering of the active site environment (ligands for metal catalysts, peripheral substituents for organocatalysts), but are harder to separate and recover from the reaction products. Product distillation away from the reaction mixture is often implemented, but this procedure is not viable when the low volatility requires use of extreme temperatures where the catalyst thermally decomposes. Furthermore, distillation is energy intensive and alternative low-temperature methods offer economic gain if equally efficient in terms of recovery.

Numerous solutions exist for combining the efficiency of homogeneous catalysis and the ease of catalyst recovery and recycling. They can be classified depending on the separation method: filtration, ultrafiltration, and decantation. The first category comprises, in addition to precipitation which requires again distillation to remove the added non-solvent, various "heterogenized" or "immobilized" protocols [3], which feature catalyst confinement in a "fluid" phase held by chemical or physical forces on the surface of a solid. If the chemical linker is sufficiently long and flexible, the single-site active catalyst operates as if it were free in solution. In "supported ionic liquid-phase" (SILP) catalysis, the active catalyst is dissolved in an ionic liquid, which is physically supported on a solid matrix [4, 5]. The solid phase should ideally be chemically inert and have a high surface area in order to optimize the mass transport kinetics. The filtration step can be avoided and replaced by magnetic sedimentation through the use of ferromagnetic solid supports [6].

In the second category, the catalyst is anchored on a soluble macromolecule (polymer, dendrimer, protein), the size of which is much greater than that of the reactants and products and can therefore be removed by ultra or nano-filtration through size-selective membranes [7, 8]. In the last category (liquid–liquid biphasic catalysis), the catalyst must be confined in a second liquid-phase, non-miscible with the reactant–product phase. This constitutes perhaps the most attractive protocol because separation does not require additional delicate hardware (filters or membranes) with associated maintenance and replacement costs. Arguably, the most attractive liquid for catalyst confinement is water (aqueous biphasic catalysis), though other catalyst confinement media such as fluorous solvents [9, 10] and ionic liquids [11] are also intensively investigated.

Aqueous biphasic catalysis can operate according to four different principles depending on the location of the catalytic act [12] (at the interface, in the bulk aqueous phase, in the bulk organic phase, or in the core of micelles) but all require eventual decantation of the organic product phase from the aqueous phase where the catalyst remains confined. In order not to physically lose any catalyst, the solubility of the latter in the organic phase should be null or negligibly small. In interfacial catalysis, the reactants are themselves completely insoluble or negligibly soluble in water, and hence the reaction mostly or solely occurs at the interface. The reaction rate increases upon increasing the interfacial area (i.e., using surfactants) but a compromise must be found with the ensuing lower rate of the emulsion breaking phase before separation.

For the reaction to occur in the bulk catalyst phase, the substrate must have sufficient solubility in it. The solubility of substrates in water can sometimes be increased by the use of appropriate additives such as cosolvents. For insufficiently water-soluble substrates, the catalytic event can be forced to take place in the bulk organic phase by the thermomorphic approach [13], anchoring the catalyst onto soluble macromolecules with LCST behavior (hydrophilic at low temperature and hydrophobic at high temperature). Thus, the catalyst-bearing polymer migrates to the organic phase at the high temperatures (>cloud point) where the reaction takes place, and then migrates back to the aqueous phase upon cooling for recovery. Catalyst leaching with this approach depends on the residual polymer solubility in the organic phase at the separation temperature.

Finally, the catalyst may also be anchored onto the hydrophobic part of surfactants or amphiphilic diblock copolymers that are able to self-assemble as micelles at relatively low concentrations (above the critical micelle concentration or CMC). This self-organization places the catalyst inside the hydrophobic micro-environment of the micellar core, where the reaction takes place. The substrates migrate to the micellar core from the organic bulk phase and the products migrate back toward the organic phase, while the micellar catalyst always remains confined in the aqueous phase. This micellar catalysis principle is not to be confused with the micellar effect in interfacial catalysis [14–16], where surfactants generate micelles but the reaction occurs only at the water/organic interface through the action of a water-soluble catalyst. Micellar catalysis presents many attractive features including versatility (wide choice of surfactant/polymers with tailored size and functionality), fast kinetics because of the high local concentration of catalyst and substrate, and a low negative impact of mass transport kinetics. However, two negative features have so far limited the widespread application of this methodology at the industrial scale: excessive core swelling and the free-arm/micelle equilibrium. The former occurs when the organic phase is highly compatible with the micellar core, leading to the generation of stable emulsions that introduce delays in the phase separation step. The second one, particularly problematic for high CMC, is associated with placement of the free surfactant molecules or polymers as a Langmuir–Blodgett film at the water/organic interface and/or with the formation of inverse micelles in the organic phase, leading to physical catalyst losses.

This chapter focuses on metal-based catalytic applications. Contrary to organocatalytic applications where the active site may be linked to the polymer by covalent bonds, the active metal needs to be anchored to the polymer scaffold through coordinative bonds; hence, the polymers of interest must be functionalized with ligands suitable for binding metals. The chapter presents the development, recently realized in our laboratories, of phosphine-functionalized unimolecular versions of micelles, achieved by core-cross-linking, where the two above-mentioned negative features of micelles should in principle be removed. Two related architectures can be distinguished (see Fig. 1). In the first one, called Core-Cross-linked Micelle (CCM), amphiphilic diblock copolymers that are ligand-functionalized in the hydrophobic part are cross-linked at the hydrophobic chain end resulting in star-block unimolecular nano-objects where the ligands are placed on flexible arms, outside of the cross-linked part. In the second architecture, called NanoGel (NG), the entire core is cross-linked and thus the catalyst binding sites are placed within the cross-linked core.

The results obtained for the Rh-catalyzed hydroformylation under aqueous biphasic micellar-type conditions will also be shown in this chapter, with focus on the relationship between the macromolecular architecture, the experimental conditions, and the observed catalyst loss. The perspectives for the further development of this methodology will be discussed.

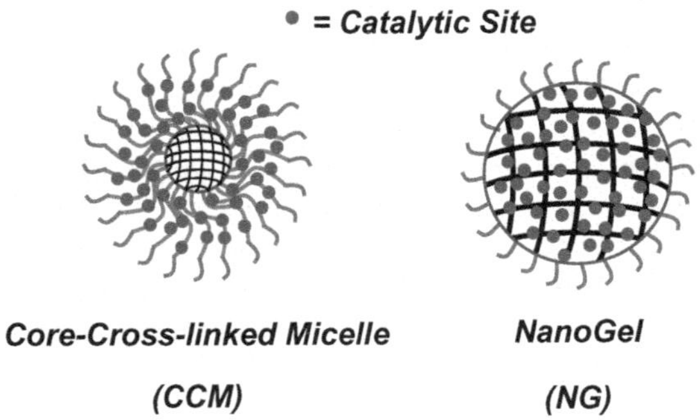

Fig. 1 Polymeric architectures of unimolecular nanoreactors used in micellar-type catalysis

2 Unimolecular Polymeric Nanoreactors for Metal-Based Catalysis: Relevant Precedents

A few unimolecular metal-containing catalytic nanoreactors have been previously presented but none of these, to the best of our knowledge, have been employed under micellar-type operating conditions prior to the development of our CCM and NG polymers. We highlight here a few of these precedents, focusing on the similarities and differences relative to our polymers. A more comprehensive coverage of polymeric nanoreactors can be found in other recent reviews [17, 18]. In terms of synthesis, enormous progress in the control of dimensions, homogeneity and architecture, has been made possible by living or controlled polymerization methods as also highlighted in other chapters of this book.

In 2008, O'Reilly et al. described shell-cross-linked micelles obtained by the two-step assembly of a diblock copolymer synthesized by nitroxide mediated polymerization (NMP) that contains a poly(*tert*-butyl acrylate) (PtBA) block and a terpy-functionalized polystyrene (PS) block [19]. The tBA ester functions of the isolated product were deprotected by trifluoroacetic acid/CH_2Cl_2 to afford an amphiphilic block copolymer with a poly(acrylic acid) (PAA) block. Then, micellization of this polymer was induced by slow water addition to a THF solution, followed by extensive dialysis. Finally, the micelles were shell-cross-linked by partial amidation with $NH_2CH_2CH_2OCH_2CH_2OCH_2CH_2NH_2$ under high dilution conditions, avoiding significant intermicellar cross-linking. The core PS-anchored terpy functions were coordinated to $RuCl_3{\cdot}6H_2O$, $FeCl_3{\cdot}6H_2O$, and $CuBr(PPh_3)_3$, to yield metal-loaded particles. The Cu-loaded nano-objects were applied to the room temperature Cu-catalyzed Huisgen cycloaddition of phenylacetylene and 3-azidocoumarin, demonstrating activity. The reaction was carried out under homogeneous reaction conditions (water/MeCN solution) and no catalyst recover/recycling was reported.

In 2011, Weck et al. developed Co-loaded nanoreactors for the substrate-selective hydrolytic kinetic resolution of epoxides [20, 21]. These nanoreactors were made by either living anionic polymerization of oxazolines [20] or ring-opening metathesis polymerization of norbornenes [21]. In both cases, a linear triblock copolymer was first produced by sequential polymerization of a hydrophilic monomer—2-methyloxazoline or a poly(ethylene glycol) methyl ether-containing norbornene, respectively—a photo-cross-linkable cinnamate-containing monomer, and a salen ligand-containing monomer. After micellar assembly by MeOH addition to a THF solution of this block copolymer (final solvent: THF/MeOH 1:2 mixture) and stabilization of the resulting nano-objects by photo-cross-linking the middle-block, cobalt was introduced in the salen-functionalized core as a MeOH solution of Co $(OAc)_2{\cdot}4H_2O$ (attained loading from 79 to 90%). The epoxide ring-opening activity was found to increase by decreasing the thickness of the hydrophilic PEG layer and by decreasing the degree of cross-linking. More hydrophobic epoxides were opened faster. The reactions were carried out under homogeneous conditions (the added epoxide and water formed a single phase together with the THF/MeOH solvent).

The recovery and recycling was investigated for the polyoxazoline-based nanoreactors by ultrafiltration, showing a slight decrease of activity over eight cycles but no detectable cobalt leaching into the filtrate.

The most closely related polymeric nanoreactors to ours are a series of triphenylphosphine-functionalized nanogels that have been developed by Terashima et al. starting in 2003 [22]. Since these are described in detail in Chap. 6, we shall only recall here their salient features. They have been assembled by a convergent strategy using $[RuCl_2(TPP)_3]$-catalyzed atom transfer radical polymerization (ATRP) (TPP = triphenylphosphine, PPh_3), with the initial polymerization of a methacrylate monomer followed by chain extension and simultaneous cross-linking by copolymerization of a ligand-functionalized monomer (4-diphenylphosphino-styrene, DPPS) and a divinyl monomer. The topology of these polymers is the same as that of NG in Fig. 1. The Ru ATRP catalyst remains entrapped in the final star polymer by coordination to the core-anchored TPP ligand, but the particles could subsequently be demetallated and the resulting metal-free particles could then be metallated again with different metals [23]. In the first contributions, the polymer outer shell was made of hydrophobic PMMA leading to fully hydrophobic nanoreactors, which were then used for Ru-catalyzed processes under homogeneous conditions [22, 24, 25]. Later, nanoreactors with an outer shell containing hydrophilic poly(ethylene oxide) methyl ether methacrylate (PEOMA) homopolymer blocks were developed [26] and used in either thermomorphic [27] or homogeneous [27, 28] transfer hydrogenation of ketones by isopropanol as the hydrogen source. With 2-octanone as substrate, the polymer lies in the water phase at room temperature and in the organic phase at the reaction temperature (100 °C), yielding a thermomorphic process [27]. At the end of the reaction after cooling, the Ru amount remaining in the organic layer was below the detection limits (<10 ppm) according to UV–visible and ICP-AES analyses. The recovered catalyst was recycled once, yielding the same activity as in the first cycle. When using shorter chain aliphatic ketones (2-butanone, cyclohexanone), on the other hand, the reaction occurred homogeneously in a single phase, whereas using aromatic ketones (*e.g.*, acetophenone), although a biphasic aqueous/organic system was generated, the catalyst remained in the organic phase even after cooling to room temperature. In those cases, the catalyst was recycled by evaporating the residual isopropanol followed by washing away the less volatile organic substances and adding fresh substrate and isopropanol [28].

Although the focus of this chapter is on catalysis by molecular metallic complexes, a couple of examples of metal nanoparticle-containing nanoreactors is also worth being illustrated. In 2003, Fréchet, Hawker et al. used a core-cross-linked star-block copolymer made by NMP and containing linear arms with an outer PS block and an inner poly(2-vinyl pyridine) (P2VP) block, cross-linked with divinyl benzene [29]. Addition of $Pd(OAc)_2$ to a toluene dispersion of these particles followed by reduction with hot ethanol yielded Pd nanoparticles held within the P2VP core, which were subsequently used in the catalyzed hydrogenation of cyclohexene in THF at 30 °C and in the Heck cross-coupling of 1-bromo-4-nitrobenzene with *n*-butyl acrylate in a xylene solution at 125 °C. In the latter case, the catalyst was recovered by precipitation with methanol and reused five times without any observable degradation of performance. In 2006, Ballauff et al. reported the generation of Ag nanoparticles by $AgNO_3$ reduction with $NaBH_4$ inside shell-thermosensitive nanoreactors made of a PS core and a poly(*N*-isopropylacrylamide) (PNIPA) shell [30]. In this case, the synthesis was not based on a controlled polymerization method but rather on conventional emulsion polymerization of styrene (S) followed by seeded emulsion polymerization of NIPA. The Ag nanoparticles are localized within the hydrophilic PNIPA shell, most probably because of the Ag^+ coordination to the N atoms. The resulting latex was used for the homogeneous catalyzed reduction of nitrobenzene by $NaBH_4$ but no detailed studies of catalyst recovery and recycling were described.

As is evident from the above examples, the previous applications of catalytic nanoreactors containing transition metals, although pioneering in several respects, were not always intended to show facile recyclability and only in one specific case this was achieved by simple decantation of a liquid/liquid biphasic reaction mixture in a thermomorphic process. In no case were catalytic metal-containing unimolecular nanoreactors applied in a micellar-type operating protocol (reaction occurring within the water-confined nanoreactor core). Also, the accurate determination of metal leaching has been rare and not correlated with the polymer structure.

3 Design and Synthesis of the CCM and NG Latexes

As stated in the introduction, our target was a unimolecular equivalent of a micelle for application in metal-based micellar-type catalysis. The first design consisted of cross-linking the flexible amphiphilic diblock copolymer arms at the hydrophobic chain end to create a nanogel core (CCM type, see Fig. 1). The core in these nanoreactors only serves to mechanically hold the micellar arms together, keeping the nanoreactors from swelling beyond the maximum chain extension in good solvents and from releasing free arms in solution, while the arm mobility should ideally be preserved like in a regular micelle and lead to rapid mass transport kinetics.

A potential technical problem in the synthesis of these objects is reaching quantitative arm cross-linking, which is notoriously difficult when carrying out a star polymer synthesis by a convergent (arm-first) method [31]. Indeed, we initially faced difficulties when attempting to generate star polymers by chain extension of TPP-functionalized PS chains with divinyl benzene by ATRP in a homogeneous medium, the highest cross-linking yields being ca. 80% [32]. Given the CCM design, a quantitative cross-linking is absolutely necessary because, upon dispersing the final product in an aqueous medium, any residual uncross-linked arm would generate regular catalytic micelles in addition to the catalytic unimolecular nanoreactors. Hence, the problems alluded to in the introduction would not be removed when applying the material in catalysis. This problem was solved by switching from polymerization performed in a homogeneous medium to emulsion polymerization, which involves the polymerization-induced self-assembly (PISA) phenomenon [33, 34] and allows the full synthesis to be carried out in one pot without the isolation of any intermediate. By this strategy, the last cross-linking step occurs within the micellar core space and results in quantitative incorporation of all chains into the eventual unimolecular nano-object [35–38]. An additional advantage of the PISA method is the generation of stable latexes with a high polymer content (>20% in weight), whereas only much more diluted latexes can be obtained by the step-by-step procedure of arm synthesis, micellization, and final cross-linking.

The successful strategy made use of the reversible addition–fragmentation chain transfer (RAFT) radical polymerization as controlling method, as summarized in Fig. 2. The choice of hydrophilic shell, a copolymer of methacrylic acid (MAA) and PEOMA (X_n = 19) (MAA/PEOMA = 50:50), was based on the previous optimization of the PISA-RAFT synthesis of non-functionalized and non-cross-linked copolymer micelles [39–43]. Using 4-cyano-4-thiothiopropylsulfanyl pentanoic acid (CTPPA, R_0-S_2CSEt) as RAFT agent, 4,4′-azobis(4-cyanopentanoic acid) (ACPA) as radical source, and water as solvent, the hydrophilic monomers (n equivalents) are polymerized in the first step to yield water-soluble R_0-(MAA$_{0.5}$-*co*-PEOMA$_{0.5}$)$_n$-S_2CSEt chains. These are extended in step 2 with m equivalents of a x:(1 − x) mixture of DPPS and S, to yield amphiphilic R_0-(MAA$_{0.5}$-*co*-PEOMA$_{0.5}$)$_n$-*b*-(DPPS$_x$-*co*-S$_{(1-x)}$)$_m$-S_2CSEt chains where the hydrophobic PS block anchors TPP ligands. This step starts slowly since no surfactant is present but ends up as a rapid emulsion polymerization after PISA. The chains are further extended and cross-linked in step 3 with l equivalents of a 10:90 mixture of diethylene glycol dimethacrylate (DEGDMA) and S [44]. The synthetic procedure is simple and rapid: quantitative monomer conversion is achieved in ca. 2 h in each step and no purification procedure between the different

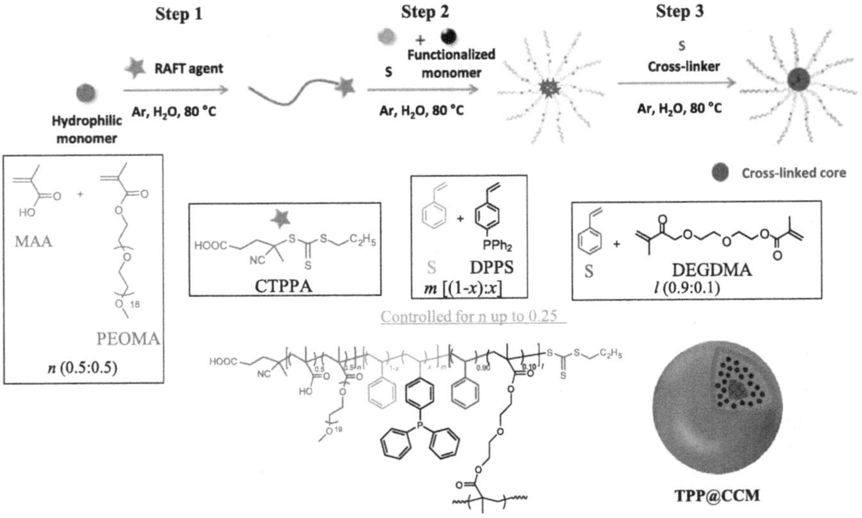

Fig. 2 Synthesis of the TPP@CCM nanoreactors by a three-step one-pot RAFT-PISA process in water

steps is needed. Typical n:m:l values are 30:(300–500):100 and the TPP content can be adjusted in the $0.05 \leq x \leq 0.25$ range [45]. The resulting low-viscosity latexes contain up to 30% in weight of well-defined TPP-functionalized core-cross-linked micelles (TPP@CCM) with controlled diameter (D_z = 80–110 nm) and low size polydispersity (PDI < 0.2). If desired, the micelle-containing latex obtained after step 2 (TPP@M) may also be recovered and used in regular micellar catalysis. Light scattering by the particles renders these latexes opaque.

The TEM image and DLS response of a representative polymer are shown in Fig. 3. THF is a good solvent for both the TPP-functionalized PS core and the P(MAA-*co*-PEOMA) shell, thus the particle diameter increases by a factor of 2–2.5 relative to water where only the outer shell is solvated. A size increase is also observed by swelling the particle core with other water-immiscible solvents (*e.g.*, toluene, chloroform). DOSY-NMR shows the absence of residual non-cross-linked chains.

The modular nature of these nanoreactors allows several modifications. As already mentioned above, the degree of polymerization (m) and the TPP density (x) of the PS block can be varied within certain ranges. Using degrees of polymerization greater than 500 led to different morphologies for the self-assembled objects that were unsuitable for cross-linking. Using >25% of DPPS in the monomer feed of step 2 led to precipitation, because of the limited solubility of solid DPPS in the styrene liquid phase. In principle, the degree of polymerization of the P(MAA-*co*-PEOMA) block (n) and of the nanogel core (l) may also be varied. Objects with longer hydrophilic blocks could be made but yielded much more viscous latexes at high polymer contents. For the cross-linked core, the 90:10 ratio of styrene and DEGDMA gave the best results, as increasing the DEGDMA content led to macrogelation. A change of the chemical nature of both the outer and the inner blocks is also possible in principle but has not yet been explored in detail. Adjusting the "solvent" properties of the particle core in such a way that the affinity of the substrate(s) of the targeted catalytic reaction is greater than that of the product(s) should positively affect the reaction rate by establishing concentration gradients in favor of substrate incorporation and product

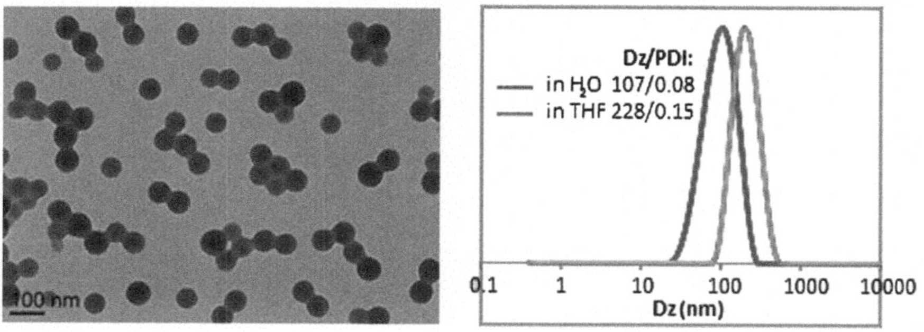

Fig. 3 TEM (**a**) and DLS (**b**) investigations of a representative TPP@CCM. The TEM image was previously published [46] and is reproduced with permission. Copyright 2015 American Chemical Society

expulsion. Finally, the chemical nature of the core-grafted ligand and the catalytic metal attached to it can be tuned to best promote the targeted catalyzed process. The TPP@CCM shown in Fig. 2 was developed with the Rh-catalyzed hydroformylation reaction in mind and was applied to this reaction (see Sect. 5).

The major problem with the ligand variation is the need to synthesize a suitable ligand monomer for incorporation in the core. In order to have a relatively homogeneous distribution of the functionalized ligand in the hydrophobic block, the comonomers should have reactivity ratios not too far from 1. For instance, the S/DPPS mixture is characterized by $r_S = 0.52$ and $r_{DPPS} = 1.43$ [47], which gives a greater initial incorporation of DPPS and a copolymer with a slight concentration gradient (experimentally verified by the relative rate of monomer consumption) [32, 48]. Another critical issue for the successful incorporation of the chosen ligand monomer is its miscibility with the other comonomer. As already alluded to above, success of PISA requires that the hydrophobic comonomer mixture constitutes a single phase, which is fully incorporated by diffusion through the water phase inside the micellar core after self-assembly. One ligand modification has already been accomplished by adding *p*-OMe substituents to the two dangling Ph groups of the anchored phosphine ligand. Thus, replacement of DPPS with the new ligand monomer [4-bis(*p*-methoxyphenyl)phosphino]styrene (BMOPPS) has yielded a CCM nanoreactor where the PS-grafted ligand is now bis(*p*-methoxyphenyl)-phenylphosphine (BMOPPP), see Fig. 4 [49]. The solubility of this ligand monomer in styrene is lower than that of DPPS, and thus BMOPPP@CCM nanoreactors could only be made with 5% of ligand monomer in the hydrophobic core ($x = 0.05$). The DLS analysis of these particles gave $D_z = 80$ nm (PDI = 0.2) in water and $D_z = 200$ nm (PDI = 0.2) in THF. As it

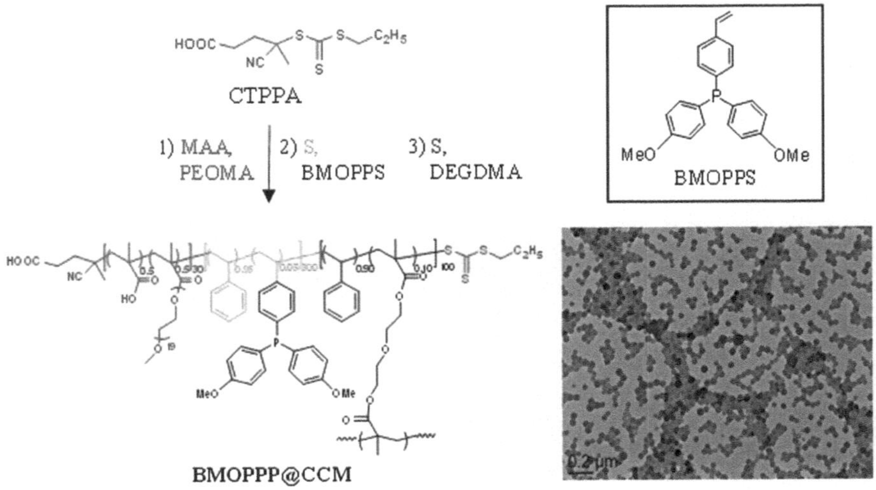

Fig. 4 Structure and TEM characterization of the BMOPPP@CCM nanoreactors. The TEM image was previously published [49] and is reproduced with permission. Copyright 2015 Elsevier Science

will be shown later, the best catalytic performances are those of the nanoreactors having a lower ligand density, thus a limited ligand incorporation is not necessarily a problem.

The nanogel (NG) nanoreactor morphology (see Fig. 1) was attained by a minor modification of the synthetic protocol: steps 2 and 3 were carried out simultaneously rather than sequentially. This synthesis is more delicate than that of the CCM, more readily leading to macrogelation. In order to better control it, a short chain extension of the R_0-($MAA_{0.5}$-$PEOMA_{0.5}$)$_3$-TTC macroRAFT agent (TTC = trithiocarbonate) with neat styrene (50 units), sufficient to induce self-assembly, was inserted in the synthetic sequence prior to addition of the mixture of functionalized monomer, cross-linking agent (DEGDMA), and additional styrene to complete the synthesis. Hence, both TPP@NG and BMOPPP@NG nanoreactors with the same total composition (n and $m + l$) as the $x = 0.05$ CCM nanoreactors (*i.e.*, 15 MAA, 15 PEOMA, 15 functionalized styrene, 10 DEGDMA, and 375 styrene molecules per initial RAFT agent) were made and characterized (Fig. 5). In one instance, however, a TPP@NG latex containing 30 functionalized monomers per initial RAFT agent (equivalent to the TPP@CCM with $x = 0.1$) could be isolated.

As stated in the introduction, these NG nanoreactors bear close similarities with those developed by Terashima, Sawamoto (TS) and their group (see Chap. 6 in this book). We note, however, a few important differences. An obvious one is the use, as controlling method, of ATRP by TS and RAFT by us. However, both strategies insure excellent control. Of greater impact on the synthetic outcome is the choice of the PISA protocol for the RAFT synthesis of the NG particles, leading to quantitative cross-linking of the hydrophilic arms and to the direct generation of a latex, whereas

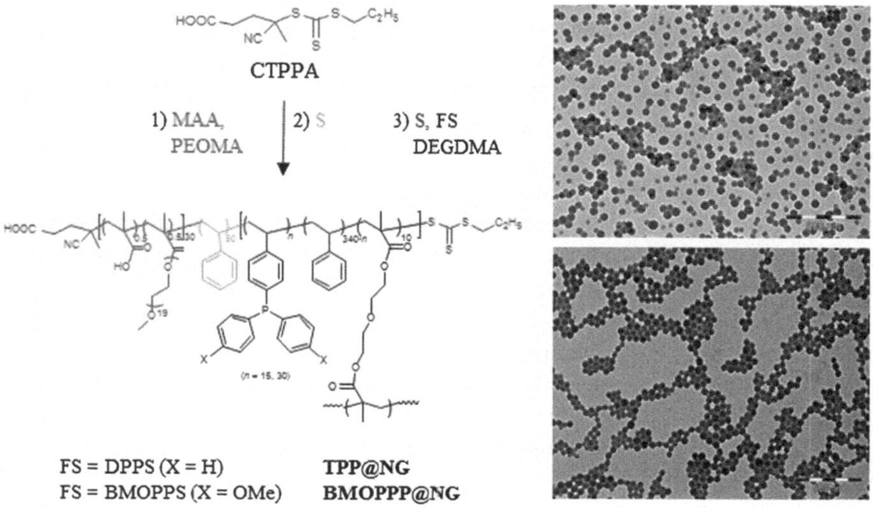

Fig. 5 Structure and TEM characterization of the TPP@NG (*above*) and BMOPPP@NG (*below*) nanoreactors

the nanogels reported by TS were made in a homogeneous medium and remained contaminated by a certain fraction of the unreacted free arms. A second important difference is the use of large amounts of styrene as a diluent in the cross-linking step for the NG synthesis, whereas the particles obtained by TS have a core solely constituted by the functionalized monomer (DPPS) and the cross-linking divinyl monomer. Consequently, the NG particles made by us have a reduced ligand density and a reduced cross-linking density (greater core flexibility): out of 40 monomers in the hydrophobic core, only 1 is a divinyl cross-linker and 1.5 is a phosphine-functionalized monomer.

4 Nanoreactor Stability, Mass Transport, Metal Coordination

The efficiency of metal coordination and mass transport for regular micelles has been amply demonstrated through their successful applications, not only in catalysis, but rarely quantified [14–16]. A less-favored transport of guest molecules to the core may be anticipated for the unimolecular objects because of the scaffold rigidification by cross-linking. The previous investigations of unimolecular nanoreactors (see Sect. 2) indirectly prove, through their positive results in catalysis, that metal coordination and mass transport take place. In some instances, nanoreactor-confined metal catalysts have yielded even higher turnover frequencies than the equivalent molecular catalyst in a homogeneous medium. However, the mass transport kinetics, swelling capacity, metal coordination chemistry, and the effect of the catalytic conditions on the polymer structure and dimension have rarely been looked at. These questions were addressed for the CCM and NG nanoreactors before analyzing their catalytic performance, using DLS and NMR spectroscopy (^{1}H, ^{31}P) as convenient tools.

The DLS study on the TPP@CCM latex revealed a slight pH dependence for D_z (Fig. 6a), initially increasing upon deprotonation and then decreasing again at higher pH. The natural pH of the polymer latex, self-regulated by the acid dissociation of the outer shell methacrylic acid functions, is in the 4.9–5.5 range depending on dilution. Prolonged heating at 90 °C at the natural pH produces an initial slight expansion with no further evolution, while the polydispersity remains narrow (Fig. 6b). No irreversible aggregation occurs under any conditions [44].

NMR has provided valuable information on the core structure and content. The pristine latex only reveals the resonances of the water-solvated hydrophilic shell. However, following core swelling by the introduction of suitable organic molecules, the chains become solvated and mobile and therefore visible in the NMR spectra. In particular, the ^{1}H NMR spectrum allows quantifying the amount of organic compounds incorporated in the core (i.e., *ca.* 2000 $CHCl_3$, 800 toluene, or 150 *n*-nonanal molecules per chain at room temperature) [44]. Visual observation of the phase volume changes, after adding excess organic solvent to the latex and stirring, shows that the latex swelling process is essentially immediate and the first recorded NMR

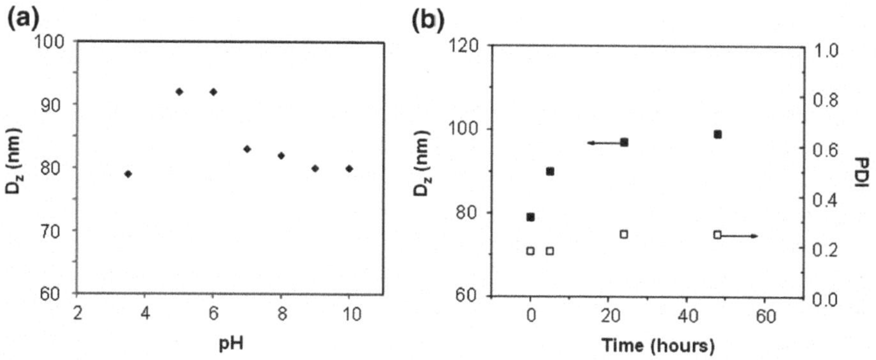

Fig. 6 Variation of the TPP@CCM particle size (D_z from DLS measurements) as a function of pH (**a**) and upon prolonged heating at 90 °C at the natural pH (**b**). Reproduced from Ref. [44] with permission. Copyright 2014 Wiley-VCH

spectrum reveals the equilibrium state. The NMR study indicates no significant core swelling by neat 1-octene (the substrate chosen for the catalytic application, see next section) at room temperature. However, a significant amount of 1-octene is drawn into the polymer core when offered to the latex at room temperature as a 1:1 mixture with toluene (ca. 100 1-octene and 500 toluene molecules per chain). In addition, a DLS study shows that neat 1-octene swells the polymer core upon heating to 90 °C (the temperature used for the catalytic transformation).

Incorporation of larger molecules is slower: stirring the latex with a small amount of a concentrated toluene solution of complex [Rh(acac)$(CO)_2$] (acac=acetylacetonato) causes immediate crystallization of the orange water-insoluble metal complex, showing that the toluene molecules migrate to the polymer core faster than the larger metal complex molecules. However, the metal complex can be slowly incorporated (ca 15 min for full loading) and coordinated to the phosphine ligands by stirring the complex solution with a toluene-pre-swollen latex at room temperature. The [Rh (acac)$(CO)_2$] complex binds by facile substitution of one CO ligand to yield [Rh (acac)(CO)(L@Pol)] (L = TPP or BMOPPP; Pol = CCM or NG), Eq. 1.

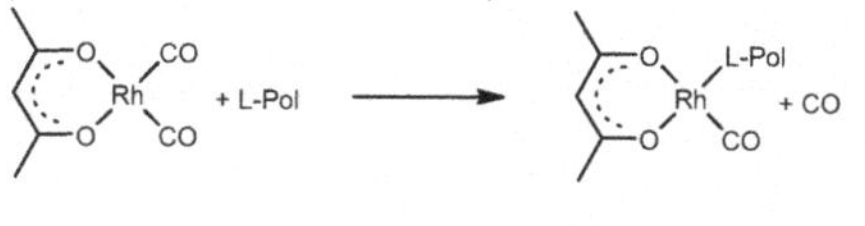

(1)

Qualitatively, the metal loading process is equally fast for the CCM and NG particles and the particle size is identical before and after metal loading as shown by DLS. The degree of metal loading can be assessed by ^{31}P NMR. For complexes [Rh (acac)(CO)(L@Pol)], a rapid degenerate ligand exchange occurs between the product and residual-free phosphine, an already known phenomenon for the molecular system with TPP [50], resulting in a single broad resonance when both are present and no resonance at all if they are in a ca. 50:50 ratio (see Fig. 7a for TPP@CCM; the behavior for the BMOPPP@CCM latex is similar). Addition of further [Rh (acac)$(CO)_2$] beyond 1 equivalent per phosphine ligand produces no further modification of the NMR resonance, indicating that the phosphine coordination to Rh is quantitative and no ligand remains free after addition of a stoichiometric amount of metal.

Incorporation of $[RhCl(COD)]_2$ (COD = 1,5-cyclooctadiene) also yields coordination by chloride bridge splitting to generate [RhCl(COD)(L@Pol)], Eq. 2. In this case the self-exchange process is slower, thus broad resonances of both the coordinated and the free phosphine ligand can be simultaneously observed and integrated to quantify the metal coordination. Figure 7b shows the $^{31}P\{^{1}H\}$ NMR spectra of the fully loaded TPP@NG and BMOPPP@NG particles with both Rh complexes. The doublets are caused by coupling with the ^{103}Rh nucleus (100% abundance). The spectra of the related metal-loaded TPP@CCM and BMOPPP@CCM are identical to these.

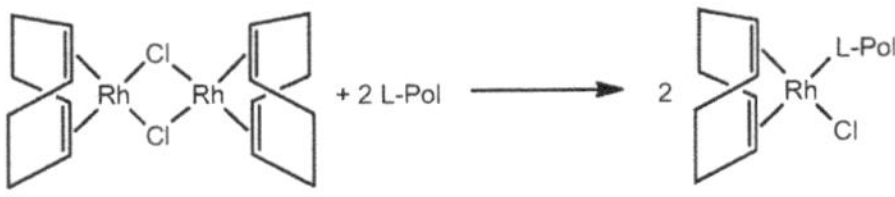

(2)

In conclusion, the nanoreactor latexes show thermal stability after an initial relaxation phenomenon and rapid transport of small molecules to the nanoreactor core, are capable of incorporating even the less compatible 1-octene at high temperatures, and the core-anchored ligands are available to easily and quantitatively coordinate suitable catalytic metal complexes.

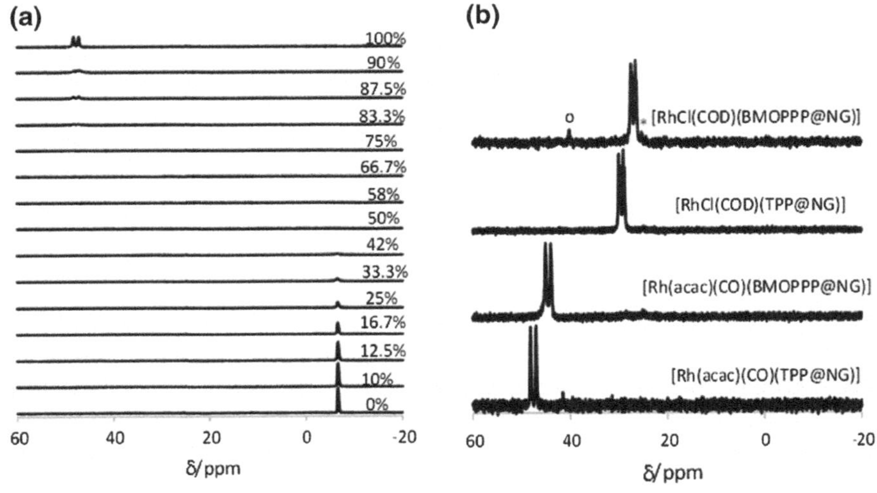

Fig. 7 **a** $^{31}P\{^{1}H\}$ NMR spectra of TPP@CCM at various levels of Rh loading. **b** $^{31}P\{^{1}H\}$ NMR spectra of different fully metal-loaded latexes. The starred resonance corresponds to a small amount of oxidized phosphine. The resonance marked with a circle belongs to an unknown impurity. All spectra were recorded on the toluene-swollen latex after dilution with D_2O for the instrument lock

5 Aqueous Biphasic Hydroformylation of 1-Octene

This reaction (Eq. 3) represents the ideal test case to probe the efficiency of the new nanoreactors in micellar-type biphasic catalysis. The hydroformylation of α-olefins, also known as oxo process [51, 52], is an industrially important homogeneously catalyzed reaction with a World production capacity of >12 Mt/a in 2008 [53]. The aldehyde products are mostly reduced to linear alcohols for use as bulk intermediates to generate plasticizers and surfactants. The process is also important in fine chemistry to generate fragrances and pharmaceutical products. Rhodium is a better catalyst because of its higher activity and especially greater regioselectivity in favor of the preferred linear aldehyde product but is much more expensive that the other commonly used metal, cobalt. Consequently, the use of rhodium is justified only if efficiently and quantitatively recovered and recycled (losses of the order of the ppb may be tolerated). Currently, rhodium is industrially used mostly to hydroformylate the light olefins, mainly propene and butene, using two processes [54, 55]. The first one is a homogeneous process with product separation from the catalyst by distillation. It cannot be extended to the higher olefins because of the insufficient vapor pressure of the resulting aldehydes, requiring extreme temperatures for distillation where the catalyst partially decomposes. The second process is aqueous biphasic catalysis, known as the Rhône-Poulenc/Ruhrchemie hydroformylation (>0.5 Mt/a production) [56], where rhodium is confined in water by coordination to triphenylphosphine trisulfonate (TPPTS). This biphasic process becomes impractical for the higher olefins because of their insufficient water solubility. Thus, the higher olefins are still hydroformylated by the less-efficient cobalt-catalyzed process, which accounts for a large fraction of today's hydroformylation capacity.

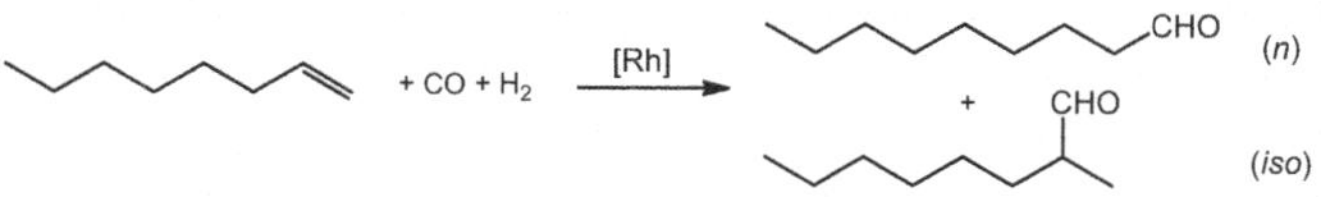

(3)

Vigorous research has been dedicated to the aqueous biphasic Rh-catalyzed hydroformylation of higher olefins, using all strategies outlined in the introduction. Reports on the use of catalytic micelles have been limited. Rhodium has been anchored through either triphenylphosphine [57] or *N*-heterocyclic carbene (NHC) [58] ligands onto an amphiphilic diblock polyoxazoline built by living cationic polymerization. It has also been anchored onto the phosphine oxide ligand in the $O{=}P[p\text{-}C_6H_4(OCH_2CH_2)_nOH]_3$ 3-arm star polymer, which also self-aggregates in water to form micelles [59, 60]. Finally, it has been anchored through an *N*,*N*-bis (2-dipyridyl) amide ligand onto an amphiphilic diblock polynorbornene built by ring-opening metathesis polymerization (ROMP) [61]. Excellent activities and recyclability were reported in all cases, though always associated to some extent of catalyst leaching in the ppm range.

The aqueous biphasic hydroformylation of 1-octene with the CCM and NG nanoreactors was carried out in a high-pressure batch reactor with a P/Rh ratio of

4:1. This is a commonly used ratio because excess free phosphine is known to improve the *n*:*iso* regioselectivity. The 1-octene substrate was diluted with *n*-decanal to mimic the conditions of a continuous stirred-tank reactor while preserving accurate yield and quantification of the *n*:*iso* selectivity by GC and NMR spectroscopy. The system turned out to be very efficient, with nearly complete conversions after 2 h (catalyst loading: 0.2% vs. substrate) and recovered yields around 90% (*n*:*iso* > 3), with only small amounts of octene isomers as byproducts. The reaction mixture was readily decanted to afford a colorless and transparent organic phase and an opaque orange aqueous layer containing the Rh-loaded nanoreactors. However, a small amount of polymer (invisible to the eye, no turbidity) is present in the organic phase as shown by the DLS studies (see next section) and by the Rh ICP-MS analysis. A selection of catalytic results is collected in Table 1.

At the end of run 1, the recovered aqueous phase was reused in a recycle run with a fresh substrate charge after prolonged exposure to air (run 1r), yielding the same activity (see Fig. 8), demonstrating not only the essentially quantitative catalyst recovery, but also a protecting effect of the polymer scaffold on the activated catalyst. In fact, the trace of run 1 shows an initially lower activity, attributed to a precatalyst activation or polymer structural reorganization phase (*i.e.*, the polymer expansion revealed by the DLS study of Fig. 6b). A series of five recycles were also carried out while always maintaining the aqueous layer under syngas protection inside the reactor (runs 2-2r4), showing minimal activity change, but always leading to Rh loss. The result of run 3 demonstrates that the reaction occurs in the nanoreactor core (micellar-type catalysis) and not at the water–organic interface (interfacial catalysis). Indeed, the water-soluble sulfoxantphos is a stronger ligand for rhodium, thus it extracts all Rh from the nanoreactor core into the aqueous phase. The conversion is dramatically retarded, whereas it should remain fast if the polymer acted merely as a surfactant to increase the water/organic interface. Note that in this case there is essentially no Rh loss.

Changing the phosphine density in the nanoreactor core (*cf.* runs 4 and 5 with run 2) shows higher activity as the phosphine density is decreased (lower catalyst concentration in the nanoconfined space). The TOF, on the other hand, should not change if the reaction rate were only under chemical control. The reason for this trend is control by the mass transport rate. When the particle core volume per catalytic site becomes smaller (and the number of substrate molecules in this volume at equilibrium gets smaller), the catalytic center remains idle for a greater fraction of time waiting for additional fresh substrate to reach the catalytic sites. A similar activity decrease upon increasing the catalyst concentration in a nanoreactor core has also been reported for the transfer hydrogenation of acetophenone catalyzed by ruthenium-incorporating microgel star polymers [22]. The view of a mass transport-limited process is confirmed by the effect of the stirring rate on the activity (*cf.* runs 6 and 7 with run 2). In the absence of mass transport limitations, the TOF should be independent on the stirring rate and this is obviously

Table 1 Results of the aqueous biphasic hydroformylation of 1-octene by $[Rh(acac)(CO)_2]$/nano-object latexes

Run	Nano-object[a]	Change versus the standard conditions[b]	Value	*n*/*iso*	TOF[c] (h^{-1})	[Rh][d] (ppm)	Ref.
1	TPP@CCM ($x = 0.1$)	V_{org}/V_{aq}	2	5.0	628	1.8	[44]
1r	TPP@CCM ($x = 0.1$)	V_{org}/V_{aq}	2	3.4	782	2.5	[44]
2	TPP@CCM ($x = 0.1$)			4.4	473	2.5	[45]
2r1	TPP@CCM ($x = 0.1$)			3.9	541	2.7	[45]
2r2	TPP@CCM ($x = 0.1$)			3.4	520	2.1	[45]
2r3	TPP@CCM ($x = 0.1$)			3.2	457	1.7	[45]
2r4	TPP@CCM ($x = 0.1$)			3.6	402	1.7	[45]
3	TPP@CCM ($x = 0.1$)	+ Sulfoxantphos	[e]	–	13	0.1	[44]
4	TPP@CCM ($x = 0.05$)			3.3	695	1.8	[45]
5	TPP@CCM ($x = 0.25$)			4.7	191	1.4	[45]
6	TPP@CCM ($x = 0.1$)	Stirring speed	1400 rpm	3.4	557	6.5	[45]
7	TPP@CCM ($x = 0.1$)	Stirring speed	1600 rpm	3.5	579	11.6	[45]
8	TPP@CCM ($x = 0.1$)	Temperature	70 °C	3.8	72	1.3	[45]
9	BMOPPP@CCM ($x = 0.05$)			2.7	742	4.5	[49]
10	TPP@M ($x = 0.1$)			3.8	560	7.2	[44]
11	TPP@NG			3.6	378	0.6	[46]

[a]For M, $n{:}m = 30{:}300$ (Fig. 2). For CCM: $n{:}m{:}l = 30{:}300{:}100$ (Fig. 2). For NG, see composition in Fig. 5. [b]Standard conditions: $[\text{1-octene}]_{org} = 1.1$ M in *n*-decanal, [P]/[Rh] = 4, $V_{org}/V_{aq.} = 3{:}1$, [1-octene]/[Rh] = 500, $T = 90$ °C, $P_{syngas} = 20$ bar ($CO/H_2 = 1{:}1$), $\omega = 1200$ rpm. [c]Turnover frequency calculated from the syngas consumption during the first 5 min following gas absorption. [d]Rh concentration in the organic phase measured by ICP/MS. [e][Sulfoxantphos]/[Rh] = 5

not the case. For this reason, it is not desirable to increase the ligand density in the nanoreactor core, the major interest being to increase the core–shell interface relative to catalyst and to optimize the kinetics of mass transport through the nanoparticle shell. One point of great interest, on which we shall return below, is the dramatic effect of the stirring rate on the Rh loss (an increase of 360% for a stirring rate increase of only 33%!). Lowering the temperature by 20 °C expectedly

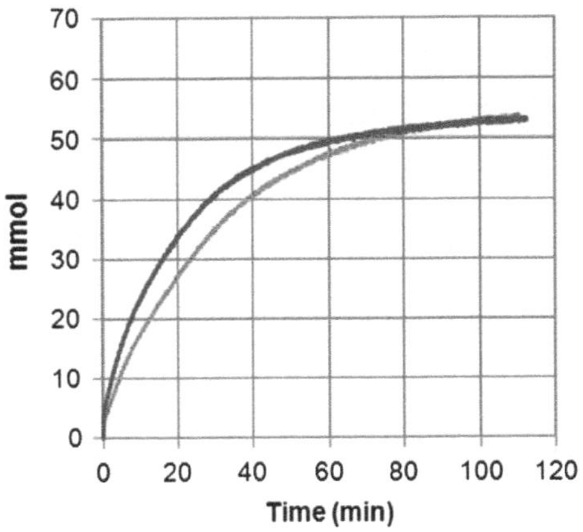

Fig. 8 Monitoring of the syngas consumption for run 1 (*red line*) and 1r (*blue line*) (Table 1). Run 1r was carried out after leaving the recovered aqueous layer in air for 5 days. Adapted with permission from Ref. [44]. Copyright 2014 Wiley-VCH

decreases the activity. A catalytic run with the BMOPPP@CCM nanoreactor (run 9) yields a slightly greater activity than the comparable TPP@CCM with the same phosphine density (*cf.* with run 4), but also a lower *n*/*iso* ratio and a much greater Rh loss. Runs 10 and 11 are interesting because they compare three different nanoreactor topologies (the micelle, M; the core-cross-linked micelle, CCM; and the nanogel, NG) with essentially the same size and phosphine content. The activity increases in the order NG < CCM < M, which may be expected, once again, on the basis of mass transport limitations: the rate is reduced upon increasing the rigidity of the scaffold. The variation is, however, relatively small (TOF reduction by only 18% on going from M to CCM and by 25% on going from CCM to NG). Conversely, the topology has a much greater effect on the Rh loss with a dramatic decrease from M (7.2 ppm) through CCM (2.5 ppm) to NG (0.6 ppm).

6 Investigations of the Causes for Catalyst Leaching

The determination of a relatively high Rh leaching, comparable to those previously reported for catalysts incorporated inside micelles, was disappointing. However, it became of interest to understand the reasons and to find, if possible, appropriate measures to minimize it or eliminate the catalyst loss. A first question is the chemical nature of the leached Rh: molecular complex (metal escape from the nanoreactor core), full metal-loaded nanoreactor, or nanoreactor fragments (*e.g.*, free arms or part or the arms such as the metal-bearing hydrophobic block) that could be present via incomplete arm cross-linking or generated during catalysis by nanoreactor shear stress. The essentially negligible Rh loss measured in the presence of sulfoxanthphos (run 3 in Table 1) may suggest that this is related to Rh binding to the polymer. However, it cannot be excluded that a more hydrophobic molecular Rh complex can be extracted from the polymer in the absence of sulfoxantphos (*e.g.*, by CO, 1-octene or other available ligands). The dramatic increase of Rh leaching at higher stirring rates may seem to suggest the generation of smaller polymer fragments by shear stress. The DLS analysis of the recovered product after catalysis and control studies of polymer solubility in *n*-decane, in the presence and absence of metal loading, gave very useful information [45].

Vigorous stirring of the pristine TPP@CCM latex, without Rh coordination, together with *n*-decanal at room temperature for 2.5 h does not result in any particle transfer to the organic phase (no significant DLS response). Stirring at 90 °C for 2.5 h (same conditions of the catalysis except for the absence of 1-octene and syngas) confirmed the transfer of particles to the organic phase (Fig. 9a), although the amount of particles was insufficient to induce significant turbidity. Interestingly, the particle size was much greater than for the pristine latex (D_z = 283 nm, PDI = 0.33) when the DLS measurement was carried out immediately after cooling, whereas the particle size distribution shrank significantly after one week standing at room temperature (D_z = 186 nm, PDI = 0.10). The size distribution measured immediately is very broad and extends to diameters beyond those expected for the maximum swelling of individual particles (*cf.* the distribution in THF solution, Fig. 3), indicating agglomeration, though it includes also non-agglomerated particles. However, the agglomerates are slowly redispersed upon standing at room temperature (Fig. 9b). Unfortunately, the DLS signal does not carry quantitative information, and thus the TPP@CCM solubility in *n*-decanal cannot be estimated and compared with the Rh loss observed during catalysis. Comparison of these results shows that, although the room temperature solubility of the CCM in *n*-decanal is negligible (or the solubilization is too slow to occur within the time of the experiment), the particles are transferred to the organic phase at high temperature, especially as agglomerates. Then, upon cooling, a significant amount of particles remains in the organic phase, though the agglomeration state is reduced. The greater dimensions after one week relative to the pristine latex (*cf.* Figs. 3 and 9a) may result from residual agglomeration and/or from particle swelling by *n*-decanal.

Repeating the same experiment after loading the TPP@CCM with [Rh(acac)$(CO)_2$] at the 25% level (P/Rh = 4), as under the catalytic conditions though still in

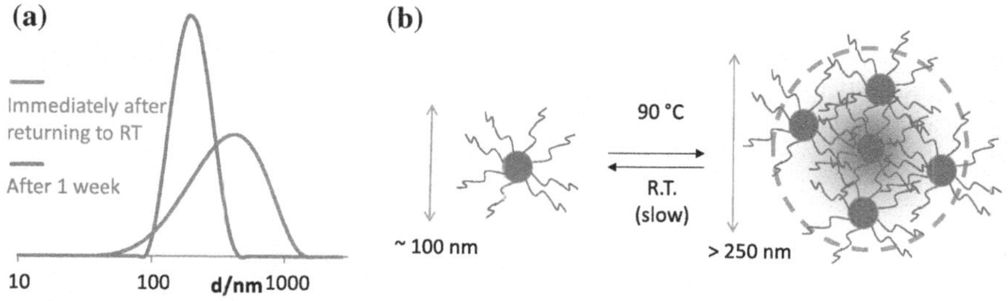

Fig. 9 **a** DLS analysis of the recovered organic phase after stirring the TPP@CCM with *n*-decanal for 2.5 h at 90 °C. Adapted with permission from Ref. [45]. Copyright 2015 Elsevier. **b** Schematic drawing of the reversible aggregation of the nano-objects

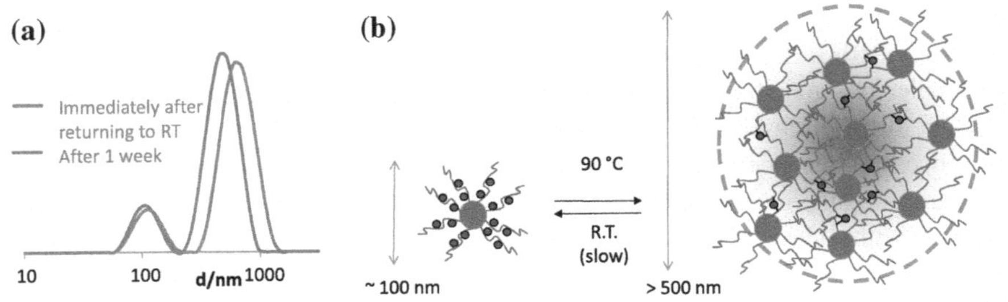

Fig. 10 **a** DLS analysis of the recovered organic phase after stirring the [Rh(acac)$(CO)_2$]/TPP@CCM solution (TPP/Rh = 4) with *n*-decanal for 2.5 h at 90 °C. Adapted with permission from Ref. [45]. Copyright 2015 Elsevier. **b** Schematic drawing of the Rh coordination induced interparticle cross-linking. Only the Rh centers involved in the cross-linking are represented, for clarity reasons

the absence of 1-octene and syngas, gave the results shown in Fig. 10a. In this case, the measurement revealed a bimodal size distribution the minor component of which, centered at ca. 100 nm, corresponds to the size of the free TPP@CCM while the major component, centered at ca. 650 nm, corresponds to agglomerated particles. After 1 week, unlike the experiment with the Rh-free particles, the distribution still reveals a major portion of agglomerated particles, though a slight shrinking indicates the slow release of swelling solvent. In addition, prolonged heating of the latex at the reflux temperature resulted in precipitation of the polymer as an orange solid, which could no longer be dispersed in the aqueous phase. Therefore, the interparticle agglomeration is irreversible in the presence of Rh. Once again, no particles were transferred to the organic phase upon prolonged stirring at room temperature. The reason for this Rh effect on the thermal behavior is attributed to the known coordination chemistry of the [Rh(acac)(CO)(TPP)]/TPP system and to the demonstrated particle interpenetration. It is in fact known that, while [Rh(acac)(CO)(TPP)] and TPP only undergo degenerate TPP exchange at room temperature (Fig. 7a), heating results in displacement of the residual CO ligand (Eq. 4) [62]. Consequently, the irreversibility of the aggregation can be attributed to the establishment of interparticle bridges by the Rh centers. The majority of the metals probably bind two TPP ligands within the same arm or of two different arms within the same polymer particle, but bridges between two arms of different particles may also be established during the interpenetration phase.

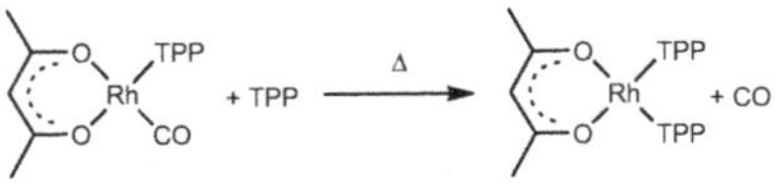

(4)

The next interesting information comes from the DLS study of the organic phases recovered from the catalytic experiments, in particular those corresponding

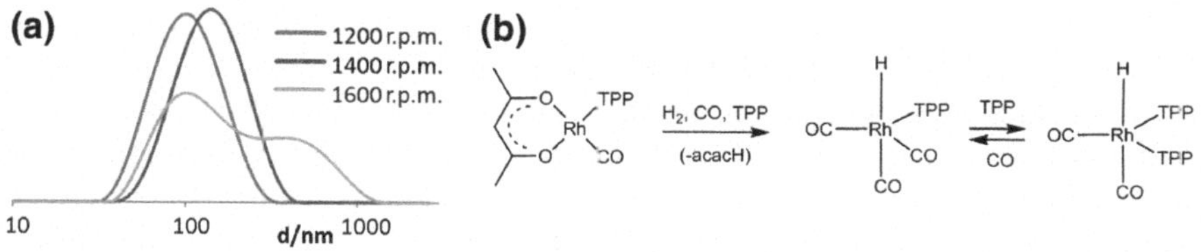

Fig. 11 **a** DLS analysis of the recovered organic phases after the catalytic runs at different stirring speeds. The measurements were carried out at room temperature ca. 1 year after the catalytic experiments. **b** Precatalyst activation upon treatment with syngas. Adapted with permission from Ref. [45]. Copyright 2015 Elsevier

to the experiments at different stirring rates (runs 2, 6, and 7 in Table 1), the results being shown in Fig. 11a. The long delay between the phase recovery after catalysis and the DLS measurement (ca. 1 year) should ensure equilibrium of the slow expulsion of the swelling molecules. A first important observation is that the particle size distribution does not reveal, in any instances (even for the control experiments discussed above), any measured scattered intensity corresponding to smaller sizes than that of the integral particles, excluding any polymer fragmentation by shear stress. The particle size distribution is displaced toward greater dimensions for higher stirring rates and a bimodal distribution is observed for the phase recovered from the experiment at the highest stirring speed. This is reminiscent of the irreversible aggregation observed for the control experiment in Fig. 10a and can be therefore assigned to interparticle cross-linking by Rh, which appears to be greater at greater stirring rates. However, the extent of aggregation is less pronounced relative to the control experiment. The difference between the two systems is in the nature of the Rh complex. During catalysis, the syngas transforms the [Rh(acac)(CO)(TPP)] precatalyst into the activated complex, which is a mixture of [RhH$(CO)_{4-n}(TPP)_n$] with different n. The distribution within this mixture depends on the TPP/Rh ratio, on the concentrations, and on the CO pressure, with the $n = 1$ species being the most active form and prevalent at high CO pressure (see Fig. 11b) [63]. Hence, it can be argued that the catalytic conditions keep the Rh complex mostly in a monophosphine form, which cannot interlink the particles, whereas heating up the precatalyst without syngas causes quantitative formation of the bis(phosphine) complex and more extensive irreversible aggregation.

Once again, the DLS signal unfortunately does not carry quantitative information for correlation with the Rh loss in the organic phase. However, the following qualitative conclusions can be drawn: (i) the polymer particles are slightly transferred to the organic phase as integral nanoreactors or as particle aggregates, with the absence of any significant fragmentation; (ii) greater stirring rates induce the formation of larger aggregates and correspondingly greater Rh loss is found by ICP-MS analysis. The logical deduction is that larger aggregates are more lipophilic and are transferred to the organic phase to a greater extent than smaller ones. This conclusion is also consistent with the trend of Rh loss observed for the M, CCM, and NG series, since interpenetration and metal-induced interlinking is expected to become less favoured less favored along the series.

We conclude from this analysis that the catalyst loss is intimately associated with transport of the nanoreactors to the organic phase and that this phenomenon can be attenuated by disfavoring particle interpenetration and cross-linking. However, the specific chemical nature of the hydrophilic shell of these nanoreactors will always yield leaching because they are transported to the organic phase even as single particles, without any interlinking (see Fig. 9a). Rh loss by dissociation of the metal center from the polymer scaffold does not appear to occur, or at least not to be the major cause of leaching. Any further efforts to develop catalytic nanoreactors with even lower leaching must therefore be oriented toward the construction of CCM and NG particles containing more hydrophilic/less lipophilic outer shells at the catalysis operational temperature.

7 Perspectives for Further Developments

The perspectives of this work are numerous. The modularity and simplicity of the PISA-RAFT synthetic protocol open access to a myriad of other ligand-functionalized CCM- and NG-type nanoreactors for aqueous biphasic micellar-type catalysis, where the core and shell chemical nature and degrees of polymerization, the chemical nature and the density of the core-anchored ligands, the cross-linking density (particularly for the NG-type particles), and the coordinated metal precatalyst can be varied nearly at will. This synthetic strategy may also be employed to prepare organocatalytic nanoreactors. In terms of industrial applications, large-scale transformations needing a quantitative recovery of expensive catalysts (e.g., Rh-catalyzed hydroformylation) require further tuning of the polymer shell in order to decrease the shell lipophilicity at high temperatures. However, many other catalyzed transformations in fine chemical production targeting products of high-added value may benefit from the already developed nanoreactors, or other similar ones in which the above-mentioned parameters can be easily optimized, allowing the implementation of simpler catalyst recovery and recycling procedures.

References

1. Cole-Hamilton DJ (2003) Homogeneous catalysis—new approaches to catalyst separation, recovery, and recycling. Science 299(5613):1702–1706
2. Gursel IV, Nol T, Wang Q, Hessel V (2015) Separation/recycling methods for homogeneous transition metal catalysts in continuous flow. Green Chem 17(4):2012–2026
3. End N, Schoning KU (2004) Immobilized catalysts in industrial research and application. Immobilized Catal 242:241–271

4. Riisager A, Fehrmann R, Haumann M, Wasserscheid P (2006) Supported ionic liquid phase (SILP) catalysis: an innovative concept for homogeneous catalysis in continuous fixed-bed reactors. Eur J Inorg Chem 4:695–706
5. Gu Y, Li G (2009) Ionic liquids-based catalysis with solids: state of the art. Adv Synth Catal 351(6):817–847
6. Mrowczynski R, Nan A, Liebscher J (2014) Magnetic nanoparticle-supported organocatalysts —an efficient way of recycling and reuse. Rsc Adv 4(12):5927–5952
7. Dijkstra HP, Van Klink GPM, Van Koten G (2002) The use of ultra- and nanofiltration techniques in homogeneous catalyst recycling. Acc Chem Res 35(9):798–810
8. Janssen M, Mueller C, Vogt D (2011) Recent advances in the recycling of homogeneous catalysts using membrane separation. Green Chem 13(9):2247–2257
9. Dobbs AP, Kimberley MR (2002) Fluorous phase chemistry: a new industrial technology. J Fluor Chem 118(1–2):3–17
10. Vincent JM (2012) Fluorous catalysis: from the origin to recent advances. In: Horvath IT (ed) Fluorous chemistry, vol 308. Topics in Current Chemistry, pp 153–174
11. Zhang Q, Zhang S, Deng Y (2011) Recent advances in ionic liquid catalysis. Green Chem 13 (10):2619–2637
12. Wachsen O, Himmler K, Cornils B (1998) Aqueous biphasic catalysis: where the reaction takes place. Catal Today 42(4):373–379
13. Behr A, Henze G, Schomaecker R (2006) Thermoregulated liquid/liquid catalyst separation and recycling. Adv Synth Catal 348 (12 + 13):1485–1495
14. Oehme G (2002) Micellar catalysis. In: Cornils B, Herrmann WA (eds) Applied homogeneous catalysis with organometallic compounds: a comprehensive handbook, vol 2. 2nd edn, pp 835–841
15. Reinsborough VC (2002) Micellar catalysis. In: Volkov AG (ed) Interfacial catalysis. Marcel Dekker, New York, NY, USA, pp 377–390
16. Khan MN (2006) Micellar catalysis. CRC Press, USA
17. Cotanda P, Petzetakis N, O'Reilly RK (2012) Catalytic polymeric nanoreactors: more than a solid supported catalyst. MRS Commun 2(4):119–126
18. Lu A, O'Reilly RK (2013) Advances in nanoreactor technology using polymeric nanostructures. Curr Opin Biotech 24(4):639–645
19. Levins AD, Wang XF, Moughton AO, Skey J, O'Reilly RK (2008) Synthesis of core functionalized polymer micelles and shell cross-linked nanoparticles. Macromolecules 41(9):2998–3006
20. Liu Y, Wang Y, Wang YF, Lu J, Pinon V, Weck M (2011) Shell cross-linked micelle-based nanoreactors for the substrate-selective hydrolytic kinetic resolution of epoxides. J Am Chem Soc 133(36):14260–14263
21. Liu Y, Pinon V, Weck M (2011) Poly(norbornene) block copolymer-based shell cross-linked micelles with Co(III)-salen cores. Polym Chem 2(9):1964–1975
22. Terashima T, Kamigaito M, Baek K-Y, Ando T, Sawamoto M (2003) Polymer catalysts from polymerization catalysts: direct encapsulation of metal catalyst into star polymer core during metal-catalyzed living radical polymerization. J Am Chem Soc 125(18):5288–5289
23. Terashima T, Nomura A, Ito M, Ouchi M, Sawamoto M (2011) Star-polymer-catalyzed living radical polymerization: microgel-core reaction vessel by tandem catalyst interchange. Angew Chem Int Ed Engl 50(34):7892–7895
24. Terashima T, Ouchi M, Ando T, Kamigaito M, Sawamoto M (2006) Metal-complex-bearing star polymers by metal-catalyzed living radical polymerization: synthesis and characterization of poly(methyl methacrylate) star polymers with Ru(II)-embedded microgel cores. J Polym Sci, Polym Chem 44(17):4966–4980
25. Terashima T, Ouchi M, Ando T, Sawamoto M (2011) Oxidation of *sec*-alcohols with Ru(II)-bearing microgel star polymer catalysts via hydrogen transfer reaction: unique microgel-core catalysis. J Polym Sci, Polym Chem 49(5):1061–1069
26. Terashima T, Ouchi M, Ando T, Kamigaito M, Sawamoto M (2007) Amphiphilic, thermosensitive ruthenium(ii)-bearing star polymer catalysts: one-pot synthesis of peg armed

star polymers with ruthenium(ii)-enclosed microgel cores via metal-catalyzed living radical polymerization. Macromolecules 40(10):3581–3588
27. Terashima T, Ouchi M, Ando T, Sawamoto M (2010) Thermoregulated phase-transfer catalysis via PEG-armed Ru(II)-bearing microgel core star polymers: efficient and reusable Ru(II) catalysts for aqueous transfer hydrogenation of ketones. J Polym Sci, Polym Chem 48(2):373–379
28. Terashima T, Ouchi M, Ando T, Sawamoto M (2011) Transfer hydrogenation of ketones catalyzed by PEG-armed ruthenium-microgel star polymers: microgel-core reaction space for active, versatile and recyclable catalysis. Polym J 43(9):770–777
29. Bosman AW, Vestberg R, Heumann A, Fréchet JMJ, Hawker CJ (2003) A modular approach toward functionalized three-dimensional macromolecules: from synthetic concepts to practical applications. J Am Chem Soc 125(3):715–728
30. Lu Y, Mei Y, Drechsler M, Ballauff M (2006) Thermosensitive core-shell particles as carriers for Ag nanoparticles: modulating the catalytic activity by a phase transition in networks. Angew Chem Int Ed Engl 45(5):813–816
31. Gao HF, Matyjaszewski K (2009) Synthesis of functional polymers with controlled architecture by CRP of monomers in the presence of cross-linkers: From stars to gels. Progr Polym Sci 34(4):317–350
32. Cardozo AF, Manoury E, Julcour C, Blanco J-F, Delmas H, Gayet F, Poli R (2013) Preparation of phosphine-functionalized polystyrene stars by metal catalyzed controlled radical copolymerization and their application to hydroformylation catalysis. Dalton Trans 42:9148–9156
33. Lansalot M, Rieger J, D'Agosto F. (2016) Polymerization-Induced Self-Assembly: the Contribution of Controlled Radical Polymerization to the Formation of Self-Stabilized Polymer Particles of Various Morphologies. In: O. Borisov, L. Billion (Eds.), Macromolecular Self-assembly, John Wiley & Sons, Inc., pp.33-82.
34. Canning S L, Smith G N, Armes S P (2016) A Critical Appraisal of RAFT-Mediated Polymerization-Induced Self-Assembly. Macromolecules 49(6):1985-2001
35. An Z, Qiu Q, Liu G (2011) Synthesis of architecturally well-defined nanogels via RAFT polymerization for potential bioapplications. Chem Commun 47(46):12424–12440
36. Liu G, Qiu Q, Shen W, An Z (2011) Aqueous dispersion polymerization of 2-Methoxyethyl acrylate for the synthesis of biocompatible nanoparticles using a hydrophilic RAFT polymer and a redox initiator. Macromolecules 44(13):5237–5245
37. Qiu Q, Liu G, An Z (2011) Efficient and versatile synthesis of star polymers in water and their use as emulsifiers. Chem Commun 47(47):12685–12687
38. Shen W, Chang Y, Liu G, Wang H, Cao A, An Z (2011) Biocompatible, antifouling, and thermosensitive core-shell nanogels synthesized by RAFT aqueous dispersion polymerization. Macromolecules 44(8):2524–2530
39. Chaduc I, Zhang W, Rieger J, Lansalot M, D'Agosto F, Charleux B (2011) Amphiphilic block copolymers from a direct and one-pot RAFT synthesis in water. Macromol Rapid Comm 32(16):1270–1276
40. Zhang X, Boisse S, Zhang W, Beaunier P, D'Agosto F, Rieger J, Charleux B (2011) Well-defined amphiphilic block copolymers and nano-objects formed in situ via RAFT-mediated aqueous emulsion polymerization. Macromolecules 44(11):4149–4158
41. Zhang WJ, D'Agosto F, Boyron O, Rieger J, Charleux B (2011) One-pot synthesis of poly (methacrylic acid-co-poly(ethylene oxide) methyl ether methacrylate)-b-polystyrene amphiphilic block copolymers and their self-assemblies in water via RAFT-mediated radical emulsion polymerization. A kinetic study. Macromolecules 44(19):7584–7593
42. Zhang WJ, D'Agosto F, Boyron O, Rieger J, Charleux B (2012) Toward better understanding of the parameters that lead to the formation of nonspherical polystyrene particles via RAFT-Mediated one-pot aqueous emulsion polymerization. Macromolecules 45(10):4075–4084

43. Zhang WJ, D'Agosto F, Dugas PY, Rieger J, Charleux B (2013) RAFT-mediated one-pot aqueous emulsion polymerization of methyl methacrylate in presence of poly(methacrylic acid-co-poly(ethylene oxide) methacrylate) trithiocarbonate macromolecular chain transfer agent. Polymer 54(8):2011–2019
44. Zhang X, Cardozo AF, Chen S, Zhang W, Julcour C, Lansalot M, Blanco J-F, Gayet F, Delmas H, Charleux B, Manoury E, D'Agosto F, Poli R (2014) Core-shell nanoreactors for efficient aqueous biphasic catalysis. Chem Eur J 20(47):15505–15517
45. Cardozo AF, Julcour C, Barthe L, Blanco J-F, Chen S, Gayet F, Manoury E, Zhang X, Lansalot M, Charleux B, D'Agosto F, Poli R, Delmas H (2015) Aqueous phase homogeneous catalysis using core-shell nano-reactors: application to rhodium catalyzed hydroformylation of 1-octene. J Catal 324:1–8
46. Poli R, Chen S, Zhang X, Cardozo A, Lansalot M, D'Agosto F, Charleux B, Manoury E, Gayet F, Julcour C, Blanco J-F, Barthe L, Delmas H (2015) One-pot RAFT synthesis of triphenylphosphine-functionalized amphiphilic core-shell polymers and application as catalytic nanoreactors in aqueous biphasic hydroformylation. ACS Symp Ser 1188:203–220
47. Rabinowitz R, Marcus R, Pellon J (1964) Synthesis, polymerization, and copolymerization of diphenyl-*p*-styrylphosphine, phosphine oxide, and phosphine sulfide. J Polym Sci, Polym Chem 2(3):1241–1249
48. Cardozo AF, Manoury E, Julcour C, Blanco J-F, Delmas H, Gayet F, Poli R (2013) Preparation of polymer supported phosphine ligands by metal catalyzed living radical copolymerization and their application to hydroformylation catalysis. Chem Cat Chem 5:1161–1169
49. Chen S, Cardozo AF, Julcour C, Blanco J-F, Barthe L, Gayet F, Charleux B, Lansalot M, D'Agosto F, Delmas H, Manoury E, Poli R (2015) Amphiphilic core-cross-linked micelles functionalized with Bis(4-methoxyphenyl)phenylphosphine as catalytic nanoreactors for biphasic hydroformylation. Polymer 72:327–335
50. Trzeciak AM, Jon M, Ziolkowski JJ (1982) Effect of free phosphine on the reactivity of phosphine-containing rhodium(I) complexes. React Kinet Catal Lett 20(3–4):383–387
51. Cornils B, Herrmann WA, Rasch M (1994) Roelen, Otto, Pioneer in Industrial Homogeneous Catalysis. Angew Chem Int Ed Engl 33(22):2348–2348
52. Beller M, Cornils B, Frohning CD, Kohlpaintner CW (1995) Progress in hydroformylation and carbonylation. J Mol Catal A 104(1):17–85
53. Steinborn D (2011) Fundamentals of organometallic catalysis. Wiley-VCH, Weinheim
54. Tudor R, Ashley M (2007) Enhancement of industrial hydroformylation processes by the adoption of rhodium-based catalyst: part II key improvements to rhodium process, and use in non-propylene applications. Platinum Met Rev 51(4):164–171
55. Tudor R, Ashley M (2007) Enhancement of industrial hydroformylation processes by the adoption of rhodium-based catalyst: part I. Development of the LP oxo process to the commercial stage. Platinum Met Rev 51(3):116–126
56. Kohlpaintner CW, Fischer RW, Cornils B (2001) Aqueous biphasic catalysis: ruhrchemie/rhone-poulenc oxo process. Appl Catal A 221(1–2):219–225
57. Nuyken O, Persigehl P, Weberskirch R (2002) Amphiphilic poly(oxazoline)s—synthesis and application for micellar catalysis. Macromol Symp 177 (Synthesis of defined polymer architectures):163–173
58. Zarka MT, Bortenschlager M, Wurst K, Nuyken O, Weberskirch R (2004) Immobilization of a rhodium carbene complex to an amphiphilic block copolymer for hydroformylation of 1-octene under aqueous two-phase conditions. Organometallics 23(21):4817–4820
59. Liu XZ, Li HM, Kong FZ (2002) Polyether phosphine oxide induced phase separable homogeneous catalysis for hydroformylation of higher olefins. J Organomet Chem 664(1–2):1–4

60. Liu X, Kong F, Zheng X, Jin Z (2003) Polyether triaryl phosphine oxides for hydroformylation of oleyl alcohol in micellar catalysis. Catal Commun 4(3):129–133
61. Pawar GM, Weckesser J, Blechert S, Buchmeiser MR (2010) Ring opening metathesis polymerization-derived block copolymers bearing chelating ligands: synthesis, metal immobilization and use in hydroformylation under micellar conditions. Beilstein J Org Chem 6(1):28
62. Mukhedkar AJ, Mukhedkar VA, Green M, Stone FGA (1970) Reactons of low-valent metal complexes with fluorocarbons. 15. Acetylacetonatobis(methyldiphenylphosphine of triphenylphosphine)rhodium. J Chem Soc A 18:3166–3171
63. van Leeuwen PWNM, Claver C (eds) (2000) Rhodium Catalyzed Hydroformylation, vol 22. Catalysis by Metal Complexes, Kluwer, Dordrecht

Catalysis Within Dendrimers

Anne-Marie Caminade, Armelle Ouali, Régis Laurent and Jean-Pierre Majoral

Abstract Dendrimers are hyperbranched macromolecules, synthesized step by step (generation after generation) in an iterative fashion, which structure is reminiscent to that of the branches of trees. Most of their properties are due to their terminal functions, which can be easily modified at will to fulfill the desired properties. In particular, many types of catalytic entities have been used as terminal groups of dendrimers. In some cases, a dendritic effect, that is the enhancement of the catalytic properties when a catalyst is linked to a dendrimer, has been observed. It is also generally possible to recover and reuse the dendritic catalysts. The internal structure of dendrimers can also play a key-role, as it manages cavities which can accommodate the catalytic entities, and enable the substrates to interact with them. Catalytic sites included inside the structure of dendrimers are rare, excepted if they constitute the core of dendrimers (or of dendrons, which are dendritic wedges). Effect of the confinement on the catalysis outcome is generally the main aim of these works. Another type of dendritic catalytic entities taking profit of the internal structure concerns metallic nanoparticles used as core of dendrimers. In this chapter, we will gather information about catalytic entities included inside dendrimers, either covalently linked, or noncovalently entrapped, and on their syntheses. The main types of reactions studied, the role of the generation (size) of the dendrimers, their recovery and reuse, and in general the effect of the confinement inside the dendritic structures on the catalytic efficiency will be discussed.

A.-M. Caminade (✉) · A. Ouali · R. Laurent · J.-P. Majoral
Laboratoire de Chimie de Coordination (LCC), UPR 8241 CNRS,
205 Route de Narbonne, Toulouse Cedex 4 31077, France
e-mail: anne-marie.caminade@lcc-toulouse.fr

A.-M. Caminade · A. Ouali · R. Laurent · J.-P. Majoral
Université de Toulouse, UPS, INPT, Toulouse Cedex 4 31077, France

R. Poli (ed.), *Effects of Nanoconfinement on Catalysis*,
Fundamental and Applied Catalysis, DOI 10.1007/978-3-319-50207-6_8

1 Introduction

Dendrimers [1] are hyperbranched macromolecules, synthesized step by step in an iterative fashion. The presence of branches has three consequences on the structure of dendrimers: (i) a bowl-shaped global structure, at least for large generations, (ii) a large number of terminal groups, generally easily accessible, and (iii) internal cavities, managed by the branches in the internal part. Two main methods of synthesis of dendrimers are known: the divergent method and the convergent method. The divergent method is the most widely used. It consists in adding branched monomers to a multifunctional core, then after activation or deprotection, adding a second layer of branched monomers, and so on. Each time a new layer is created, a new generation is obtained. The principle of the divergent method is shown in Fig. 1. The main problem associated with this method is the possible non-reaction of a few terminal functions, in particular for high generation dendrimers.

F. Vögtle proposed in 1978 the very first example of dendrimers, which he called "cascade molecules" [2]. These branched polyamines were synthesized by a divergent process, but only up to the second generation (two layers of branching points), due to difficulties in the synthesis and purification. Several years later, in 1985 Tomalia et al. [3] described the first "dendrimers" (including the well-known polyamidoamines, i.e., PAMAM), up to the tenth generation. The same year, Newkome et al. [4] published "arborols", having also a branched structure. The

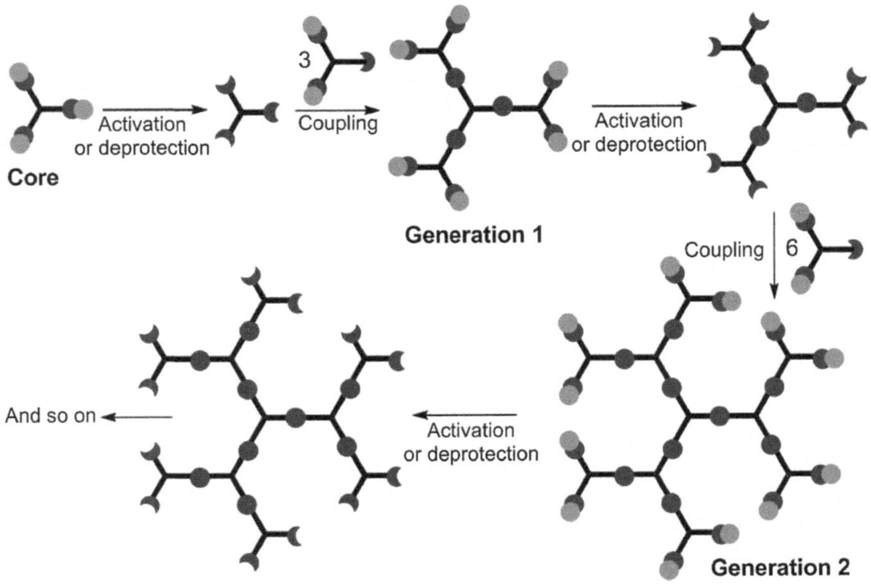

Fig. 1 The principle of the divergent synthesis of dendrimers

name “dendrimers”, proposed by D. Tomalia, is now widely used for naming such compounds.

In 1990, J.M.J. Fréchet proposed a different way for synthesizing dendrimers, a convergent process [5]. A branched monomer, in which only one of the three functions is reactive, is coupled with another branched monomer having two reactive functions, affording a dendron (a dendron is defined as a dendritic wedge). The core of this dendron can be reacted with a branched monomer, or used for the grafting to a core, to afford a dendrimer, as shown in Fig. 2. The convergent process affords dendrimers with a better defined structure than the divergent process, as only a few sites are concerned by the reactions at each step (one or two for the growing, three for grafting to a trifunctional core in the last step), against several tenths, hundredths, or even thousandths (depending on the generation) for the divergent process. However, the grafting to a core becomes difficult when the generation of the dendron increases, due to steric hindrance at the core of the dendron, and it is often difficult to activate the surface functions of dendrimers synthesized by a convergent process, as they need to be very stable to survive to the entire synthetic process. The convergent process is used frequently for the synthesis of dendritic compounds having a single catalytic group at the core.

Also in 1990, T.M. Miller and T.X. Neenan presented the synthesis of small polyphenylene dendrimers [6] (more developed later on by K. Müllen [7]). In 1993,

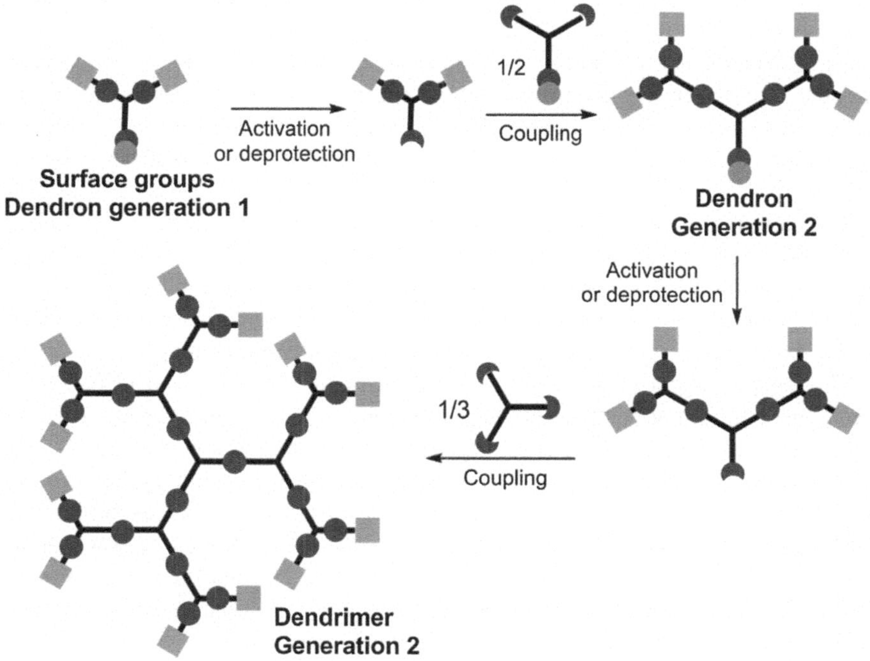

Fig. 2 The principle of the convergent synthesis of dendrimers

the groups of R. Mulhaupt [8] and E.W. Meijer [9] proposed independently improved methods for the synthesis of the Vögtle's dendrimers, now named **PPI** (polypropyleneimine). Also in 1993, J. Roovers presented the first examples of carbosilane dendrimers [10]. In 1994, our group proposed the synthesis of phosphorus-containing dendrimers, more precisely of poly(phosphorhydrazone) dendrimers [11], up to generation 12 [12]. Later on, E. Simanek presented triazine-based dendrimers in 2000 [13]. In most cases, divergent processes are used, eventually accompanied by convergent processes at some steps of the synthesis. Nowadays, emphasis is on the uses of dendrimers, rather than on their synthesis. Main applications concern catalysis, materials and biology/nanomedicine [14].

The use of dendrimers as catalysts was proposed for the first time in 1994, in two papers. The first one (published in November 1994) [15] concerned palladium complexes of small poly(phosphine) dendrimers used for the catalytic electrochemical reduction of CO_2; the second one (published in December 1994) [16] concerned polysilane dendrimers ended by nickel catalytic sites suitable for the regiospecific catalysis of the Kharasch addition of polyhalogenoalkanes to carbon–carbon double bonds. These early examples spawned a multitude of publications concerning the use of dendrimers as catalysts, as shown by many reviews in the field, for early work [17–22], and recently [23–25]. In most cases, the catalytic entities constitute the terminal groups of dendrimers; it means that a large number of catalytic entities are gathered in a single molecule. The proximity of many catalytic entities may induce a positive dendritic effect [26, 27] (an increase of the efficiency of the catalysis, even when considering a same number of catalytic sites), or no special effect, or a detrimental effect (a decrease of the catalytic efficiency), but it is difficult to rationalize and predict such effect. However, it is generally possible to recover the dendritic catalyst and to reuse it, due to its large size compared to the reagents and products. Precipitation by adding solvent or filtration by membrane filters are the most widely used methods for recovering dendritic catalysts.

The presence of catalytic entities inside the structure of dendrimers, at the core or in the branches, is less developed than for catalytic entities as terminal groups. Research in this field is generally carried out with the aim to taking profit of the encapsulation of the catalytic site [28], as is known for the active site of enzymes. A few reviews have gathered some examples of such compounds [29–31].

This chapter about catalysis inside dendrimers will be organized in two main parts, concerning first catalytic entities (organometallic, metallic nanoparticles, or organic entities) located at the core of dendrimers (or of dendrons); this part is the most developed to date. Then the case of catalytic entities constituting the branches of dendrimers, at one or several layers will be displayed. Many examples will illustrate each part, but this chapter will not be exhaustive. Only cases in which the catalytic entities are precisely located in the structure of dendrimers will be considered. The cases of nanoparticles entrapped somewhere in the dendritic structure and/or inside networks of dendrimers will not be considered, as well as the case of organic or organometallic entities entrapped in an undefined way inside dendrimers [32].

2 Catalytic Entities Located at the Core of Dendritic Structures

The very first example of catalysis inside a very small dendrimer was given by H. Brunner who synthesized a series of compounds bearing 4 or 8 chiral arms, in particular linked to a pyridine diimine core. The corresponding copper complexes were used for catalyzing enantioselective cyclopropanations. Slight enantiomeric excesses were obtained (<10%), not very different from those obtained with simpler ligands [33]. "Dendrizymes", i.e., a diphosphine chelating core, surrounded by 4 small dendritic branches were reported by the same group [34]. The corresponding rhodium complex (Fig. 3) was used for catalyzing the hydrogenation of acetamidocinnamic acid. A slight increase of the efficiency was observed on the rate of the reaction, compared with the parent tetraphenyl diphosphine (diphos). The reaction was complete after 3 s with diphos, after 2 s with the "dendrizyme". The 1,3,5-branched ligand led to this rate acceleration, whereas the 1,2,5-isomer showed a rate retardation.

2.1 Polyarylether Dendrons

In view of these early encouraging results, many other attempts were proposed. A large number of them concern dendritic structures in which the branches are constituted of Fréchet's type dendrons (polyarylether) [5]. The main reason is that these dendrons are synthesized in a convergent way, i.e., providing a reactive core at each step. Furthermore, the chemical structure of these dendrons is stable and

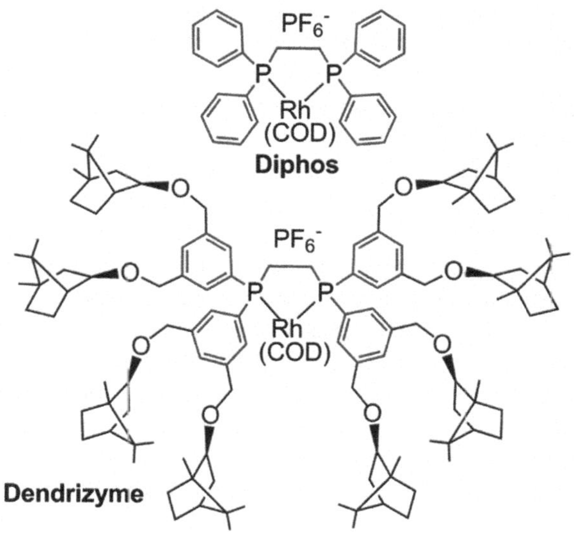

Fig. 3 Chemical structure of the Rh complexes of diphos and "dendrizyme" used for catalyzing the hydrogenation of acetamidocinnamic acid

compatible with many catalytic conditions. On the other hand, phosphines are widely used as versatile ligands for catalysis. Many examples of phosphine complexes located at the core of dendrons have been reported for catalysis experiments, essentially for asymmetric hydrogenations. A recent paper has reviewed the field [31]. Numerous types of phosphine ligands linked through various linkers to a variable number of polyarylether dendrons of different generations have been already proposed, as shown in Fig. 4. This work was carried out with the aim of avoiding the main drawbacks of classical catalyzed asymmetric hydrogenations. Phosphines are generally easily oxidized; the dendritic wedges may prevent their oxidation. Catalysts deactivation processes can occur by formation of nonreactive multimetallic species; the site isolation effect induced by the branches can prevent such deactivation process. Sophisticated chiral phosphine ligands/complexes are often expensive, thus their recovery and reuse are highly desirable; the presence of the dendritic wedges should facilitate the recovery by precipitation or ultrafiltration. The stereoselectivity of asymmetric hydrogenations has often to be improved; the specific microenvironment at the core, around the catalytic site, may influence favorably the catalytic performance, in particular the enantioselectivity.

The first example of increased efficiency concerned the BINAP (2,2′-bis (diphenylphosphino)-1,1′-binaphthyl) ligand with two polyarylether dendrons introduced in the 5,5′ positions through amide linkers. These dendritic ligands were used for Ru-catalyzed asymmetric hydrogenations of 2-(4-isobutylphenyl) acrylic acid. The reactivity increased on moving from generation 2 to generation 3, illustrating a positive dendritic effect, and the second generation could be reused at least three times [35]. The presence of two dendritic wedges is necessary to improve the catalytic performances. Indeed, the ligand with a single wedge used for the same catalytic reaction did not show any dendritic effect [36]. The two-wedge BINAP ligands were also used for Ru-catalyzed hydrogenation of ketones, with good to high enantioselectivities [37]. The same dendrimers were also used for Ir-catalyzed asymmetric hydrogenation of quinolones. They were highly effective, with TOF and TON as high as 3450 h^{-1} and 43,000, respectively. The catalytic activity increased with the dendrimer generation, presumably due to a decreased catalyst dimerization [38].

The chiral PHOX (phosphinooxazoline) surrounded by polyarylether dendrons was used for Ir-catalyzed asymmetric hydrogenation of 2,4-diaryl-1,5-benzodiazepine derivatives, displaying considerable enhancement in activity compared to monomeric catalysts. In this case, the complex was stabilized by the site isolation effect, which prevents the formation of unreactive trinuclear iridium species [39]. The PyrPhos (3,4-bis(diphenylphosphino)pyrrolidine) ligand was used as core of one or two polyarylether dendrons, and the corresponding Rh complexes were applied for hydrogenation of α-acetamido cinnamic acid. A negative dendritic effect was observed, as the rate of catalysis decreased as the number of the generations increased. Generation 4 with a single wedge was almost non-active, and for each generation, the two-wedge ligands were less efficient than the single wedge compound, indicating an encapsulation of the core [40].

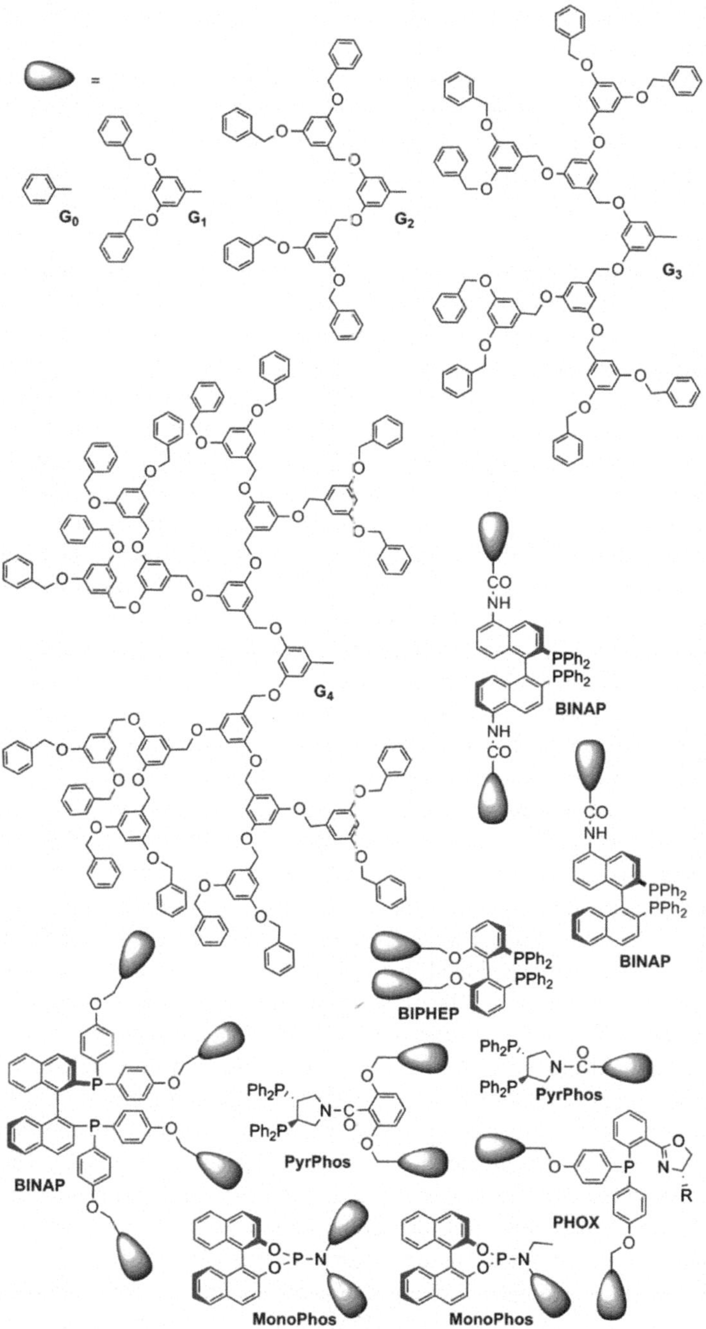

Fig. 4 Chemical structure of polyarylether dendrons, from generation 0 to generation 4, and types of phosphine ligands used as cores. The petal structures symbolize the dendrons

Encapsulation of the catalytic site may have either positive or negative effects on the rate and yield of the catalysis as shown above, but it can also influence the enantioselectivity. The Ru-complexes of the BIPHEP (2,2′-bis(diphenylphosphino)-1,1′-biphenyl) ligand used as core of different generations of dendritic structures was applied in the catalyzed hydrogenation of β-ketoesters. A negative dendritic effect on the enantioselectivity was observed, which could be correlated to an increase of the dihedral angle [41]. The Ru-complexes of the BINAP ligand modified by four dendritic wedges, attached onto the four P-phenyl groups, instead of two classically attached to the naphthyl group, were also applied to the Ru-catalyzed asymmetric hydrogenation of β-ketoesters. Excellent enantiomeric excesses were obtained using high generation dendrons, when compared to the corresponding monomeric catalytic species [42]. These catalysts were also used with α-ketoesters and α-ketoamides, affording also high enantioselectivities.

The monophosphoramidite (MonoPhos) ligand was used as core of two series of dendritic catalysts obtained by grafting one [43] or two [44] polyarylether dendrons. The one-wedge ligand was applied in the Rh-catalyzed asymmetric hydrogenation of α-dehydroamino acid esters and dimethyl itaconate, with excellent enantioselectivities, slightly higher than with monomeric phosphoramidite. The bulky dendritic substituents also decreased the decomposition of the Rh-catalysts induced by hydrolysis in protic solvents [43]. The MonoPhos ligand bearing two dendritic wedges was also applied in the Rh-catalyzed hydrogenation of α-dehydroamino acid derivatives and enamides. A very positive dendritic effect was observed on enantioselectivity in both cases, but also a decrease on the catalytic activity. It was proposed that the bulky dendritic wedges caused a steric repulsion increasing the chirality, but also decreasing the accessibility to the catalytic site [44].

Recycling of these catalysts has been carried out several times by solvent precipitation. For instance, the second generation of the MonoPhos ligand was precipitated and reused five times with similar enantioselectivities [44]. However, efforts have been made to facilitate the catalyst recycling. For this purpose, the end groups of the dendritic wedges were modified. "One-phase catalysis and two-phase separation" system was developed using dendritic catalysts with core–shell structures. In particular, dendritic BINAP ligands bearing a variable number of alkyl chains on the 3,5 or 3,4,5 positions of the terminal aryl groups were used in *n*-hexane/ethanol. Both solvents are totally miscible, but separate in a biphasic solution when a small amount of water (2.5% v:v) is added (Fig. 5). The non-polar catalyst is preferably soluble in *n*-hexane, whereas the products of the asymmetric hydrogenation of 2-arylacrylic acids are preferably soluble in ethanol. The hexane layer containing the dendritic catalyst was easily separated, and reused in the next run. The second generation dendrimer could be reused four times, with practically no difference in the yield and the enantioselectivity [45].

All the previous examples contain chiral phosphines as core of polyarylether dendrons, but some examples of non-chiral dendritic phosphines have been proposed. Dendritic phosphines functionalized by TEG (tetraethylene glycol) or n-C_{12} moieties as substituents on the terminal aryl groups of polyarylether dendrons (Fig. 6) were used for Pd-catalyzed Suzuki–Miyaura coupling reactions. A positive

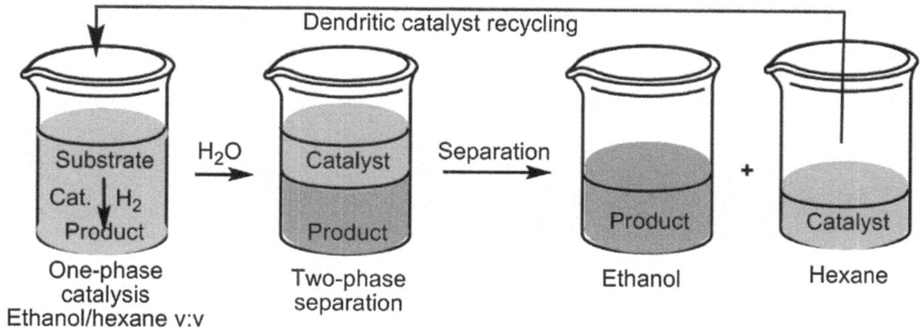

Fig. 5 The principle of the "one-phase catalysis and two-phase separation" system

dendritic effect was observed on the efficiency of coupling of 4-chlorotoluene with phenylboronic acid with the TEG dendrimers, on going from generation 1 (36% yield) to generation 2 (93% yield). On the contrary, the n-C_{12} dendrimers were practically non-active in this reaction, demonstrating the major role played by the surface, even for the catalyst at the core [46]. The role of the internal structure on the efficiency of the same type of reaction was also assessed by introducing ammonium groups relatively close to the catalytic center (Fig. 6). The presence of six permanent cationic charges in the backbone of these ligands was proposed to result in a significant interligand Coulombic repulsion. Nonactivated aryl bromides and activated aryl chlorides can be coupled at a low Pd loading (0.1 mol%), showing high catalytic activity for these systems. A positive dendritic kinetic effect was observed among the different generations, in the coupling reactions of activated aryl chlorides [47].

Beside phosphines, other types of ligands have been used as core of polyarylether dendrimers. Several examples are gathered in Fig. 7, depending on the type of donor atom (O, N, or C). Titanium complexes of optically active binaphthol dendritic derivatives were used as catalysts for C–C bond forming reactions of aldehydes with allyl stannane. No modification in the selectivity was observed, compared to binaphthol [48]. The versatile chiral ligand TADDOL ((R,R)-$\alpha,\alpha,\alpha',\alpha'$-tetraaryl-1,3-dioxolane-4,5-dimethanol) was incorporated as core of polyarylether dendrimers up to the fourth generation. The Ti-complexes were employed as catalysts for the addition of Et_2Zn to PhCHO. Comparison of the stereoselectivity with that obtained with simple Ti TADDOLate shows that there was no detectable decrease of selectivity up to the second generation [49]. A third-generation polyarylether dendron-appended terphenyl carboxylate ligand and ended by alkyl chains was used to complex iron. More precisely, four dendrons were bridging two iron atoms, and this complex was used as catalyst for oxidation of 9,10-Dihydroanthracene and Anthrone, in the presence of an excess of oxygen [50].

Chiral bis(oxazoline) copper complexes linked to the core of polyarylether dendrons were used as Lewis acid catalysts in enantioselective aldol reactions. No influence on the yield and the enantioselectivity was observed between the different generations of the dendrons, and when compared to the corresponding small

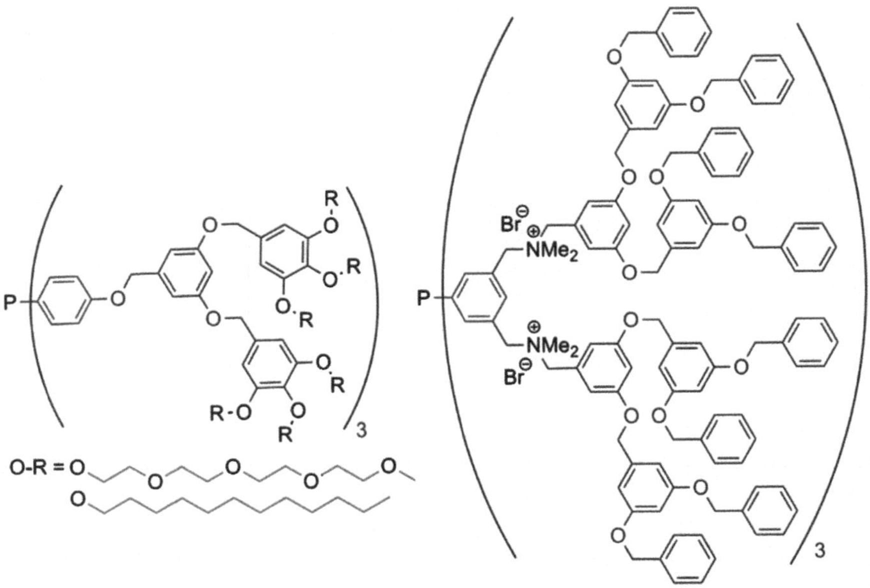

Fig. 6 Polyarylether dendrimers with phosphine as core, and used as ligands for Pd-catalyzed Suzuki-Miyaura coupling reactions

catalysts [51]. The same type of bis(oxazoline) Cu-complexes, but surrounded by elongated poly(arylether) branches (O–$(CH_2)_3$–O linkages between the aryl groups instead of O–CH_2), were used for catalyzing Diels–Alder reactions [52]. The reaction rate dropped abruptly for the third generation, probably as a consequence of the back-folding of the dendritic wedges. However, a slight but noticeable substrate selectivity was observed with this third generation, in competitive experiments involving two dienophiles of different steric sizes [53]. The osmium complexes of very small dendrons with alkyl chains as terminations and either DMAP (dimethylaminopyridine) or a tetrahydrophthalyl derivative were used as phase selective and easily recyclable catalysts for olefin dihydroxylation. The principle of the "one-phase catalysis and two-phase separation" system already shown in Fig. 5 was also applied in these cases. Up to 10 runs could be carried out in the case of the tetrahydrophthalyl derivative [54]. Polyarylether dendrons bearing a pyridyl alcohol as core were used for catalyzing the enantioselective addition of diethylzinc to benzaldehyde. The corresponding secondary alcohol was obtained in high yield, and with good enantioselectivity (up to 86% ee) [55]. Polyarylether dendrons having either one or two rhodium carbene complexes as core were used for catalyzing the hydrosilylation of ketones. The yields gradually increased with increasing dendrimer generation, in the case of one rhodium carbene complexes [56], whereas no effect was observed in the case of the di-carbene complexes [57]. Very recently, bis(carbene)palladium complexes containing a poly(benzyl ether) dendron as *N*-substituent of the NHC ligands were used as catalysts for Heck reactions, and could be recovered by nanofiltration techniques. Conformational

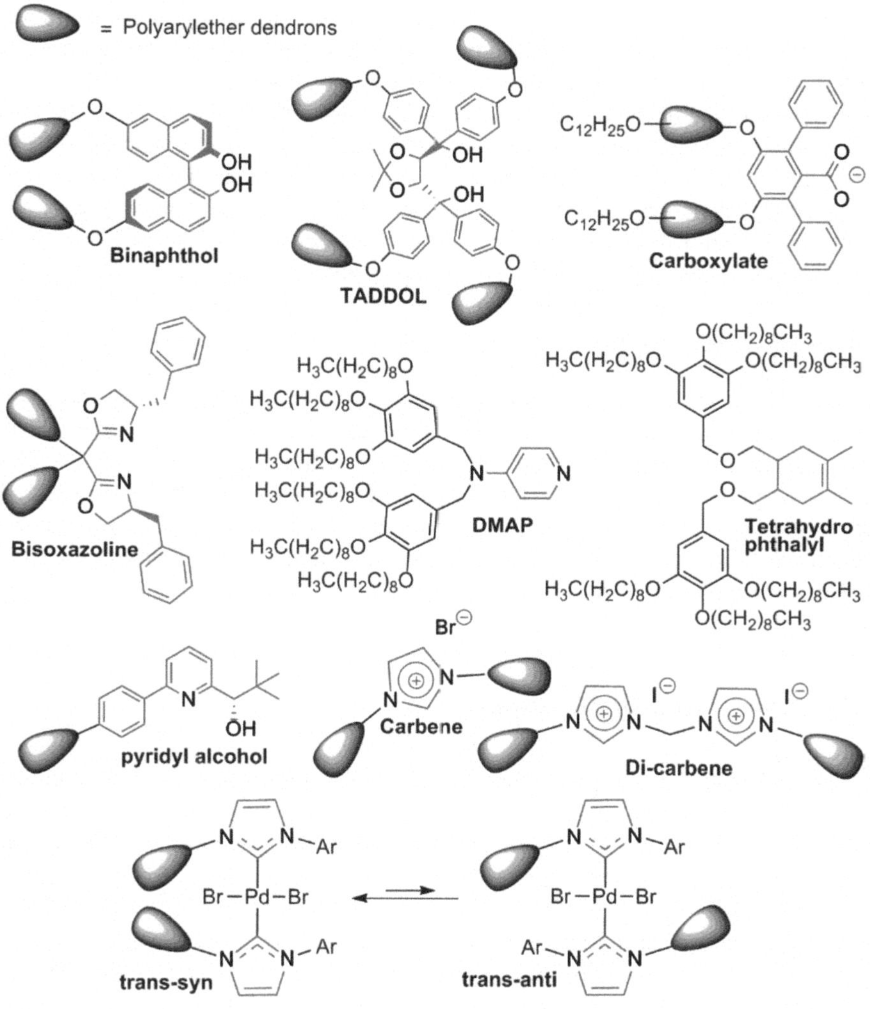

Fig. 7 Non-phosphine ligands as core of polyarylether dendrons

semiflexibility of the benzylic link between the poly(benzyl ether) dendrons and the *N*-heterocyclic ligands prevent steric hindrance, even for the third generation (*trans-syn* and *trans-anti* in Fig. 7) [58].

2.2 *Polycarbosilane Dendrons*

Besides polyarylether dendrons, polycarbosilane dendrons [10] have been used also several times for functionalizing ligands, most generally of type phosphines, as shown in Fig. 8. The dendritic 1,1′-bis(diphenylphosphino)ferrocene (dppf) palladium complex was used for catalyzing the allylic alkylation of 3-phenylallylacetate with diethyl 2-sodio-2-methylmalonate. This reaction affords a mixture of the linear *trans* product, and of the branched product. The reaction rate decreased as the size of the dendrimer increased, whereas the percentage of branched product increased. This effect might be due to the increasing steric bulk of the dendrimer hindering the attack of the nucleophile on the Pd–allyl [59]. This work was expanded with the synthesis of carbosilane dendrons having as core either xantphos (9,9-dimethyl-4,6-bis(diphenylphosphino)xanthene) or tpp (triphenylphosphine) ligands. The corresponding rhodium complexes were applied for catalyzing the hydroformylation of oct-1-ene, and for hydrogenations. In most cases, no difference with the parent small ligands was observed. However, only the dendrimers could be used successfully in a batch reactor and in a continuous-flow membrane reactor [60]. The dendritic phosphoramidite ligand was also used for Rh-catalyzed asymmetric hydrogenation of methyl 2-acetamidocinnamate, with a high effectivity [61]. The nickel complex of the dendritic P,O ligands was used for the oligomerization of ethylene. The dendritic P,O ligand suppresses the formation of inactive bis-(P,O)Ni complexes in toluene, and thus outperforms the parent ligand in catalysis in this solvent [62].

2.3 Other Types of Dendrons

Besides polyarylether and polycarbosilane dendrons, a few other types of dendrons have been used as supports for ligand cores, as shown in Fig. 9. Polyphosphorhydrazone dendrimers [11] have been functionalized very often with catalytic entities as terminal groups [25], but only one time with a catalytic entity as core. The ruthenium diphosphine complex located at the core is an efficient and recoverable catalyst in diastereoselective Michael additions. No difference in efficiency or selectivity was observed, when compared with the phosphorhydrazone dendrimer having the same catalytic entities as terminal functions [63]. Other examples of diphosphine rhodium complexes as core of specially engineered chiral branches were proposed for the asymmetric hydrogenation of (Z)-Methyl-2-acetamido-3-phenylacrylate. Remote chirality within a dendritic catalyst was relayed over 14 bonds from the chiral surface groups to the core, via the helical secondary structure of the dendron (B in Fig. 9), to control the enantioselectivity of this reaction [64]. Generations 1 and 2 of dendrimers bearing copper(I) (hexabenzyl) tren complexes (tren = tris(2-aminoethyl)amine) as core and carbosilane triazole wedges ended by TEG terminal groups were applied as catalysts for 1,3-cycloaddition between azides and alkynes (CuAAC reactions). These metallodendrimers catalyze the CuAAC reaction between azidomethylbenzene and phenylacetylene more rapidly than the non-dendritic complexes [65]. Two peptide dendrons of type (Ac-Glu-Ser)$_8$(Dap-Glu-Ala)$_4$(Dap-Amb-Tyr)$_2$Dap-Cys-Asp-NH (Dap = branching 2,3-diaminopropanoic acid, Amb = 4-aminomethyl-benzoic acid, Ac = acetyl,

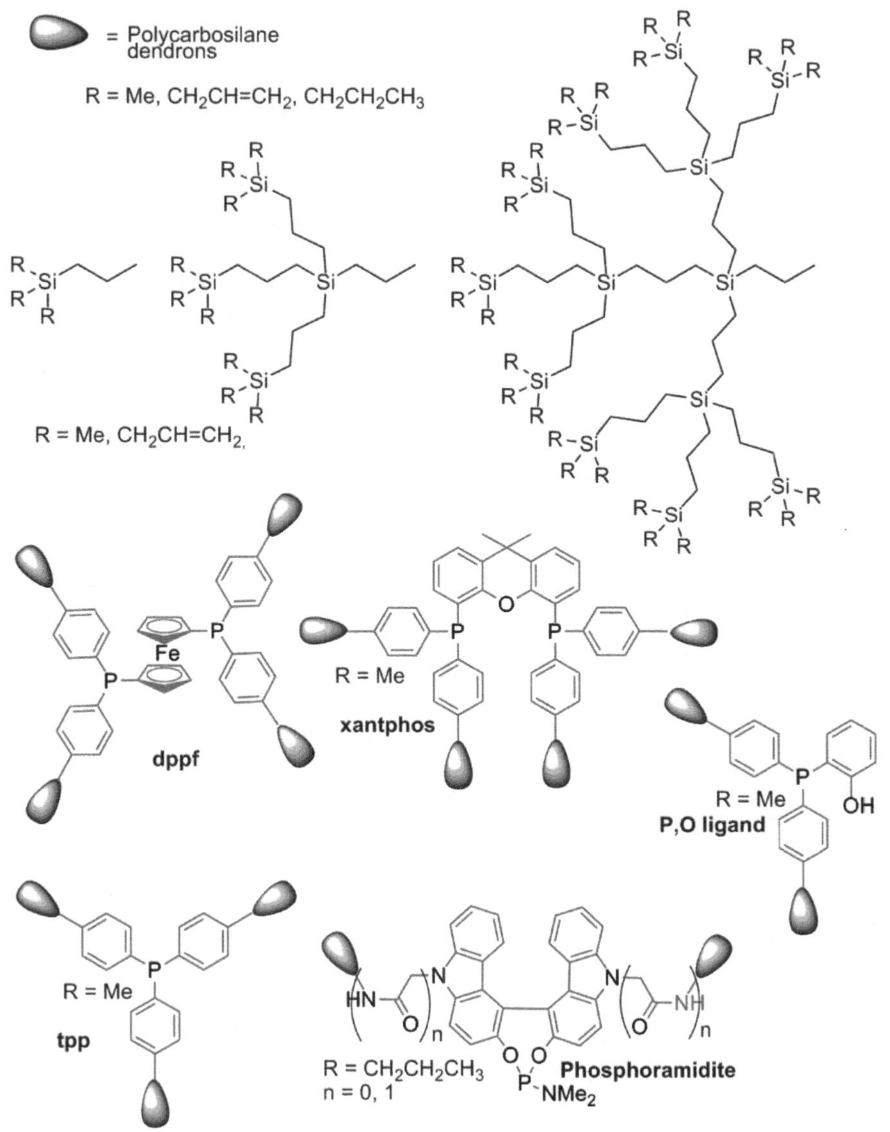

Fig. 8 Polycarbosilane dendrons and types of ligands used as cores

Glu = glutamate, Ser = serine, Ala = alanine, Tyr = tyrosine) linked to a bipyridine as core were associated around Fe(II) to form a di-coordinated complex at pH 6.5, as electrostatic repulsion between the polyanionic dendrimer branches precludes larger associations. This complex catalyzed the oxidation of *o*-phenylenediamine with H_2O_2, with enzyme-like kinetics (k_{cat}/k_{uncat} = 90,000), while Fe^{2+} or $[Fe(bipy)^3]^{2+}$ were inactive, emphasizing the role of the branches [66].

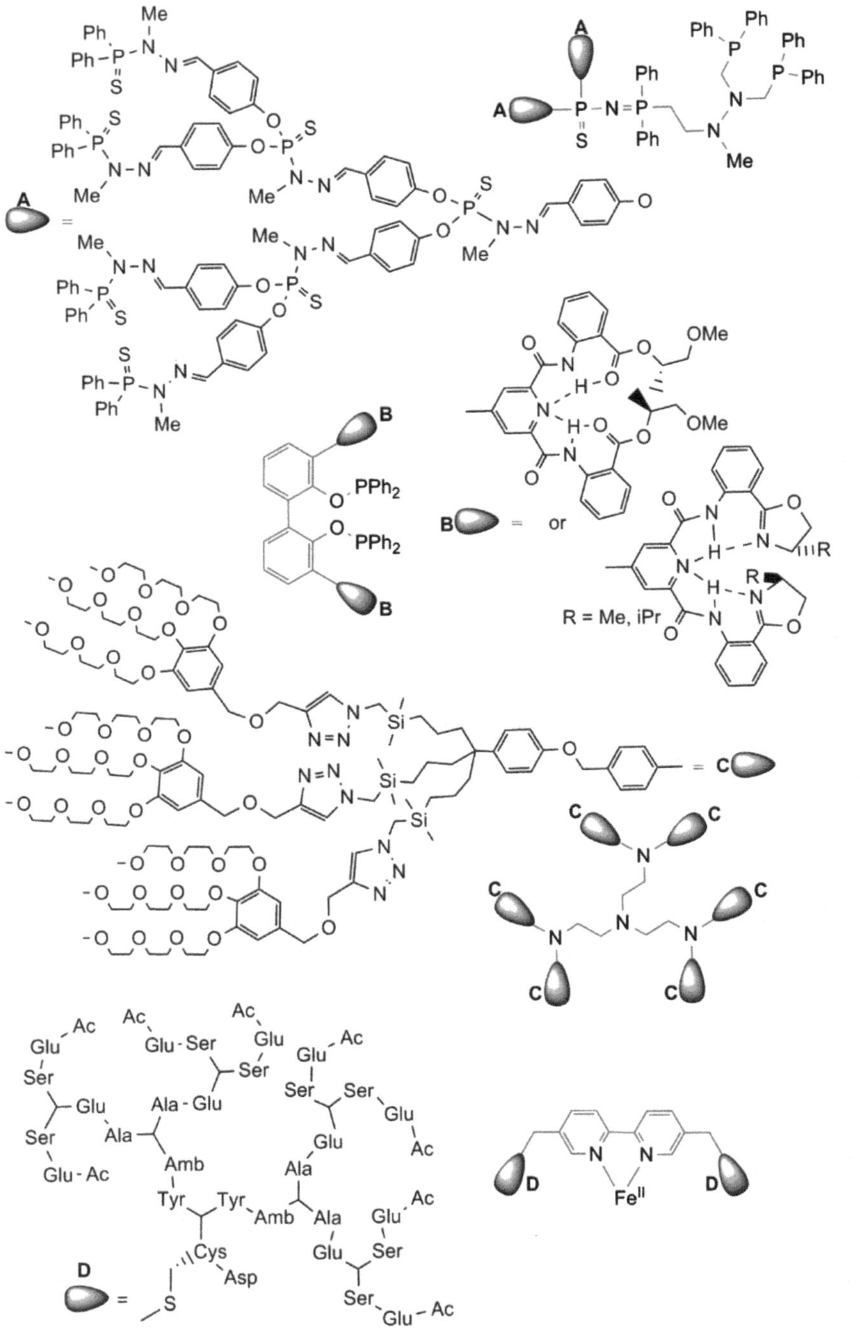

Fig. 9 Miscellaneous dendritic wedges and ligands used for catalyzed reactions

2.4 Macrocyclic Complexes as Cores of Dendrimers

Macrocyclic derivatives, mainly porphyrin derivatives and to a lesser extent phthalocyanine derivatives, have been used as core of dendrimers (Fig. 10), and some of their complexes were applied as catalysts. The first example concerned 8 poly (phenylester) branches ended by tBu groups, appended to the *meta*-positions of the 5,10,15,20-tetrakis(3′,5′-hydroxyphenyl)-porphinatomanganese(III) chloride core. These complexes have been examined for epoxidation of non-conjugated dienes and alkene mixtures of 1-alkene and cyclooctene, using iodosylbenzene as the oxygen donor. Significantly greater regioselectivity was observed with the dendrimer metalloporphyrins, relative to the corresponding Mn-porphyrine parent [67, 68]. The same poly(phenylester) branches, but ended by acetoxyphenyl groups, were attached to a porphyrin as core complexing iron. Hyperbranched polymers were also used instead of the dendritic branches. Both types of compounds were used for catalyzing the epoxidation of a linear alkene and a cyclic alkene, in isolation and together in equal ratios, in order to make a valid comparison between dendrimers and hyperbranched polymers. However, no change in product distribution was observed when four dendrons were grafted in the 4-position of the phenyl rings; in that case, branching is directed away from the core, resulting in a relatively open core environment and a more densely packed external structure. Grafting eight dendrons in the 3,5-position induced a more sterically crowded environment, that was more sensitive to ligand size and geometry. The relative yields of oxidation product obtained from the larger alkene were smaller than those obtained using the smaller alkene. 2,6-disubstitutions in porphyrin unit was used to impose steric demand at the core. In that case, only hyperbranched polymers were synthesized. Catalytic studies carried out on the two alkenes revealed high yields in both cases, which suggests that the polymer environment (around the core) might be helping the catalytic process. When the two alkenes were reacted together with the catalysts, a marked switch in relative selectivity was observed, in favor of the linear alkene. It was concluded that hyperbranched polymers can be applied as genuine alternatives to dendrimers in applications requiring site isolation or encapsulation [69].

Attachment of poly(benzyl ether) dendrons to carbonylruthenium(II) *meso*-tetraphenylporphyrin was carried out either in the *meta*-position (eight wedges) or in the *para*-position (4 wedges). All of these dendritic ruthenium porphyrins are highly selective catalysts for epoxidation of alkenes with 2,6-dichloropyridine *N*-oxide. The chemo- or diastereoselectivity increased with the generation number of the dendron and with the number of dendrons. A variety of alkene substrates was studied; the cyclopropanation of styrene and its *para*-substituted derivatives with ethyl diazoacetate was highly *trans* selective [70]. Iron porphyrins with mono-imidazole axial ligation, and 4 TEG-ended dendrons in *ortho*-position were synthesized as mimics of heme monooxygenases. Epoxidation of olefins and oxidation of sulfides to sulfoxides, catalyzed by these dendritic iron porphyrins, were investigated with iodosylbenzene as the oxidant. The total turnover numbers were

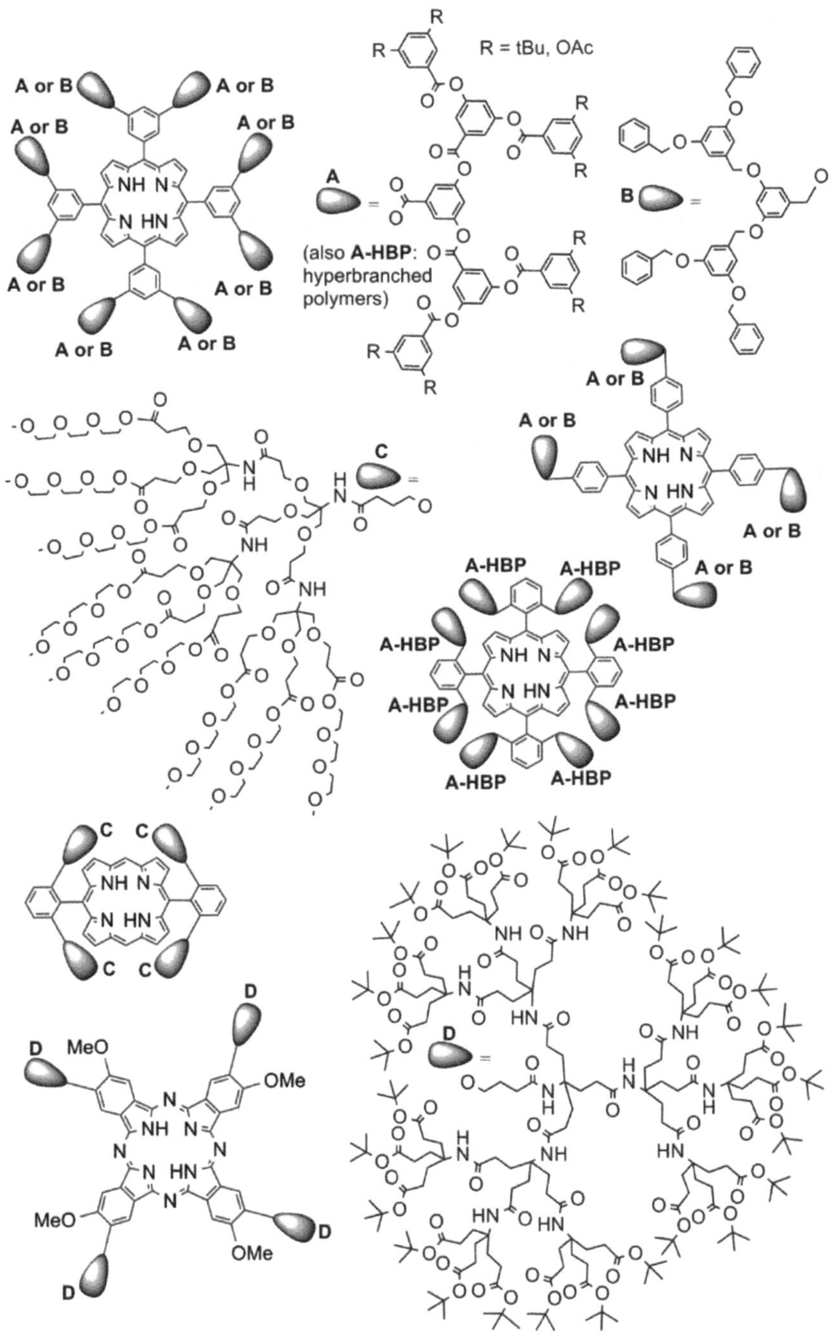

Fig. 10 Macrocyclic cores (porphyrins and phthalocyanine) of various dendritic structures used as ligands for catalyzed reactions

found to increase with the size of the dendrimer, due to improved catalyst stability. Very high product selectivities were observed in all oxidation reactions, regardless of dendrimer generation [71]. Very recently, Fe(porphyrin)s surrounded by 4 poly (benzyl ether) dendrons (generations 0–4) were applied as catalyst for cyclopropanation reactions, the (2 + 1) cycloaddition between a model alkene and diazomethane. Cooperative effects, generated via aggregation of dendritic units, enhanced the catalytic efficiency [72].

Sterically hindered dendritic cobalt-phthalocyanines bearing four "arborol" [73] wedges were applied for thiol oxidation in the presence of dioxygen. The dendritic structure around the cobalt phthalocyanine prevented the molecular aggregation among phthalocyanine moieties, and enhanced the catalysts stability [74].

2.5 *Multimetallic Species as Cores of Dendrimers*

In the previous examples, a single metal was located at the core of dendrons or dendrimers. However, it is also possible to have several metals when using polyoxometallates (POM). In that case, the metal cluster bears negative charges, and the interaction occurs by electrostatic interactions with dendrons having an ammonium as core. The number of dendrons surrounding the core depends on the number of charges of the core. With the trianionic tungsten cluster $[PO_4\{WO(O_2)_2\}_4]^{3-}$, known for its catalytic efficiency in alkene epoxidation and alcohol oxidation with hydrogen, three dendrons with an ammonium as the core and 9–27 terminal functions of various types have been electrostatically attached (Fig. 11). In the case of epoxide terminal groups, the complexes were used as air-stable, recoverable catalysts for oxidation reactions (epoxidation of cyclooctene) using hydrogen peroxide [75]. These compounds and others, ended by propyl or aryl thioether functions, were used also for the selective oxidation of alkenes to epoxides, sulfides to sulfones, and alcohols to ketones, in a biphasic system (aqueous/$CDCl_3$) with hydrogen peroxide as the primary oxidant. The dendritic structure increased the stability of the POM species and facilitated the recovery of the catalyst up to the eighth cycle. However, increasing the bulkiness around the POM center on going from 9 to 27 terminal functions led to a negative kinetic dendritic effect [76]. Well-defined enantiopure tripodal propyl dendritic structures bearing three amine groups have been used also for the complexation of the same POM; a single dendrimer is necessary in this case. The hybrid compounds selectively oxidize sulfides to the corresponding chiral sulfoxides, with up to 13% enantiomeric excess (ee) [77].

With the POM $[Zr_2(O_2)_2(SiW_{11}O_{39})_2]^{12-}$, bearing 12 negative charges, 12 dendrons having an ammonium as core interact electrostatically. These compounds are also stable, efficient, recoverable, and reusable catalysts for the oxidation of sulfides in aqueous/$CDCl_3$ biphasic media, with hydrogen peroxide as the oxidant. The reaction kinetics and selectivity were influenced by the structure of the counter-cation used [78].

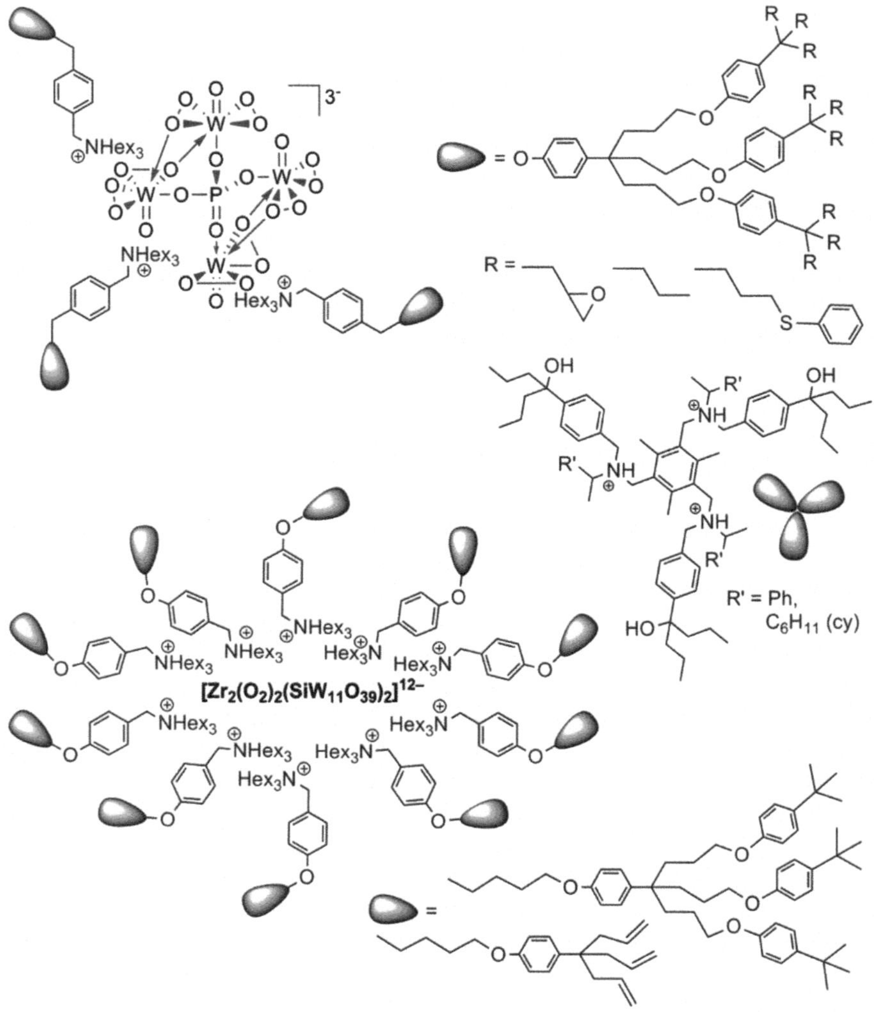

Fig. 11 Polyoxometallates (POM) as core of dendrons

Metallic nanoparticles entrapped within dendrimers have been applied very early for catalysis. For instance, platinum nanoparticles were obtained by using the fourth-generation (G_4) polyamidoamine (PAMAM) dendrimers as both template and stabilizer. The sequestered nanoparticles were applied for electrocatalytic reduction of O_2 [79]. In another early example, Pd and Pt nanoparticles, also obtained using PAMAM dendrimers as both template and stabilizer, were used for catalyzing the hydrogenation of alkenes in water, exhibiting high catalytic activity [80]. Such approach is largely developed, but in all cases the location of the nanoparticle is undefined, as illustrated in cases A and B in Fig. 12. In a related recent example, the small catalyst Cu(hexabenzyl triaminoethylamine)Br entrapped

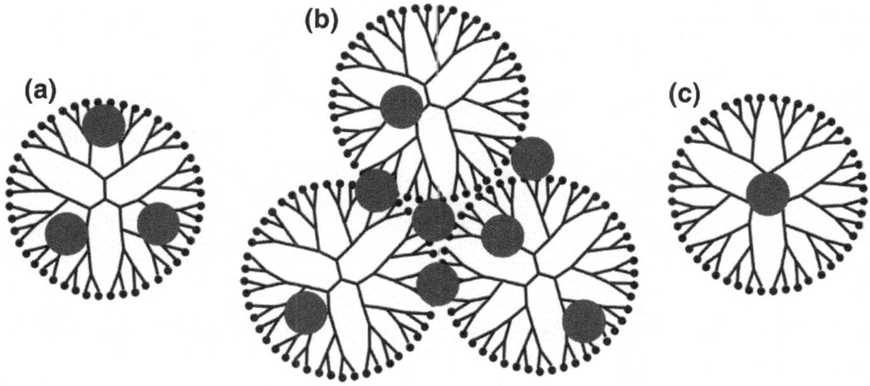

Fig. 12 Various types of interactions between metallic nanoparticles and dendritic structures

within an amphiphilic dendrimer containing 27 triethylene glycol termini and 9 triazole rings inside was used for CuI-catalyzed alkyne–azide cycloaddition (CuAAC) "click" reactions. Exceptional TONs up to 510,000 were measured, not attainable under the same conditions without the dendrimer [81]. However, it is difficult to ascertain the location of the catalytic entities, and the real role played by the confinement in these cases, and we will not consider this approach in this chapter. There exists another possibility in the case of nanoparticles, which consists in surrounding it by several dendrons, as illustrated in the case C in Fig. 12. The number of dendrons is not exactly defined in general, and there is a poor control of the nanoparticle core dimension, but the location of the nanoparticle is clearly defined.

Unprotected nanometer-sized metal particles have a tendency to aggregate, which leads to a reduction of the observed catalytic activity. The use of dendrons to protect them should preclude any aggregation. These systems should combine the high efficiency of homogeneous catalysis with the ease of separation associated with heterogeneous catalysis. As already reported for the cases of single catalytic species at the core of dendrons, polyarylether dendrons have been used for the protection of nanoparticles. The first example concerned palladium nanoparticles complexed by thiol functions located at the core of benzylether dendrons (Fig. 13; see Fig. 4 for the structure of the benzylether dendrons). These protected nanoparticles, having approximately 300 Pd atoms in the metallic core, to which are attached about fourteen G-3 benzylether dendrons, were applied for efficiently catalyzing Heck and Suzuki reactions. Nearly 90% of the metal nanoparticle surface is unpassivated and available for catalysis [82]. Pd nanoparticles were also protected by carbosilane dendrons (see their structure in Fig. 8) with a thiol at the core. In mixtures with Al_2O_3, an improved stability of the Pd/Al_2O_3 catalyst was observed in the dimethyl ether steam reforming reaction, to generate hydrogen with good performance. The reason for this improvement is the formation of SiO_2 (from the carbosilane dendron upon calcination) at the Pd–Al_2O_3 interface that acted as a

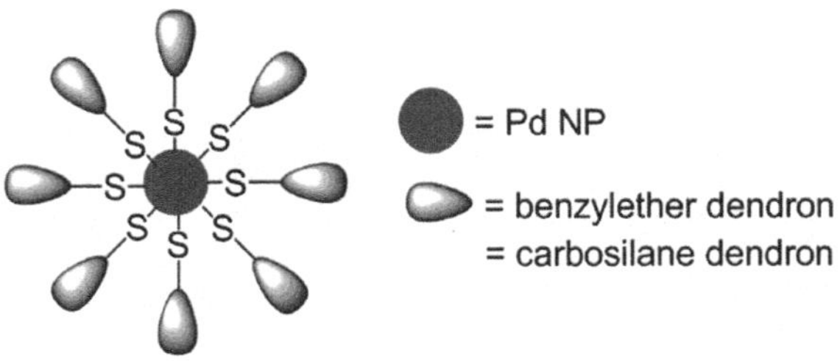

Fig. 13 Dendrons used for protecting nanoparticles at their core

pinning center and prevented Pd migration under the reaction conditions as well as the accumulation of carbon [83].

2.6 Organocatalysis

The interest in organocatalysis has increased spectacularly in the last few years [84], and this field has been reviewed recently, concerning the use of dendrimers [23]. One of the variants of the nitroaldol (Henry) reaction consists in the use of amines as catalysts. Dendritic molecules with a single triethylene amine core surrounded by polyether dendrons from generation 0 to generation 3 were synthesized for such purpose (Fig. 14). The nitroaldol reaction between aromatic aldehydes and nitroalkanes was chosen as test reaction. The activity of the dendritic catalysts decreased when the generation number increased. No significant change in stereocontrol was observed on passing from lower to higher generations of dendrimers [85]. This work was expanded by the synthesis of more dense dendrons, in which the branching points multiply by 3 the number of functions, instead of two for more classical dendrimers, and also less dense star compounds. It was shown that the catalytic power for the nitroaldol reaction decreased as the generation number and/or the degree of branching increased [86]. Cinchonidine derivative was used as core of generations 1–3 of benzylether dendrons. These chiral dendritic molecules was used as phase transfer catalysts in the alkylation of a glycine imine ester, reaching a moderate level of asymmetric induction [87].

Three generations of poly(arylether) dendrimers with a diselenide core were synthesized for mimicking glutathione peroxidase activity (GPx, a mammalian antioxidant seleno-enzyme, which catalyzes the reduction of a variety of hydroperoxides). The initial rates for the reduction of H_2O_2 by PhSH in the presence of dendrimer mimics of GPx were determined. Dendrimers G_1 and G_2 show relatively low GPx activity, whereas dendrimer G_3 shows a high activity, with a remarkable rate enhancement of 1400-fold compared to classical mimics of GPx [88]. Optically active dendritic amino alcohols were used as catalysts for the enantioselective borane reduction of prochiral ketones, leading to the desired products with up to 96% ee. The branches were constituted of polyarylethers, either derived from 3,5-dihydroxybenzyl alcohol, or with longer linkers derived from

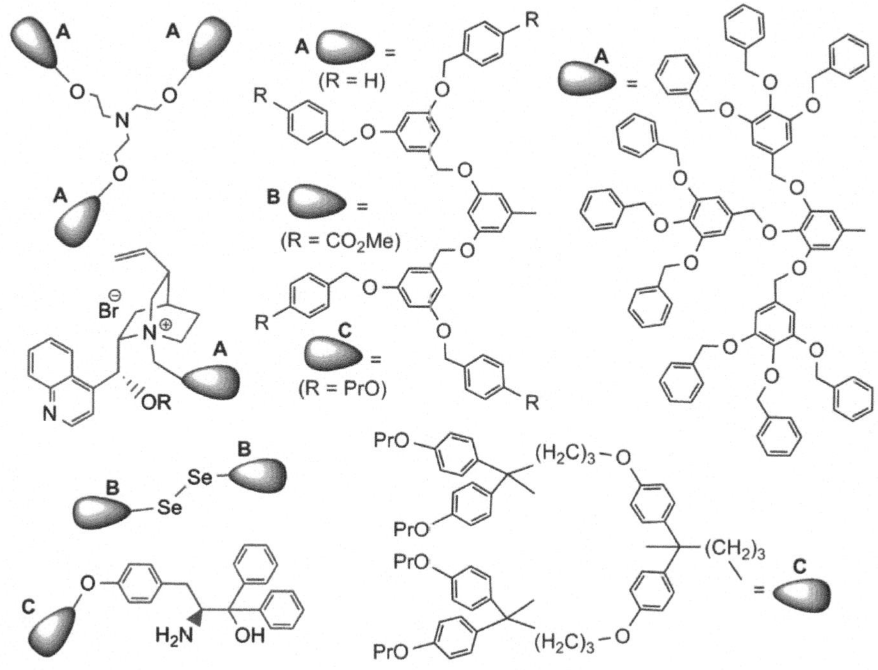

Fig. 14 Organocatalysts as cores of polyarylether dendritic structures

4,4-bis(4′-hydroxyphenyl) pentanol, but no influence of this parameter on the catalytic efficiency was observed [89]. The small DMAP derivative (shown in Fig. 7) was used as catalyst for Baylis–Hillman reactions of various aryl aldehydes with methyl vinyl ketone and acrylonitrile. The recyclability of this catalytic system has been demonstrated in the binary solvent system (DMF–cyclohexane, 1:1, v/v) [90].

Dendrimeric pyridoxamine-based on PAMAM dendrimers up to the sixth generation were synthesized with the aim of mimicking transaminases (Fig. 15). The rate of the transamination reactions from the ketoacids, forming racemic alanine and phenylalanine, was increased by 1000-fold with the sixth generation terminated by NMe_2 groups, compared to simple pyridoxamine. The NHAc terminations induced a lower efficiency than the NMe_2 terminations, suggesting that the terminal NMe_2 groups provided better general acid–base catalysis than the internal tertiary amines [91]. An *o,o′,p,p′*-tetraalkoxybenzophenone function surrounded by aliphatic polyester backbones based on 2,2-bis(hydroxymethyl)propionic acid was chosen as a sensitizing core for the generation of singlet oxygen ($^1O_2$). This system having a hydrophobic interior and hydrophilic surface was elaborated for [4 + 2] cycloaddition of singlet oxygen to dienes with subsequent reduction to the allylic diol, owing to the large polarity difference between substrate and product in this reaction. The model reaction involved formation of *cis*-2-cyclopentene-1,4-diol from cyclopentadiene. It has been clearly seen that higher generation dendrimers lead to

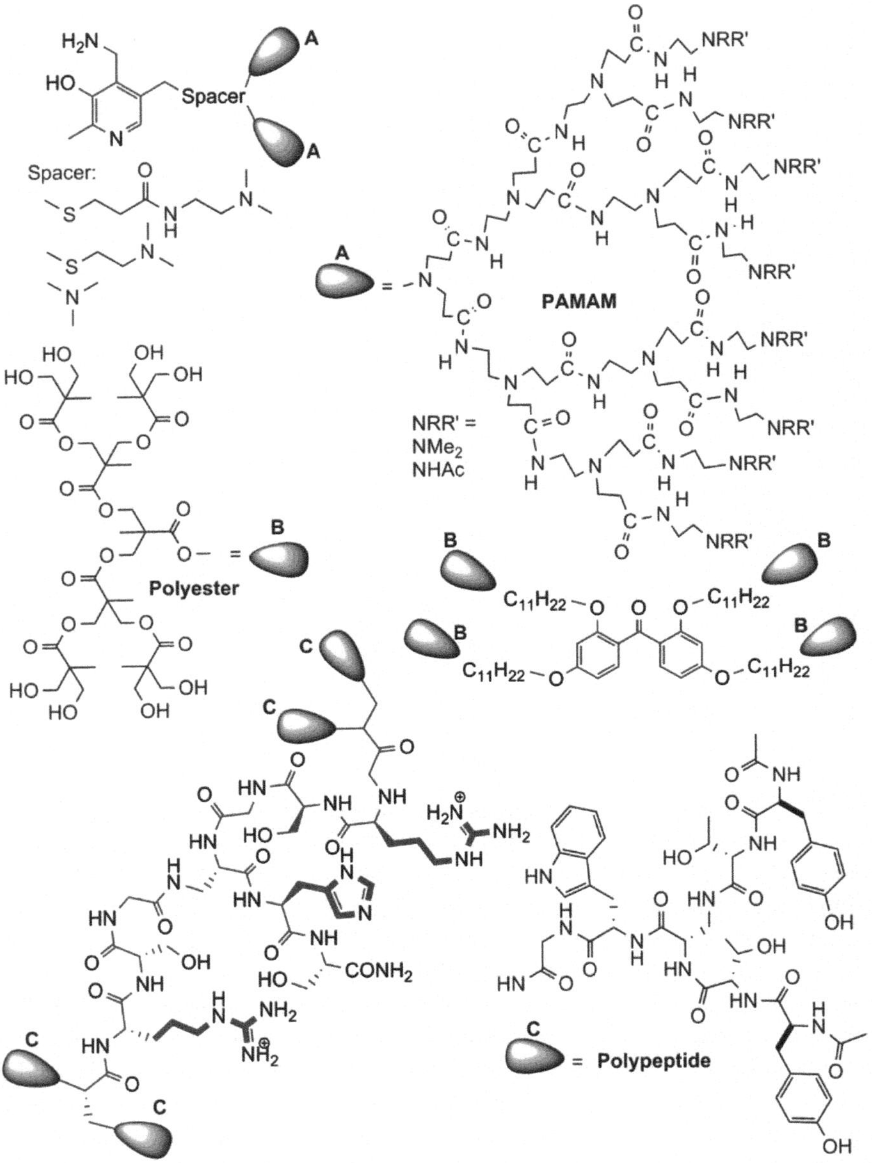

Fig. 15 Organocatalysts as cores of other dendritic structures

faster reactions and higher levels of conversion. The observed behavior is primarily due to the increased local concentrations of substrate as well as of $^{1}O_2$ [92].

Screening of a 65,536-member combinatorial library of third-generation peptide dendrimers using fluorogenic 1-acyloxypyrene-3,6,8-trisulfonates as substrates led to the discovery of a catalytic esterase peptide dendrimer with a core active site

(Fig. 15). Ester hydrolysis was catalyzed by a single histidine residue at the peptide dendrimer core, whereas a pair of arginine residues in the first-generation branches assisted substrate binding. Catalysis was also enhanced by the outer dendritic branches consisting of aromatic amino acids [93]. Other peptide dendrimers have shown an impressive increase of the rate of ester hydrolysis, which will be illustrated in Sect. 3.

3 Catalytic Entities Located at One or Several Layers of Dendritic Structures

This field is less developed than the previous one, as the synthesis of such compounds is highly demanding. In particular, catalytic entities precisely located at a single layer of the internal structure of dendrimers are rare. Amphiphilic benzyl ether and aliphatic ester dendrimers were prepared in a convergent manner from a common trivalent expanded core containing three 4-(dialkylamino)pyridine groups (Fig. 16). Catalysis experiments of acylation reactions employing sterically demanding tertiary alcohols as substrates clearly indicates that the nanoenvironment played the dominant role in determining the activity of the catalysts, with the polyester platform being superior to the benzyl ether one [94].

Catalytically active arylpalladium complexes, which bear a tethered sulfato group, were noncovalently attached to ionic core-shell benzylether dendrimers, having 8 cationic groups. The octacationic dendrimers of generations 1–3 form discrete 1:8 assemblies with the arylpalladium complexes, strongly associated in solution (Fig. 16). The catalytic performance of these assemblies was tested in the aldol condensation reaction between benzaldehyde and methyl isocyanoacetate in dichloromethane. The catalytic performances of the metallodendritic assemblies were comparable, but slightly lower than those of the unsupported catalyst [95].

Catalytic entities located at several or all layers inside the dendritic structures have been presented for the first time in 1994. A second generation dendrimer containing 15 phosphines in its structure was used for complexing 5 Pd (3 phosphines per Pd) (Fig. 17). These complexes were used for catalyzing the electrochemical reduction of CO_2 to CO, with a rate and selectivity similar to those of monomeric catalysts [15]. Larger dendritic phosphines, up to generation 4, having also a phosphine at each branching point were used for the complexation of rhodium (one phosphine per Rh). These complexes were successfully employed as efficient catalysts for olefin hydrogenation, using decene as the substrate. The catalytic activity (turnover number (TON) = $mol_{prod}/mol_{cat} \sim 200$, turnover frequency = $mol_{prod}/mol_{cat}/h \sim 400$) of the Rh(I) dendrimers was found similar to that of the monomeric complex. The generation 4 could be recovered and reused one time [96].

A generation 4 benzylether dendrimer including 21 alcohol groups in the internal structure and 48 tetradecyl units as terminal groups was synthesized by a convergent way. These compounds were used as organocatalysts for elimination reactions

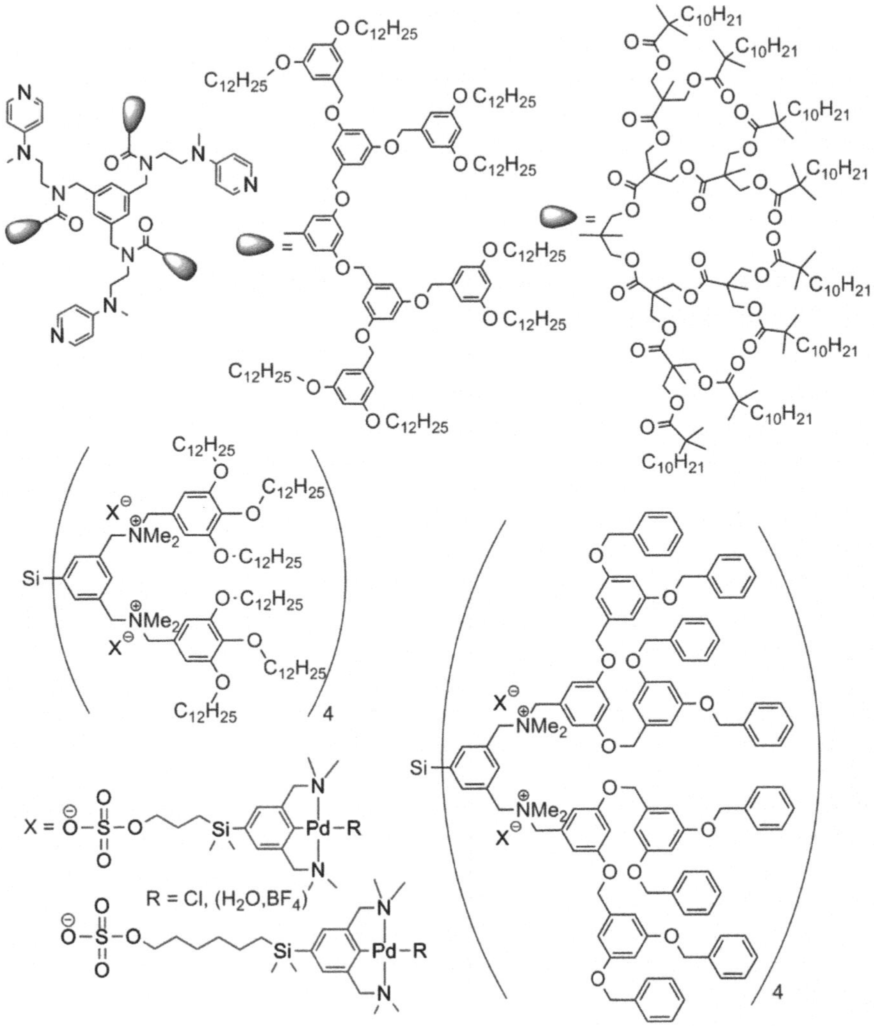

Fig. 16 Catalytic entities at a single layer inside dendrimers

from tertiary alkyl halides, using an inorganic base to trap the acidic hydrogen halide byproduct. Complete conversions (turnover number of approximately 17,400) were obtained for both 2-iodo-2-methylpropane and 2-iodo-2-methylheptane, in the presence of 0.57 mol% of generation 4 dendrimer, upon heating in cyclohexane at 70 °C for 24 h. Little or no reaction could be observed for control experiments carried out under identical conditions but without dendrimer. Decreasing the size of the dendrimer from generation 4 to generation 3 resulted in a 15–20% reduction in both rate of reaction and turnover number. The SN_2 alkylation of pyridine with CH_3I was also catalyzed by these dendrimers.

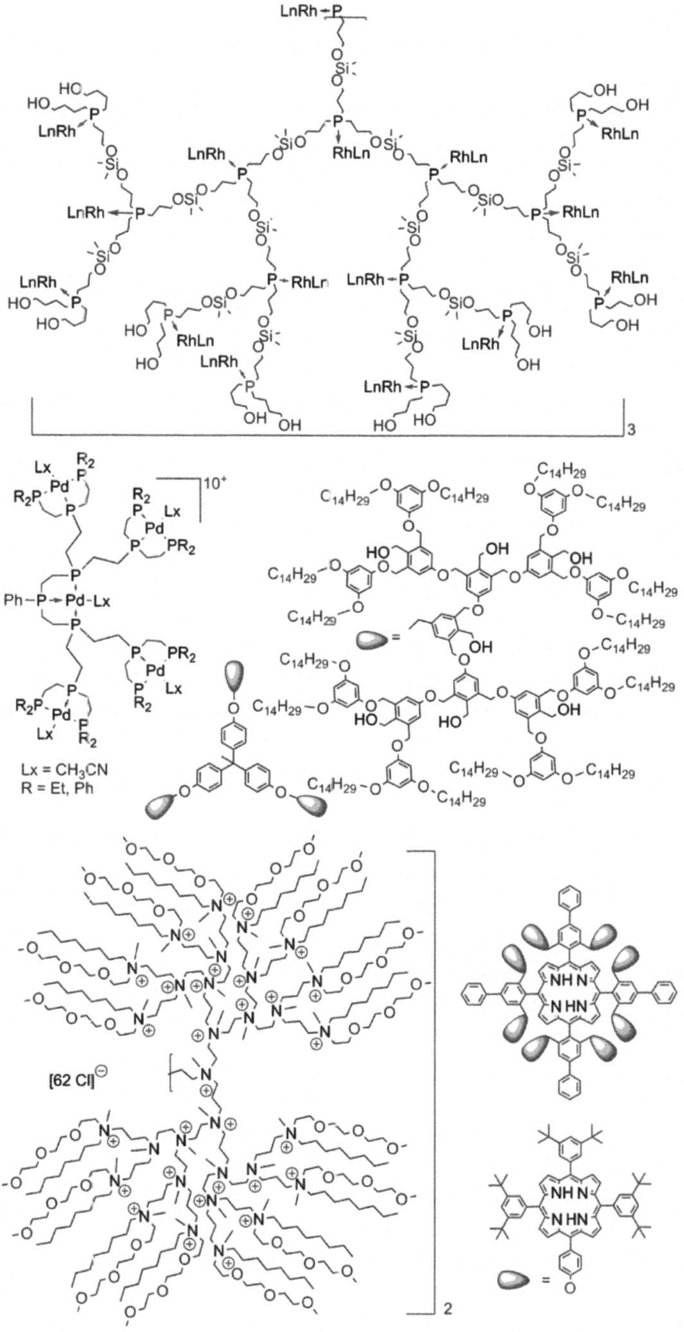

Fig. 17 Catalytic entities at several layers inside dendrimers

Catalytic activity was observed for all the dendrimers studied; the highest catalytic ability was again observed for the 4th generation dendrimer [97].

Tertiary amine dendrimers with both hydrophilic and hydrophobic chains on every end, synthesized from poly(propyleneimine) dendrimers, were quaternized completely with methyl iodide and converted to quaternary ammonium chlorides. These fully polycationic dendrimers of generations 2, 4 and 5, were able to solubilize lipophilic compounds, and were used for catalyzing the decarboxylation of 6-nitrobenzisoxazole-3-carboxylic acid. The rate of the decarboxylation in aqueous solutions of the generation 5 cationic dendrimer was up to 500 times faster than in water alone [98]. Neutral poly(propyleneimine) dendrimers from generation 2 to generation 5, ended by perfluorooctanoyl chains, were used as phase transfer catalysts (water/CO_2) for two types of reactions, i.e., the halogen exchange reaction of benzyl chloride into benzyl bromide and the esterification of oxalic acid with pentafluorobenzyl bromide. It was found that the rate of reaction for both cases depends on the generation of the dendrimer. For the higher generation dendrimers, it is more difficult for the substrate to migrate from the bulk carbon dioxide phase to the interior of the dendrimer [99].

Several multiporphyrinic compounds, including a porphyrin core surrounded by 8 porphyrins, were active catalysts for the light-induced generation of singlet oxygen ($^1O_2$) from ground-state oxygen. These compounds were useful photosensitizers for the oxidation of various olefinic compounds to the corresponding allylic hydroperoxides. High TONs of 10,000 were observed after 8 h of reaction. The regio- and stereoselectivities of the products were identical to the known selectivity of singlet oxygen reactions. Recycling of these dendrimers was possible by nanofiltration technology, but the peripheral porphyrin units were prone to photodegradation [100].

A series of papers has reported the synthesis of peptide dendrimers of various composition having catalytic properties for the hydrolysis of esters. In the first example, the symmetrical, achiral diamino acid (1,3-diaminoisopropyloxy)acetic acid was chosen as the branching unit B which provides the dendrimeric architecture, and 3-dimethylaminoisophthaloyl group as terminal functions (Cap). Permutations of the catalytic triad of the amino acids histidine (His), aspartate (Asp), and serine (Ser) were chosen as the variable positions A^1, A^2, and A^3 to generate a family of esterolytic peptide dendrimers, of structure $[((CapCONH\text{-}A^3)_2BA^2)_2B\text{-}Cys\text{-}(A^1)\text{-}S]_2$, resulting in a family of 21 different peptide dendrimers (Fig. 18). The dendrimers were assayed for the catalysis of ester hydrolysis using chromogenic and fluorogenic acetyl esters and acetoxymethyl ethers of umbelliferone, 8-acetoxypyrene-1,3,6-trisulfonate, and 7-acetoxy-1-methylquinolinium sulfate [101]. The activity pattern indicated that catalysis was correlated with the presence of histidine (or His-Ser) residues at the outermost positions A^3 of the dendrimers [102]. A combinatorial approach based on split-and-mix synthesis and on-bead screening was exemplified by the discovery of catalytic and binding peptide dendrimers in a 65,536-membered library. The split-and-mix library of peptide dendrimers was designed by distributing sixteen proteinogenic amino acids (histidine (His), glutamic acid (Glu), tryptophan (Trp),

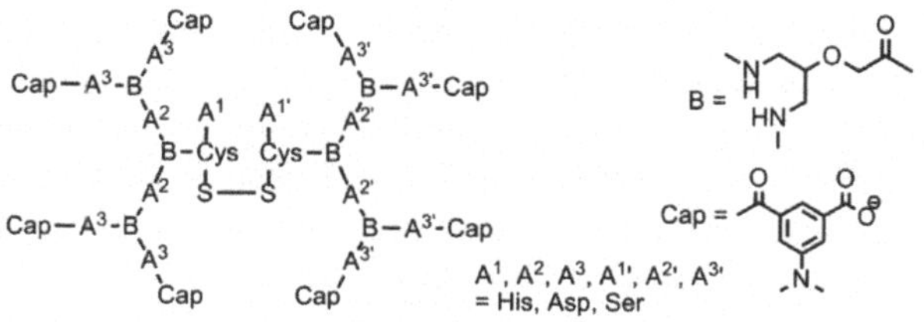

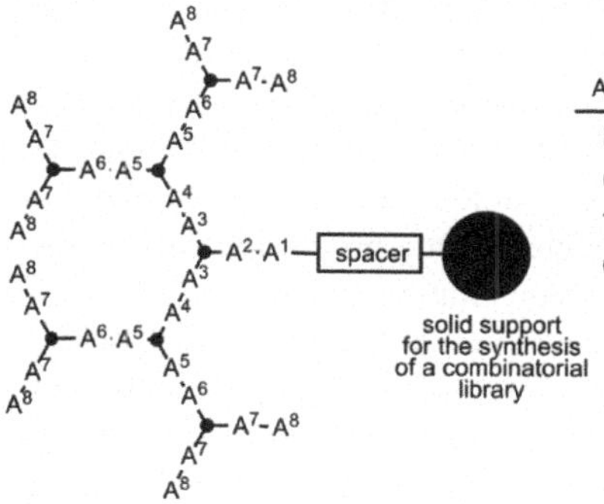

A8, A6	A7, A5	A4, A2	A3, A1
His	Pro	Lys	Asp
Glu	Ser	Phe	Ala
Trp	Arg	Ile	Thr
Gly	Leu	Tyr	Val

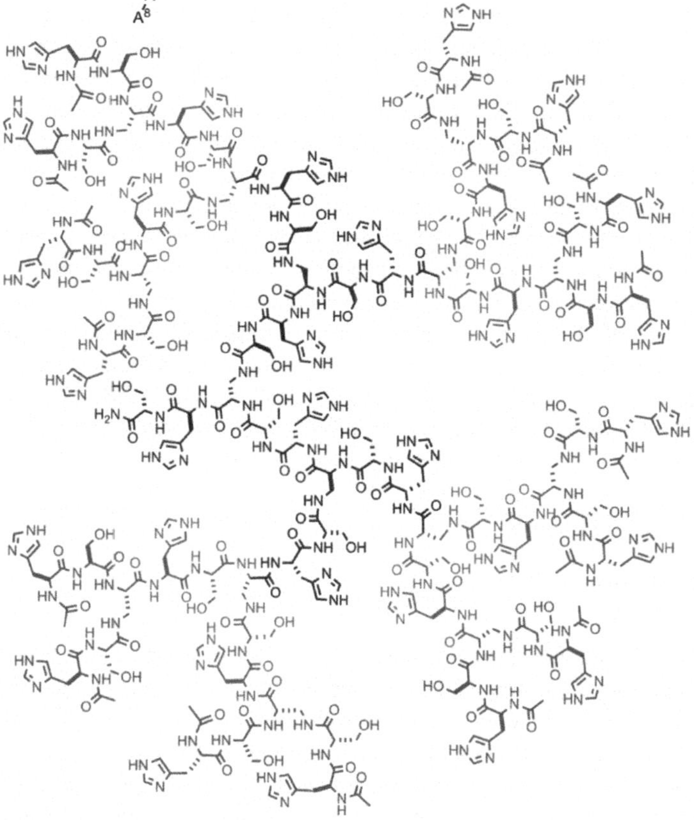

Fig. 18 Schematized peptide dendrimers, and full structure of a generation 4 of a peptide dendrimer

glycine (Gly), proline (Pro), serine (Ser), arginine (Arg), leucine (Leu), lysine (Lys), phenylalanine (Phe), isoleucine (Ile), tyrosine (Tyr), aspartic acid (Asp), alanine (Ala), threonine (Thr), and valine (Val)) into four groups of four amino acids each. The peptide dendrimer library was screened to find catalytic peptide dendrimers for an ester hydrolysis reaction. In this case also, catalytically active dendrimers contained histidine at the outermost position A^8 [103].

In view of the role played by histidine, a series of peptide dendrimers having His-Ser residues at each layer, from generation 1 to generation 4 was synthesized (see full structure in Fig. 18). All dendrimers were catalytically active with multiple turnover, for hydrolysis of pyrene trisulfonate esters. Kinetic parameters showed that catalysis increased with increasing generation number. The catalytic rate constants and the substrate binding constants were directly proportional to the total number of histidines per dendrimer, resulting in a quadratic increase in catalytic efficiency, as a function of dendrimer size. The generation 4 dendrimer was 140,000-fold more efficient than 4-methyl-imidazole as a reference catalyst, amounting to a 4500-fold acceleration per histidine function [104]. Thirty-two mutants of the original third generation of these peptide dendrimers were synthesized later. The fastest catalyst in this series was the threonine mutant ((Ac-His-Thr)$_8$(Dap-His-Thr)$_4$(Dap-His-Thr)$_2$Dap-His-Thr-NH_2), which catalyzed ester hydrolysis with a fivefold improvement over the original sequence [105]. Comparison of the dendritic versus linear structure indicated that the catalytic efficiency per histidine residue is by one order of magnitude higher with the dendrimers [106].

For obtaining higher generations, peptide dendrimers bearing four or eight chloroacetyl groups at their N-termini underwent multiple thioether ligation with generations 2 and 3 of peptide dendrons with a cysteine residue at their focal point, to give generations 5 and 6 dendrimers containing up to 341 amino acids. The efficiency of the esterase catalysts was comparable to that of their lower generation analogs. The generations 4–6 of these dendrimers with multiple N-terminal prolines were investigated for catalyzing the aldol addition of acetone to 4-nitrobenzaldehyde. A remarkable reactivity increase was observed with the highest generations (G_5 and G_6) for this reaction [107].

4 Conclusion

The synthesis of dendrimers and dendritic structures having a functional interior [108] suitable for performing catalysis experiments is still a challenge. Most of the efforts already carried out concerns the introduction of a single catalytic entity at the core of dendrons or dendrimers. Such work was performed with different aims. The main reasons were for protecting the ligand (case of easily oxidized phosphines), for preventing catalyst deactivation (for avoiding the formation of multimetallic species), for increasing the substrate concentration (more substrate concentrated around the catalytic sites, induced by the branches), or for increasing the regio- or enantioselectivity (a specific microenvironment may modify the access of substrates to the catalytic site). In all cases, it is highly desirable to recover and reuse the dendritic catalysts, as they are generally expensive; this is carried out most often by precipitation or by ultrafiltration. A large part of the work carried out with dendritic structures having a single catalytic entity at the core concerns polyarylether structures. There are two main reasons for that: first, these dendrimers/dendrons are synthesized by a convergent method, in which there is always a reactive function at the core, suitable for anchoring a catalytic entity; and second, the polyarylether structures are stable enough to be compatible with many different types of catalytic reactions and conditions. However, a large variety of other types of structures was used also successfully.

The case of dendrimers having catalytic entities at several or all layers of the structure is less explored than the case of a single catalytic entity at the core. Indeed, the synthesis of such compounds is generally difficult, and by no way straightforward. However, such compounds have the largest number of catalytic entities gathered in a single dendrimer structure, of course much larger than for a single catalyst at the core, but also larger than for dendrimers having catalytic entities as terminal groups. Despite the potential difficulties for the elaboration of such dendrimers fully constituted of catalytic entities, dramatic outcomes could be obtained, as already illustrated by a generation 4 dendrimer fully constituted of histidine residues, which was 140,000-fold more efficient than the reference catalyst for the hydrolysis of esters [103].

Despite the quantity of work already carried out, it is difficult to rationalize the effects observed on the catalytic efficiency (rate of reaction, regio- or enantioselectivity). In some cases the expected positive dendritic effects [26–28] are observed, in other cases not, or even detrimental effects. The subtle balance between the protection of the catalytic site and a forbidden access to it, depending on the degree of steric hindrance provided by the dendritic branches, is difficult to predict. No doubts that progresses in the modeling by molecular dynamics of dendritic structures [109] should be very helpful for understanding the observed effects, and may be also for predicting them. The nanoscale confinement of catalytic entities is of fundamental interest, and a recent development concerns PAMAM dendrimers entrapped in inorganic nanoporous support scaffolds, affording a potential double entrapment of the catalytic site [110]. No doubts that the imagination of chemists should continue to afford new series of aesthetically pleasant, while highly performant, dendritic catalysts.

References

1. (a) Frechet JMJ, Tomalia DA (eds) (2001) Dendrimers and other dendritic polymers. Wiley, Chichester; (b) Newkome GR, Moorefield CN, Vögtle F (eds) (2001) Dendrimers and dendrons. Concepts, syntheses, applications. Wiley VCH, Weinheim
2. Buhleier E, Wehner F, Vögtle F (1978) "Cascade-" and: "nonskid-chain-like" syntheses of molecular cavity topologies. Synthesis 78:155–158

3. Tomalia DA, Baker H, Dewald J, Hall M, Kallos G, Martin S, Roeck J, Ryder J, Smith P (1985) A new class of polymers—starburst-dendritic macromolecules. Polym J 17:117–132
4. Newkome GR, Yao ZQ, Baker GR, Gupta VK (1985) Micelles. 1. Cascade molecules—a new approach to micelles—A [27]-arborol. J Org Chem 50(11):2003–2004
5. Hawker CJ, Frechet JMJ (1990) Preparation of polymers with controlled molecular architecture—a new convergent approach to dendritic macromolecules. J Am Chem Soc 112 (21):7638–7647
6. Miller TM, Neenan TX (1990) Convergent synthesis of monodisperse dendrimers based upon 1,3,5-trisubstituted benzenes. Chem Mater 2(4):346–349
7. Morgenroth F, Reuther E, Mullen K (1997) Polyphenylene dendrimers: from three-dimensional to two-dimensional structures. Angew Chem-Int Ed Engl 36(6):631–634
8. Worner C, Mulhaupt R (1993) Polynitrile-functional and polyamine-functional poly (trimethylene imine) dendrimers. Angew Chem-Int Ed Engl 32(9):1306–1308
9. de Brabander van den Berg EMM, Meijer EW (1993) Poly(propylene imine) dendrimers—large-scale synthesis by hetereogeneously catalyzed hydrogenations. Angew Chem-Int Ed Engl 32(9):1308–1311
10. Zhou LL, Roovers J (1993) Synthesis of novel carbosilane dendritic macromolecules. Macromolecules 26(5):963–968
11. Launay N, Caminade AM, Lahana R, Majoral JP (1994) A general synthetic strategy for neutral phosphorus-containing dendrimers. Angew Chem-Int Ed Engl 33(15–16):1589–1592
12. Lartigue ML, Donnadieu B, Galliot C, Caminade AM, Majoral JP, Fayet JP (1997) Large dipole moments of phosphorus-containing dendrimers. Macromolecules 30(23):7335–7337
13. Zhang W, Simanek EE (2000) Dendrimers based on melamine. Divergent and orthogonal, convergent syntheses of a G3 dendrimer. Org Lett 2(6):843–845
14. Caminade AM, Turrin CO, Laurent R, Ouali A, Delavaux-Nicot B (eds) (2011) Dendrimers: towards catalytic, material and biomedical uses. Wiley, Chichester, UK
15. Miedaner A, Curtis CJ, Barkley RM, Dubois DL (1994) Electrochemical reduction of CO_2 catalyzed by small organophosphine dendrimers containing palladium. Inorg Chem 33 (24):5482–5490
16. Knapen JWJ, Van der Made AW, De Wilde JC, Van Leeuwen PWNM, Wijkens P, Grove DM, Van Koten G (1994) Homogeneous catalysts based on silane dendrimers functionalized with arylnickel(Ii) complexes. Nature 372(6507):659–663
17. Astruc D, Chardac F (2001) Dendritic catalysts and dendrimers in catalysis. Chem Rev 101 (9):2991–3023
18. Kreiter R, Kleij AW, Gebbink R, van Koten G (2001) Dendritic catalysts. In: Topics in current chemistry. Dendrimers IV, vol 217. Springer, Berlin, pp 163–199
19. Oosterom GE, Reek JNH, Kamer PCJ, van Leeuwen P (2001) Transition metal catalysis using functionalized dendrimers. Angew Chem Int Edit 40(10):1828–1849
20. van Heerbeek R, Kamer PCJ, van Leeuwen P, Reek JNH (2002) Dendrimers as support for recoverable catalysts and reagents. Chem Rev 102(10):3717–3756
21. Mery D, Astruc D (2006) Dendritic catalysis: major concepts and recent progress. Coord Chem Rev 250(15–16):1965–1979
22. Caminade AM, Servin P, Laurent R, Majoral JP (2008) Dendrimeric phosphines in asymmetric catalysis. Chem Soc Rev 37(1):56–67
23. Caminade AM, Ouali A, Keller M, Majoral JP (2012) Organocatalysis with dendrimers. Chem Soc Rev 41(11):4113–4125
24. Wang D, Astruc D (2013) Dendritic catalysis-basic concepts and recent trends. Coord Chem Rev 257(15–16):2317–2334
25. Caminade AM, Ouali A, Laurent R, Turrin CO, Majoral JP (2016) Coordination chemistry with phosphorus dendrimers. Applications as catalysts, for materials, and in biology. Coord Chem Rev 308(2):478–497
26. Helms B, Frechet JMJ (2006) The dendrimer effect in homogeneous catalysis. Adv Synth Catal 348(10–11):1125–1148

27. Caminade AM, Ouali A, Laurent R, Turrin CO, Majoral JP (2015) The dendritic effect illustrated with phosphorus dendrimers. Chem Soc Rev 44(12):3890–3899
28. Hecht S, Frechet JMJ (2001) Dendritic encapsulation of function: applying nature's site isolation principle from biomimetics to materials science. Angew Chem Int Edit 40(1):74–91
29. Twyman LJ, King ASH, Martin IK (2002) Catalysis inside dendrimers. Chem Soc Rev 31 (2):69–82
30. Ouali A, Caminade AM (2011) Catalytic sites inside the dendrimeric structure for homogeneous catalysis. In: Caminade AM, Turrin CO, Laurent R, Ouali A, Delavaux-Nicot B (eds) Dendrimers: towards catalytic, material and biomedical uses. Wiley, Chichester, UK, pp 183–195
31. He YM, Feng Y, Fan QH (2014) Asymmetric hydrogenation in the core of dendrimers. Acc Chem Res 47(10):2894–2906
32. Astruc D, Wang D, Deraedt C, Liang LY, Ciganda R, Ruiz J (2015) Catalysis inside dendrimers. Synthesis 47(14):2017–2031 (Stuttgart)
33. Brunner H, Altmann S (1994) Enantioselective catalysis, 90. Optically-active nitrogen ligands with dendrimeric structure. Chem Ber 127(11):2285–2296
34. Brunner H (1995) Dendrizymes—expanded ligands for enantioselective catalysis. J Organomet Chem 500(1–2):39–46
35. Fan QH, Chen YM, Chen XM, Jiang DZ, Xi F, Chan ASC (2000) Highly effective and recyclable dendritic BINAP ligands for asymmetric hydrogenation. Chem Commun 9:789–790
36. Deng GJ, Fan QH, Chen XM (2002) Synthesis of dendritic BINAP ligands and their applications in asymmetric hydrogenation. Chin J Chem 20(11):1139–1141
37. Deng GJ, Fan QH, Chen XM, Liu GH (2003) Dendritic BINAP based system for asymmetric hydrogenation of simple aryl ketones. J Mol Catal A-Chem 193(1–2):21–25
38. Wang ZJ, Deng GJ, Li Y, He YM, Tang WJ, Fan QH (2007) Enantioselective hydrogenation of quinolines catalyzed by Ir(BINAP)-cored dendrimers: dramatic enhancement of catalytic activity. Org Lett 9(7):1243–1246
39. Ma BD, Ding ZY, Liu J, He YM, Fan QH (2013) Highly enantioselective hydrogenation of 2,4-diaryl-1,5-benzodiazepines catalyzed by dendritic phosphinooxazoline Iridium complexes. Chem Asian J 8(6):1101–1104
40. Yi B, Fan QH, Deng GJ, Li YM, Qiu LQ, Chan ASC (2004) Novel chiral dendritic diphosphine ligands for Rh(I)-catalyzed asymmetric hydrogenation: remarkable structural effects on catalytic properties. Org Lett 6(9):1361–1364
41. Deng GJ, Li GR, Zhu LY, Zhou HF, He YM, Fan QH, Shuai ZG (2006) Dendritic BIPHEP: synthesis and application in asymmetric hydrogenation of beta-ketoesters. J Mol Catal A-Chem 244(1–2):118–123
42. Ma BD, Miao TT, Sun YH, He YM, Liu J, Feng Y, Chen H, Fan QH (2014) A new class of tunable dendritic diphosphine ligands: synthesis and applications in the Ru-catalyzed asymmetric hydrogenation of functionalized ketones. Chem-Eur J 20(32):9969–9978
43. Tang WJ, Huang YY, He YM, Fan QH (2006) Dendritic MonoPhos: synthesis and application in Rh-catalyzed asymmetric hydrogenation. Tetrahedron-Asymmetry 17(4): 536–543
44. Zhang F, Li Y, Li ZW, He YM, Zhu SF, Fan QH, Zhou QL (2008) Modular chiral dendritic monodentate phosphoramidite ligands for Rh(II)-catalyzed asymmetric hydrogenation: unprecedented enhancement of enantioselectivity. Chem Commun 45:6048–6050
45. Deng GJ, Fan QH, Chen XM, Liu DS, Chan ASC (2002) A novel system consisting of easily recyclable dendritic Ru-BINAP catalyst for asymmetric hydrogenation. Chem Commun 15:1570–1571
46. Fujihara T, Yoshida S, Ohta H, Tsuji Y (2008) Triarylphosphanes with dendritically arranged tetraethylene glycol moieties at the periphery: an efficient ligand for the Palladium-catalyzed Suzuki-Miyaura coupling reaction. Angew Chem Int Edit 47 (43):8310–8314

47. Snelders DJM, van Koten G, Gebbink R (2009) Hexacationic Dendriphos ligands in the Pd-catalyzed Suzuki-Miyaura cross-coupling reaction: scope and mechanistic studies. J Am Chem Soc 131(32):11407–11416
48. Yamago S, Furukawa M, Azuma A, Yoshida J (1998) Synthesis of optically active dendritic binaphthols and their metal complexes for asymmetric catalysis. Tetrahedron Lett 39 (22):3783–3786
49. Rheiner PB, Seebach D (1999) Dendritic TADDOLs: synthesis, characterization and use in the catalytic enantioselective addition of Et_2Zn to benzaldehyde. Chem-Eur J 5(11):3221–3236
50. Zhao M, Helms B, Slonkina E, Friedle S, Lee D, DuBois J, Hedman B, Hodgson KO, Frechet JMJ, Lippard SJ (2008) Iron complexes of dendrimer-appended carboxylates for activating dioxygen and oxidizing hydrocarbons. J Am Chem Soc 130(13):4352–4363
51. Yang BY, Chen XM, Deng GJ, Zhang YL, Fan QH (2003) Chiral dendritic bis(oxazoline) copper(II) complexes as Lewis acid catalysts for enantioselective aldol reactions in aqueous media. Tetrahedron Lett 44(17):3535–3538
52. Mak CC, Chow HF (1997) Dendritic catalysts: reactivity and mechanism of the dendritic bis (oxazoline)metal complex catalyzed Diels-Alder reaction. Macromolecules 30(4): 1228–1230
53. Chow HF, Mak CC (1997) Dendritic bis(oxazoline)copper(II) catalysts. 2.1 Synthesis, reactivity, and substrate selectivity. J Org Chem 62(15):5116–5127
54. Tang WJ, Yang NF, Yi B, Deng GJ, Huang YY, Fan QH (2004) Phase selectively soluble dendrimer-bound osmium complex: a highly effective and easily recyclable catalyst for olefin dihydroxylation. Chem Commun 12:1378–1379
55. Bolm C, Derrien N, Seger A (1996) Hyperbranched macromolecules in asymmetric catalysis. Synlett 4:387–388
56. Fujihara T, Obora Y, Tokunaga M, Sato H, Tsuji Y (2005) Dendrimer N-heterocyclic carbene complexes with rhodium(I) at the core. Chem Commun 36:4526–4528
57. Fujihara T, Obora Y, Tokunaga M, Tsuji Y (2007) Rhodium(III) complexes with a bidentate N-heterocyclic carbene ligand bearing flexible dendritic frameworks. Dalton Trans 16:1567–1569
58. Ortiz AM, Sanchez-Mendez A, de Jesus E, Flores JC, Gomez-Sal P, Mendicuti F (2016) Poly(benzyl ether) dendrimers functionalized at the core with palladium bis(N-heterocyclic carbene) complexes as catalysts for the Heck coupling reaction. Inorg Chem 55(3):1304–1314
59. Oosterom GE, van Haaren RJ, Reek JNH, Kamer PCJ, van Leeuwen P (1999) Catalysis in the core of a carbosilane dendrimer. Chem Commun 12:1119–1120
60. Oosterom GE, Steffens S, Reek JNH, Kamer PCJ, van Leeuwen P (2002) Core-functionalized dendrimeric mono- and diphosphine rhodium complexes; application in hydroformylation and hydrogenation. Top Catal 19(1):61–73
61. Botman PNM, Amore A, van Heerbeek R, Back JW, Hiemstra H, Reek JNH, van Maarseveen JH (2004) Dendritic phosphoramidite ligands in Rh-catalysed asymmetric hydrogenations. Tetrahedron Lett 45(31):5999–6002
62. Muller C, Ackerman LJ, Reek JNH, Kamer PCJ, van Leeuwen P (2004) Site-isolation effects in a dendritic nickel catalyst for the oligomerization of ethylene. J Am Chem Soc 126 (45):14960–14963
63. Maraval V, Laurent R, Caminade AM, Majoral JP (2000) Phosphorus-containing dendrimers and their transition metal complexes as efficient recoverable multicenter homogeneous catalysts in organic synthesis. Organometallics 19(20):4025–4029
64. Yu JF, RajanBabu TV, Parquette JR (2008) Conformationally driven asymmetric induction of a catalytic dendrimer. J Am Chem Soc 130(25):7845–7847
65. Liang LY, Ruiz J, Astruc D (2011) The efficient Copper(I) (hexabenzyl)tren catalyst and dendritic analogues for green "click" reactions between azides and alkynes in organic solvent and in water: positive dendritic effects and monometallic mechanism. Adv Synth Catal 353(18):3434–3450

66. Geotti-Bianchini P, Darbre T, Reymond JL (2013) pH-tuned metal coordination and peroxidase activity of a peptide dendrimer enzyme model with a Fe(II)bipyridine at its core. Org Biomol Chem 11(2):344–352
67. Bhyrappa P, Young JK, Moore JS, Suslick KS (1996) Dendrimer-metalloporphyrins: synthesis and catalysis. J Am Chem Soc 118(24):5708–5711
68. Bhyrappa P, Young JK, Moore JS, Suslick KS (1996) Shape selective epoxidation of alkenes by metalloporphyrin-dendrimers. J Mol Catal A-Chem 113(1–2):109–116
69. Ellis A, Twyman LJ (2013) Probing dense packed limits of a hyperbranched polymer through ligand binding and size selective catalysis. Macromolecules 46(17):7055–7074
70. Zhang JL, Zhou HB, Huang JS, Che CM (2002) Dendritic ruthenium porphyrins: a new class of highly selective catalysts for alkene epoxidation and cyclopropanation. Chem-Eur J 8(7):1554–1562
71. Weyermann P, Diederich F (2002) Dendritic iron porphyrins with a tethered axial ligand as new model compounds for heme monooxygenases. Helv Chim Acta 85(2):599–617
72. Vins P, de Cozar A, Rivilla I, Novakova K, Zangi R, Cvacka J, Arrastia I, Arrieta A, Drasar P, Miranda JI, Cossio FP (2016) Cyclopropanation reactions catalysed by dendrimers possessing one metalloporphyrin active site at the core: linear and sigmoidal kinetic behaviour for different dendrimer generations. Tetrahedron 72(8):1120–1131
73. Newkome GR, Behera RK, Moorefield CN, Baker GR (1991) Chemistry of micelles series. 18. Cascade polymers—syntheses and characterization of one-directional arborols based on adamantane. J Org Chem 56(25):7162–7167
74. Kimura M, Sugihara Y, Muto T, Hanabusa K, Shirai H, Kobayashi N (1999) Dendritic metallophthalocyanines—synthesis, electrochemical properties, and catalytic activities. Chem-Eur J 5(12):3495–3500
75. Nlate S, Astruc D, Neumann R (2004) Synthesis, catalytic activity in oxidation reactions, and recyclability of stable polyoxometalate-centred dendrimers. Adv Synth Catal 346 (12):1445–1448
76. Nlate S, Plault L, Astruc D (2006) Synthesis of 9-and 27-armed tetrakis (diperoxotungsto) phosphate-cored dendrimers and their use as recoverable and reusable catalysts in the oxidation of alkenes, sulfides, and alcohols with hydrogen peroxide. Chem-Eur J 12(3): 903–914
77. Jahier C, Coustou MF, Cantuel M, McClenaghan ND, Buffeteau T, Cavagnat D, Carraro M, Nlate S (2011) Optically active tripodal dendritic polyoxometalates: synthesis, characterization and their use in asymmetric sulfide oxidation with hydrogen peroxide. Eur J Inorg Chem 5:727–738
78. Jahier C, Mal SS, Kortz U, Nlate S (2010) Dendritic Zirconium-peroxotungstosilicate hybrids: synthesis, characterization, and use as recoverable and reusable sulfide oxidation catalysts. Eur J Inorg Chem 10:1559–1566
79. Zhao MQ, Crooks RM (1999) Dendrimer-encapsulated Pt nanoparticles: synthesis, characterization, and applications to catalysis. Adv Mater 11(3):217–220
80. Zhao MQ, Crooks RM (1999) Homogeneous hydrogenation catalysis with monodisperse, dendrimer-encapsulated Pd and Pt nanoparticles. Angew Chem Int Edit 38(3):364–366
81. Deraedt C, Pinaud N, Astruc D (2014) Recyclable catalytic dendrimer nanoreactor for part-per-million CuI catalysis of "click" chemistry in water. J Am Chem Soc 136 (34):12092–12098
82. Gopidas KR, Whitesell JK, Fox MA (2003) Synthesis, characterization, and catalytic applications of a palladium-nanoparticle-cored dendrimer. Nano Lett 3(12):1757–1760
83. Ramos E, Davin L, Angurell I, Ledesma C, Llorca J (2015) Improved stability of Pd/Al_2O_3 prepared from Palladium nanoparticles protected with carbosilane dendrons in the dimethyl ether steam reforming reaction. ChemCatChem 7(14):2179–2187
84. Dalko PI, Moisan L (2004) In the golden age of organocatalysis. Angew Chem-Int Edit 43 (39):5138–5175
85. Morao I, Cossio FP (1997) Dendritic catalysts for the nitroaldol (Henry) reaction. Tetrahedron Lett 38(36):6461–6464

86. Zubia A, Cossio FP, Morao LA, Rieumont M, Lopez X (2004) Quantitative evaluation of the catalytic activity of dendrimers with only one active center at the core: Application to the nitroaldol (Henry) reaction. J Am Chem Soc 126(16):5243–5252
87. Guillena G, Kreiter R, Van de Coevering R, Gebbink RJMK, Van Koten G, Mazon P, Chinchilla R, Najera C (2003) Chiroptical properties and applications in PTC of new dendritic cinchonidine-derived ammonium salts. Tetrahedron Asymmetry 14(23):3705–3712
88. Zhang X, Xu HP, Dong ZY, Wang YP, Liu JQ, Shen JC (2004) Highly efficient dendrimer-based mimic of glutathione peroxidase. J Am Chem Soc 126(34):10556–10557
89. Bolm C, Derrien N, Seger A (1999) Hyperbranched chiral catalysts for the asymmetric reduction of ketones with borane. Chem Commun 20:2087–2088
90. Yang NF, Gong H, Tang WJ, Fan QH, Cai CQ, Yang LW (2005) Phase selectively soluble dendritic derivative of 4-(N, N-dimethylamino)pyridine: an easily recyclable catalyst for Baylis-Hillman reactions. J Mol Catal A-Chem 233(1–2):55–59
91. Liu L, Breslow R (2003) Dendrimeric pyridoxamine enzyme mimics. J Am Chem Soc 125 (40):12110–12111
92. Hecht S, Frechet JMJ (2001) Light-driven catalysis within dendrimers: designing amphiphilic singlet oxygen sensitizers. J Am Chem Soc 123(28):6959–6960
93. Javor S, Delort E, Darbre T, Reymond JL (2007) A peptide dendrimer enzyme model with a single catalytic site at the core. J Am Chem Soc 129(43):13238–13246
94. Helms B, Liang CO, Hawker CJ, Frechet JMJ (2005) Effects of polymer architecture and nanoenvironment in acylation reactions employing dendritic (dialkylamino)pyridine catalysts. Macromolecules 38(13):5411–5415
95. van de Coevering R, Alfers AP, Meeldijk JD, Martinez-Viviente E, Pregosin PS, Gebbink RJMK, van Koten G (2006) Ionic core-shell dendrimers with an octacationic core as noncovalent supports for homogeneous catalysts. J Am Chem Soc 128(39):12700–12713
96. Petrucci-Samija M, Guillemette V, Dasgupta M, Kakkar AK (1999) A new divergent route to the synthesis of organophosphine and metallodendrimers via simple acid-base hydrolytic chemistry. J Am Chem Soc 121(9):1968–1969
97. Piotti ME, Rivera F, Bond R, Hawker CJ, Frechet JMJ (1999) Synthesis and catalytic activity of unimolecular dendritic reverse micelles with "internal" functional groups. J Am Chem Soc 121(40):9471–9472
98. Pan YJ, Ford WT (2000) Amphiphilic dendrimers with both octyl and triethylenoxy methyl ether chain ends. Macromolecules 33(10):3731–3738
99. Goetheer ELV, Baars MWPL, van den Broeke LJP, Meijer EW, Keurentjes JTF (2000) Functionalized poly(propylene imine) dendrimers as novel phase transfer catalysts in supercritical carbon dioxide. Ind Eng Chem Res 39(12):4634–4640
100. Chavan SA, Maes W, Gevers LEM, Wahlen J, Vankelecom IFJ, Jacobs PA, Dehaen W, De Vos DE (2005) Porphyrin-functionalized dendrimers: synthesis and application as recyclable photocatalysts in a nanofiltration membrane reactor. Chem-Eur J 11(22): 6754–6762
101. Esposito A, Delort E, Lagnoux D, Djojo F, Reymond JL (2003) Catalytic peptide dendrimers. Angew Chem Int Edit 42(12):1381–1383
102. Douat-Casassus C, Darbre T, Reymond JL (2004) Selective catalysis with peptide dendrimers. J Am Chem Soc 126(25):7817–7826
103. Clouet A, Darbre T, Reymond JL (2004) A combinatorial approach to catalytic peptide dendrimers. Angew Chem Int Edit 43(35):4612–4615
104. Delort E, Darbre T, Reymond JL (2004) A strong positive dendritic effect in a peptide dendrimer-catalyzed ester hydrolysis reaction. J Am Chem Soc 126(48):15642–15643
105. Delort E, Nguyen-Trung NQ, Darbre T, Reymond JL (2006) Synthesis and activity of histidine-containing catalytic peptide dendrimers. J Org Chem 71(12):4468–4480
106. Biswas R, Maillard N, Kofoed J, Reymond JL (2010) Comparing dendritic with linear esterase peptides by screening SPOT arrays for catalysis. Chem Commun 46(46):8746–8748

107. Uhlich NA, Darbre T, Reymond JL (2011) Peptide dendrimer enzyme models for ester hydrolysis and aldolization prepared by convergent thioether ligation. Org Biomol Chem 9 (20):7071–7084
108. Hecht S (2003) Functionalizing the interior of dendrimers: synthetic challenges and applications. J Polym Sci Part A-Polym Chem 41(8):1047–1058
109. Caminade AM, Fruchon S, Turrin CO, Poupot M, Ouali A, Maraval A, Garzoni M, Maly M, Furer V, Kovalenko V, Majoral JP, Pavan GM, Poupot R (2015) The key role of the scaffold on the efficiency of dendrimer nanodrugs. Nat Commun 6:7722
110. Ficici E, Andricioaei I, Howorka S (2015) Dendrimers in nanoscale confinement: the interplay between conformational change and nanopore entrance. Nano Lett 15(7): 4822–4828

Site Isolation for Non-orthogonal Tandem Catalysis in Confined Nanospaces

Rinaldo Poli

Abstract Catalytic site isolation in nanoreactors is a desirable feature because it permits complex transformations with more than one catalyzed step, each step needing a different catalyst, to be run in a single vessel. Various types of one-pot multiple step processes may be distinguished. The present chapter focuses on those where two or more steps are chemically incompatible by the nature of the catalysts and/or the reagents, but the process is rendered possible by compartmentalization of the mutually destructive components. A requirement to accomplish this goal is the absence of interpenetration between the compartments holding the incompatible functions. This can more easily be achieved by anchoring the different components on rigid scaffolds, but successful processes may also result when using non-miscible flexible scaffolds and even when using physically compatible nanogels that cannot interpenetrate because of a sufficiently dense cross-linking network. This chapter provides a short history of site isolation for catalytic or stoichiometric one-pot tandem process and then reviews the recently developed nanostructured materials able to promote non-orthogonal tandem catalytic processes.

1 Introduction

Catalyst site isolation provides multiple advantages. It allows the active species to be chemically protected through suitable location in an environment that stops unwanted degradation processes [1, 2]. It may also result in improved efficiency and modified selectivities [3–5], sometimes approaching those of natural enzymes [6],

---

R. Poli (✉)
Laboratoire de Chimie de Coordination (LCC), UPR 8241 CNRS,
205 Route de Narbonne, BP 44099, 31077 Toulouse Cedex 4, France
e-mail: rinaldo.poli@lcc-toulouse.fr

R. Poli
Institut Universitaire de France, 1, rue Descartes, 75231 Paris Cedex 05, France

R. Poli (ed.), *Effects of Nanoconfinement on Catalysis*,
Fundamental and Applied Catalysis, DOI 10.1007/978-3-319-50207-6_9

as also highlighted in several other chapters of this book. Furthermore, isolation by confinement can also serve the purpose of more easily recovering and recycling the catalyst.

Catalyst degradation may result from unwanted reactions of the active site with compounds different than the reaction substrate, whether they are small amounts of impurities or compounds present in large amounts in solution. If the catalyst is embedded in an environment that is physically incompatible with the harmful reagents but compatible with the substrates to be transformed, such degradative processes may be reduced or suppressed without negatively affecting the catalytic efficiency. This is achieved, for instance, by anchoring the catalyst in a hydrophobic pocket for the transformation of hydrophobic reagents while keeping away polar reagents such as water, as achieved for instance by the micellar catalysis approach [7–9]. Catalyst degradation may also result from dimerization/oligomerization, when the active species is a highly unsaturated monometallic species. This phenomenon may be avoided by anchoring the active sites on a rigid scaffold such as an activated silica surface, as shown for instance for the [($\equiv$SiO)$_3$Zr-H] species for which no equivalent molecular species is known [10]. To this end, it is necessary to confine the active sites in distant scaffold locations, avoiding their mutual interaction. This scenario is quite different than that of cooperative catalysis, where on the contrary the close proximity of (usually identical) catalytic sites is beneficial [11], as detailed in another chapter for the example of densely grafted catalytic sites on a dendrimer surface (the so-called dendridic effect).

A more specific type of site isolation pertains to those systems that contain two or more different catalytic sites with the goal of realizing tandem processes in one pot [12]. There are various possible scenarios. In the first one, the transformation consists of *orthogonal tandem catalytic processes*, namely not interfering with each other [13–15]. In this case, the relative location of the different catalysts is not an important issue and the process can occur equally well upon mixing the required catalysts under homogeneous conditions, thus site isolation is not required. However, placement of the two catalytic functions in close proximity usually results in reaction rate enhancement because of mass transport effects or better selectivities because the product of the first step can be trapped more efficiently in the second step before leading to degradative side reactions. A second scenario is that of *cooperative catalysis*, where two different functions act together to promote a single chemical transformation via a dual activation mechanism [16]. In this case, having the two cooperating functions in close proximity is a requirement to benefit from the positive effect. There are many examples of this phenomenon, especially by combining a catalytic function grafted on a solid surface with other functions that are already present on the same surface, such as silanol groups on silica or Al^{3+} centers on silica-alumina [17–20]. Cooperative effects may be achieved via two different compatible (e.g., non-annihilating) organic functional groups, such as acid–base, acid–thiol, amine–urea, imidazole–alcohol–carboxylate groups, etc. [21], but it is also possible to have cooperating functions that are in principle incompatible, when steric impediments block the mutual quenching (e.g., a frustrated Lewis acid–base pair). Single molecules possessing these characteristics are termed bifunctional catalysts [22]. In this chapter, tandem orthogonal systems and cooperative or bifunctional catalytic systems will only be highlighted for their historical relevance. Although those areas of research are flourishing, we wish to restrict our attention to the third scenario: that of tandem transformations where different steps are catalyzed by chemically incompatible functions. We define these as *non-orthogonal tandem processes*, a term that does not appear to have previously been used in the context of tandem reactions, to distinguish them from the first category described above. In this case, the relative location of the catalytic sites is critical; it is essential to avoid that the different catalysts and/or reagents required to accomplish the different steps of the tandem process come into contact with each other. For the situation of two mutually destructive catalysts, this is similar to the "frustration" concept, except that the mutual quenching is not avoided by the ligand steric bulk but rather by the site confinement. An example that can be easily imagined, and indeed has attracted much attention, is that of a transformation involving one acid-catalyzed and one base-catalyzed step when the strength of the acidic and basic sites is sufficient for mutual neutralization in a homogeneous system. Another important example, of great current interest for harvesting solar energy and transforming it into chemical energy, is that of combining a photosensitizer for the separation of electrons and holes with one or more catalysts for the transformation of substrates into products with higher energy content. While the photosensitizer and the catalyst are mutually compatible and act cooperatively, the subsequent use of electrons and holes for separate chemical transformations, each of which requires a different catalytic function, needs site isolation.

It is necessary to distinguish two general situations of site isolation. One concerns systems where the two incompatible functions are separately confined on different bulk phases (usually different solids) that, through lack of mixing, avoid physical contact of the two incompatible functions while interface crossing is allowed for the reagents and products. The second one concerns situations where the different sites are installed spatially separated at the nanoscale within the same support. The first type of dual catalytic systems is more easily constructed but the second one presents the clear advantage of faster mass transport kinetics and the possibility, as already mentioned above, of optimizing selectivities when an intermediate of the cascade transformation has limited stability.

Nature is a great source of inspiration [23–25], as complex living objects are able to carry out several, often incompatible catalytic transformations by compartmentalizing them, thereby avoiding interference and also allowing the simultaneous reactions to run in a variety of environments that may differ by pH, ionic strength, hydro/lipophilicity, and so forth. Site isolation for use in non-orthogonal tandem catalysis can in principle be accomplished more easily by catalyst confinement on a rigid scaffold such as the surface of a bulk solid or the inner surface of a microporous material [26], but the use of mobile supports (supramolecular systems, polymers, artificial enzymes) is also of growing interest [27–30], although the scaffold flexibility makes the site confinement issue less straightforward. The development of complex bionanostructures and their catalytic application in cascade processes has experienced an explosive growth in recent years [31].

We will first review the historical development of catalytic site confinement and then focus our attention on dual catalytic systems that are spatially separated at the nanoscale.

2 Evolution of Site Confinement for Stoichiometric and Catalytic Processes

Although we are specifically interested in this book on catalytic transformations, the development of site confinement strategies for tandem transformations was pioneered with stoichiometric reagents and the most relevant contributions will be highlighted here. We have elected to include in this section various different approaches such as the use of different non-mutually destructive reagents or catalysts on the same support, or mutually destructive ones on different supports. Processes run either consecutively or simultaneously are included. The common strive of all these contributions was to achieve complex reaction sequences in one pot, simplifying synthetic, purification, and/or catalyst recovery operations.

2.1 The Early Days with Cross-Linked Polymers

The use of polymer-supported catalysts, using both soluble and insoluble polymers, has long been practiced [32]. Indeed, the site isolation and the multifunctional catalysis concepts have first been introduced using cross-linked polymer supports. Use of insoluble resins is quite widespread given the commercial availability of ion exchange materials, where charged reagents or catalysts can be readily anchored, and of functionalized cross-linked polymers such as Merrifield resins in which covalently linked functions can be easily installed. All contributions on non-orthogonal tandem processes using cross-linked polymers have made use, to the best of our knowledge, of a combination of separately grafted stoichiometric or catalytic reagents anchored on different supports, while site isolation at the nanoscale level has not been developed on this type of support, with the exception of more recently developed porous organic frameworks (see Sect. 3.1.3).

In a pioneering 1975 study, Pittman and Smith described the simultaneous use of two polymeric organometallic catalysts for sequential cyclooligomerization-hydrogenation or cyclooligomerization-hydroformylation of butadiene [33]. The catalysts were anchored either separately or together on $-PPh_2$ functionalized styrene-divinylbenzene resins, see Scheme 1.

Anchored $Ni(CO)_2(PPh_3)_2$ catalyzed the cyclooligomerization reaction, whereas $RhCl(PPh_3)_3$ or $RuCl_2(CO)_2(PPh_3)_2$ catalyzed the hydrogenation process and RhH $(CO)(PPh_3)_3$ catalyzed the hydroformylation process. The cyclooligomerization catalyzed by the Ni complex (either molecular $[Ni(CO)_2(PPh_3)_2]$ or

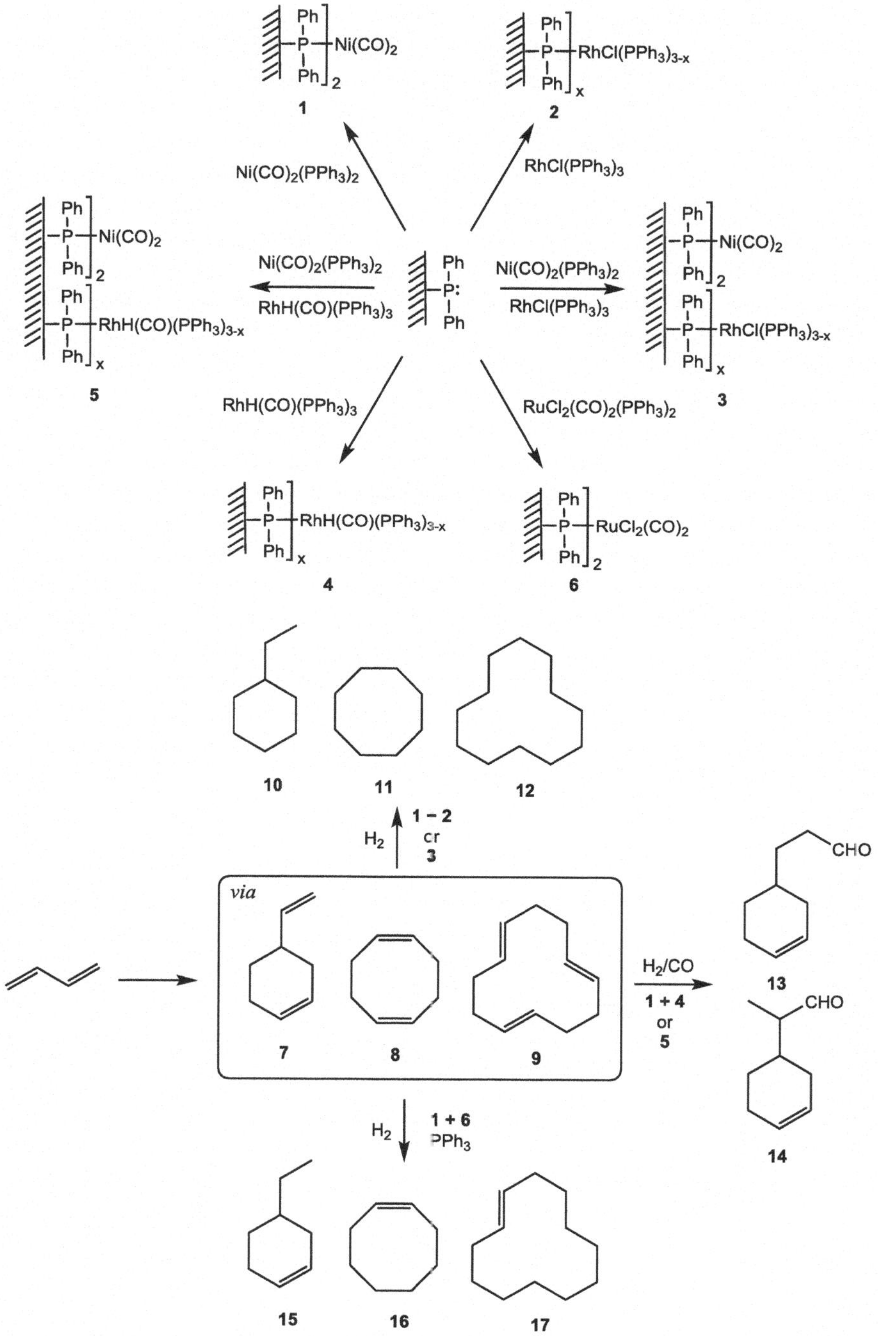

Scheme 1 Sequential processes catalyzed by dual metal catalysts anchored on styrene-divinylbenzene resins [33]

polymer-anchored **1**) yields a mixture of products **7**, **8,** and **9**. In the presence of the Wilkinson catalyst (either molecular [$RhCl(PPh_3)_3$] or polymer-anchored **2**), these are hydrogenated to **10**, **11,** and **12**, whereas in the presence of the ruthenium catalyst [$RuCl_2(CO)_2(PPh_3)_2$] they are converted into the monoene products **15**, **16,** and **17**. In the presence of [$RhH(CO)(PPh_3)_3$] (either molecular or as polymer-anchored **4** or **5**), intermediate **7** was selectively hydroformylated to a mixture of linear and branched products **13** and **14**. In all cases, the two steps were carried out sequentially, not simultaneously (H_2 for hydrogenation or syngas for hydroformylation were introduced after completion of the cyclooligomerization process). These catalysts do not interfere with each other, thus site isolation is not needed to insure success for these processes. However, the study illustrated for the first time the principle of one-pot tandem catalysis on a polymer support. The polymer-anchored systems allowed facile catalyst recovery but yielded results overall quite similar to those of the mixed-catalyst homogeneous systems.

Use of polymer supports for the spatial localization of incompatible reagents and application to non-orthogonal tandem transformations has apparently been shown for the first time in 1977 by Cohen and coworkers, who termed the processes "wolf and lamb" reactions, by making use of polymer-supported stoichiometric reagents [34–36]. They used separate beads of styrene-divinylbenzene resins (PS), one functionalized with the strongly basic trityllithium functions, PS-C_6H_4-CPh_2Li, and the other one with the acylating *o*-nitrophenyl ester functions, PS-C_6H_3-$NO_{2\text{-}4}$-OCOX (X = Ph, OEt). In this fashion, a sequential deprotonation/acylation of nitriles and carbonyl compounds could be carried out in one pot as shown in Scheme 2. Other one-pot stoichiometric processes have been published later [37–40], even involving more than two steps [41].

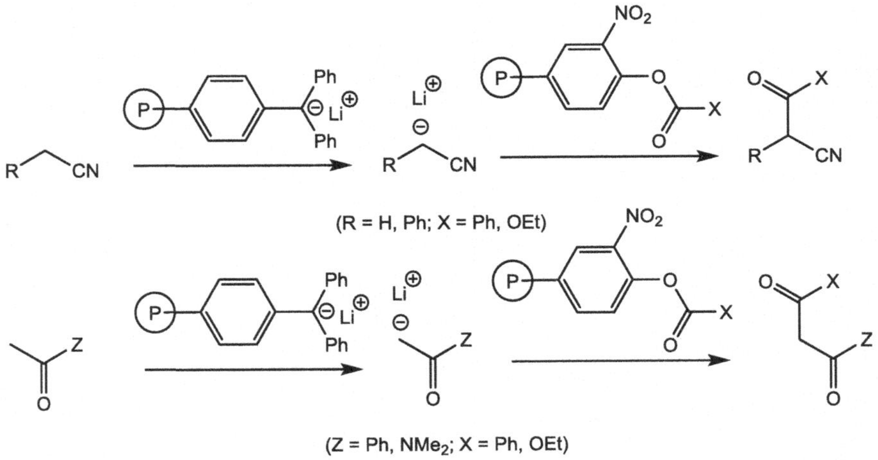

Scheme 2 "Wolf and lamb" reactions accomplished with incompatible acid- and base-functionalized polystyrene resins [35]

In contrast to the above fully stoichiometric processes, Cainelli and coworkers were first to report a tandem process with site isolation involving one catalytic species. In a little-cited 1980 contribution [42], they used anionic phosphonate reagents supported on an anion exchange resin (Amberlyst A-26) to accomplish a variety of aldehyde and ketone olefination reactions (Scheme 3a), then coupled this process to a first step consisting of an acid-catalyzed carbonyl compound release by deprotection of the dioxolane using Amberlyst 15 H (Scheme 3b). The stoichiometric reagent was prepared by ion exchange from the polymeric ammonium hydroxide and the phosphonate $XCH_2P(P)(OEt)_2$. The reaction is efficient, particularly for X=COOMe, making it suitable for the column technique, and is equally applicable to the dioxolanes of ketones and of aromatic and α,β-unsaturated aldehydes. The procedure would be impossible in solution because of the immediate neutralization of the acidic catalyst by the basic phosphonate reagent.

Other tandem processes with incompatible catalysts separately anchored on different cross-linked polymers have later been presented and a comprehensive coverage is out of the scope of this chapter. One contribution worth highlighting is the extension of the principle of site isolation of mutually self-destructive species to "kinetic isolation" by Bergbreiter and Chandran in 1985 [43]. Only one species is segregated in an insoluble polymer whereas the other one is in solution, but mutual annihilation does not occur on the timescale of the catalytic experiment. The principle was demonstrated using a stoichiometric Cr^{VI} oxidant bonded to a cross-linked poly(vinyl pyridine) resin (PVPCC), in combination with a soluble polyethylene-linked phosphine, PE-PPh_2, which was used as ligand for a rhodium hydrogenation catalyst, (PE-PPh_2)$_3$RhCl. This functionalized polymer, produced by anionic oligomerization of ethylene to a MW ≥ 1200 g mol^{-1} followed by quenching by Ph_2PCl, was insoluble at room temperature but sufficiently soluble in toluene at 100 °C and was further entrapped into regular high density polyethylene (HDPE) by keeping the two polymers together under these conditions. The two reagents were then used together in the one-pot alkene reduction/alcohol oxidation for a variety of unsaturated primary and secondary alcohols as shown in Scheme 4.

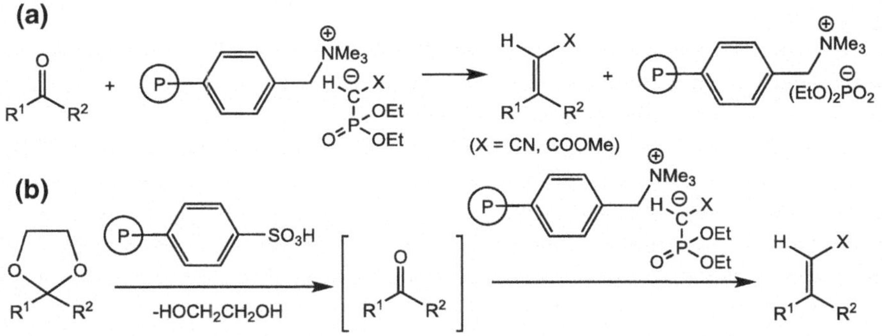

Scheme 3 Carbonyl compound olefination (**a**) and tandem dioxolane deprotection/olefination (**b**) with a combination of acidic and basic Amberlyst-type resins [42]

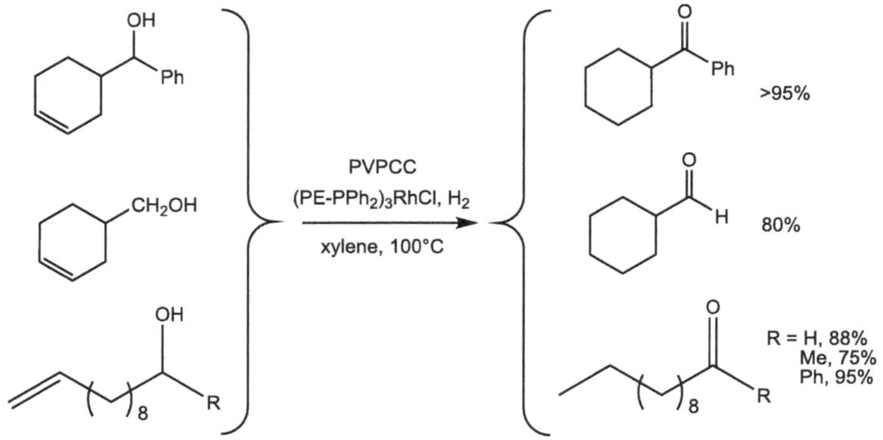

Scheme 4 One-pot alkene reduction and alcohol oxidation for a variety of unsaturated alcohols

The Rh catalyst could be recovered and reused three times, but it was no longer active after exposure to PVPCC at 100 °C for 40 h. However, replacing the PE-supported Rh catalyst with molecular [$RhCl(PPh_3)_3$] showed only alcohol oxidation and no H_2 uptake, while the PPh_3 ligand was oxidized to $Ph_3P{=}O$.

2.2 *Development of Stoichiometric and Catalytic Non-orthogonal Tandem Processes with Inorganic Solids*

The use of stoichiometric reagents or catalysts confined on solids for application to one-pot tandem processes has been introduced after the above described work with resins. It involved all sorts of solid supports, including various forms of silica and alumino-silicates (sol-gel systems, zeolites and mesoporous silica), double layered hydroxides, etc.

In a series of influential papers in the 2001–3 period [44–48], Blum and Avnir demonstrated the principle of site isolation for sequential reactions carried out in one pot using non-compatible reagents embedded in silica matrices generated by the sol-gel method. They separately prepared silica supported Brønsted or Lewis acids, bases, transition metal catalysts, or enzymes (Scheme 5). Three different methods of anchoring are involved: physical entrapment (e.g., for the Nafion polymer, $\mathbf{a_1}$, and for a poly(ammonium hydroxide), $\mathbf{b_1}$), physical entrapment with strong association with the matrix silanol groups (e.g. for inorganic MoO_3, $\mathbf{a_2}$) and covalent anchoring (e.g. for the sulfonic acid $\mathbf{a_3}$ and for the bases $\mathbf{b_2}$ and $\mathbf{b_3}$) in all cases leading to leach-proof materials. All these materials, independent on the anchoring method, will be referred to using the same symbolism, e.g., $\mathbf{a_1}$@s-g (s-g is an abbreviation for sol-gel).

Scheme 5 Functions entrapped in sol-gel systems by Blum, Avnir, and coworkers [44–48]

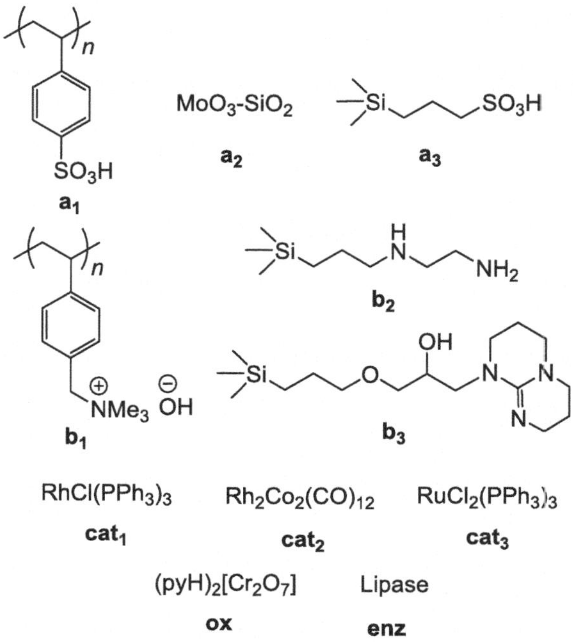

Combinations of various reagent or catalyst pairs were used for the one-pot tandem reactions shown in Schemes 6 and 7. In all cases, the reagents were deactivated, either in their homogeneous or immobilized form, in the presence of a soluble version of the other reagent, proving the importance of the immobilization and site isolation of both reagents.

A mixture of $\mathbf{b_2}$@s-g and $\mathbf{cat_1}$@s-g converts 2-iodo-1-phenylethane into ethylbenzene under hydrogen by a one-pot dehydroiodination-hydrogenation sequence (Scheme 6a) [44]. In this process, the first site-isolated reagent (the base) is stoichiometric whereas the second one (the rhodium complex) is catalytic. In Scheme 6b, the same stoichiometric base-promoted dehydroiodination of 2-iodo-1-phenylethane of Scheme 6a is followed by an acid-catalyzed aromatic alkylation of anisole [45]. On the other hand, both steps are catalytic in the process described in Scheme 6c: an acid-catalyzed pinacol rearrangement followed by a base-catalyzed Knoevenagel condensation with malonitrile of the ketone intermediate [45]. In the process shown in Scheme 6d, acid-catalyzed dehydration of 1-tetralol is combined with the catalyzed hydrogenation of the 1,2-dihydronaphthalene intermediate which, depending on conditions, yields tetralin or a mixture of *cis* and *trans*-decalin [46]. In the example of Scheme 6e, another pair of incompatible reactions, an oxidation and a reduction, were carried out in one-pot with use of **ox**@s-g as a stoichiometric oxidant and $\mathbf{cat_1}$@s-g as a hydrogenation catalyst to accomplish nitrobenzene reduction to aniline and oxidation of benzylic alcohol to benzaldehyde, followed by spontaneous condensation of the two products [47]. In this case, since the aniline intermediate is prone to

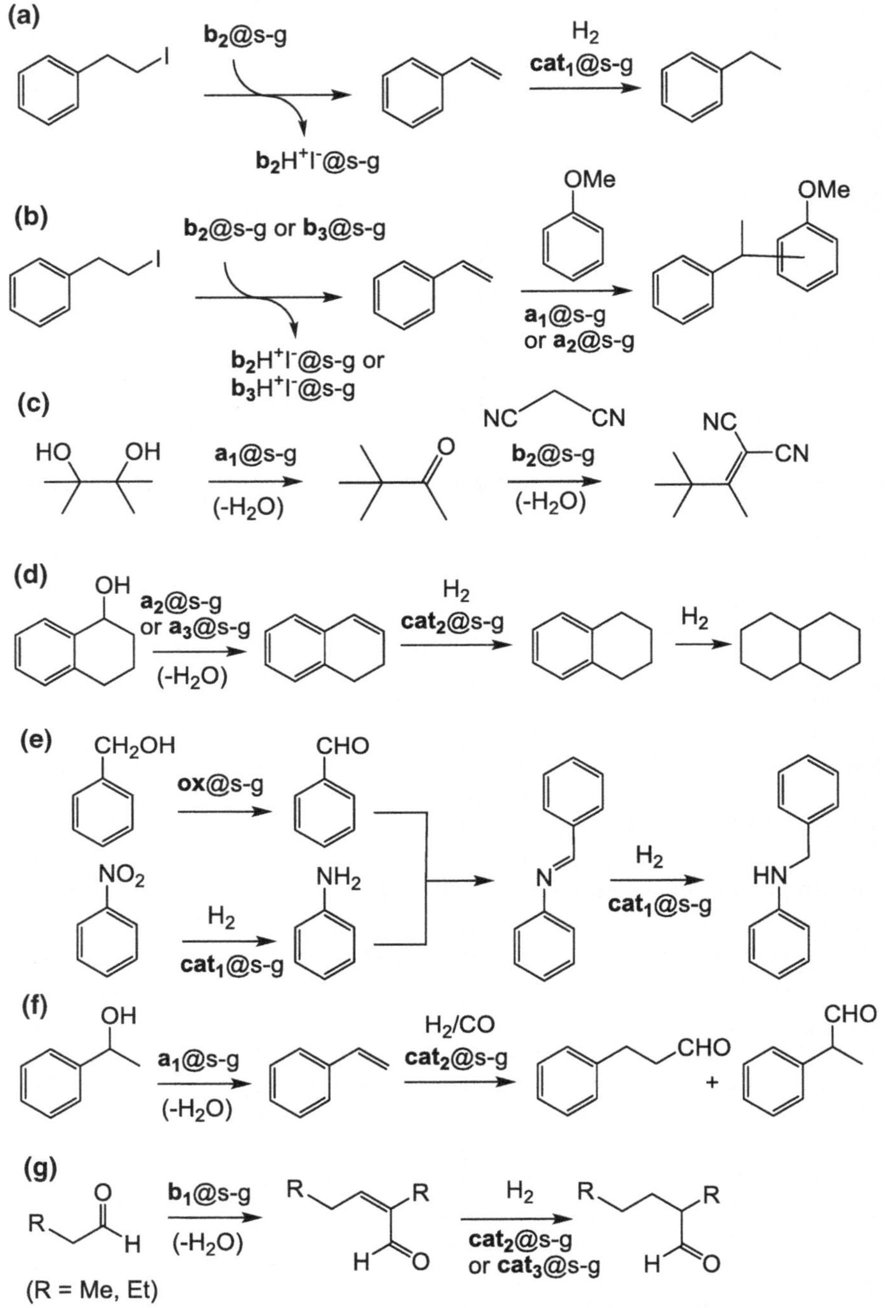

Scheme 6 One-pot-tandem processes carried out with sol-gel entrapped reagents as defined in Scheme 5

oxidation by **ox**@s-g, the H_2 addition was delayed for 1 h to allow the alcohol oxidation to reach completion, while the dichromate oxidant is not reduced by H_2 under these conditions. The subsequent nitrobenzene reduction was followed by immediate trapping of aniline, faster than reoxidation, to form the Schiff base in 91% yield. The remainder 9% of the conversion was due to further imine reduction to benzylaniline. In the reaction of Scheme 6f, an acid-catalyzed dehydration is combined with catalyzed hydroformylation of the resulting C=C unsaturation, while those of Scheme 6g involve sequential base-catalyzed aldol condensation and selective C=C hydrogenation [46].

The method has also been extended to processes where one step is enzymatic. Thus, sol-gel entrapped lipase (**enz@s-g**) and **a$_1$@s-g** or **a$_2$@s-g** were used for one-pot esterification and C–C double bond hydrogenation reactions, leading to saturated esters in good yields. In this case too, poisoning and activity quenching are observed when direct contact between the two catalysts is not blocked by segregation [48]. The system works equally well for three different situations, as outlined in Scheme 7: from an unsaturated acid (R^uCOOH) and a saturated alcohol (R^sOH), or from a saturated acid (R^sCOOH) and an unsaturated alcohol (R^uOH), or when both reagents are unsaturated. In each case, the reaction may proceed by two pathways, with the esterification and hydrogenation processes in either order. The anchored lipase induced moderate asymmetry (ee = 19%) during esterification with the prochiral alcohol 2-methyl-3-butenol. Recycling without loss of activity was

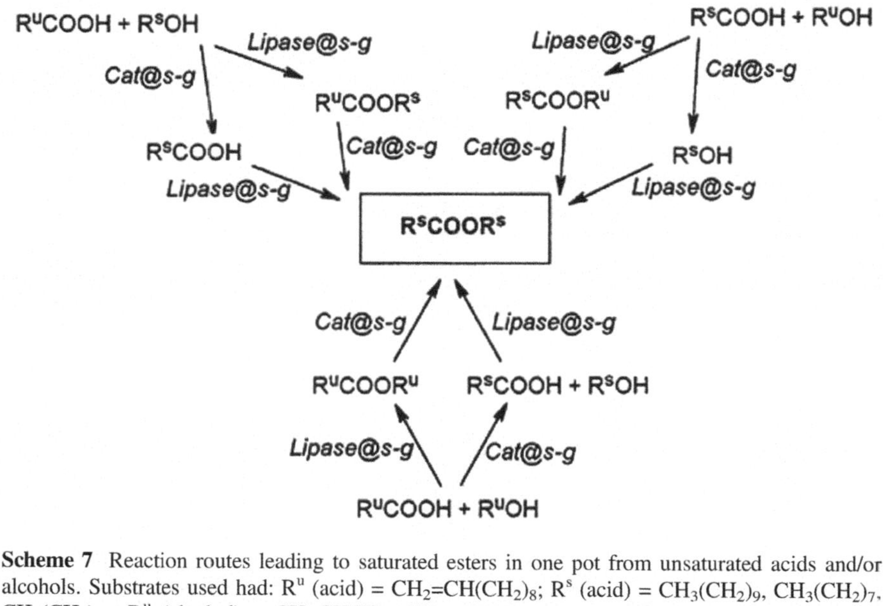

Scheme 7 Reaction routes leading to saturated esters in one pot from unsaturated acids and/or alcohols. Substrates used had: R^u (acid) = CH_2=$CH(CH_2)_8$; R^s (acid) = $CH_3(CH_2)_9$, $CH_3(CH_2)_7$, $CH_3(CH_2)_{10}$: R^u (alcohol) = CH=$CHCH_2$, CH_2=$CHCH(CH_3)CH_2$: R^s (alcohol) = $CH_3(CH_2)_2$, $CH_3(CH_2)_4$, (*S*)-$CH_3CH_2CH(CH_3)CH_2$. Reproduced with permission from Ref. [48]. Copyright 2002 American Chemical Society

demonstrated. Numerous other applications of sol-gel materials in tandem processes have been reported and reviewed elsewhere [49, 50].

In 2002, Choudary and coworkers described the preparation and use of a tri-functional $PdCl_4^{2-}$–OsO_4^{2-}–WO_4^{2-} catalyst, fixed by ion exchange on the surface of a layered double hydroxide (LDH), for a one-pot Heck-asymmetric dihydroxylation in the presence of the Sharpless chiral ligand that uses H_2O_2 as a terminal oxidant and *N*-morpholine as an electron transfer mediator (see Scheme 8) [51]. The Heck coupling and the epoxidation had to be run consecutively, not simultaneously. The use of the Na_2PdCl_4–K_2OsO_4–Na_2WO_4 catalyst combination in homogeneous medium gave analogous yields (89%) and selectivity (99% ee) for the final product, but was much more time consuming than the reaction on the LDH-PdOsW material.

Pârvulescu, De Vos and coworkers reported in 2004 the preparation of bifunctional Ir/H-beta and Ir/H-Mor catalysts by impregnation of $Ir(acac)_3$ into H-beta or H-mordenite zeolites, respectively, and subsequent calcination and reduction. These materials were used for the one-pot tandem isomerization/hydrogenation of citronellal to menthol, see Scheme 9. The first step of the reaction, carried out in the

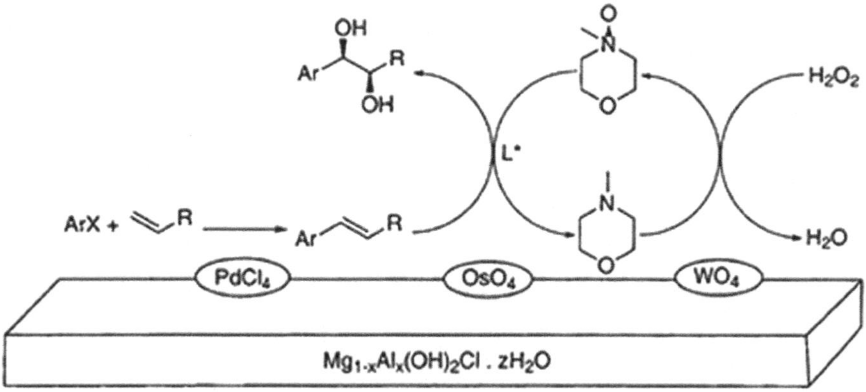

Scheme 8 Catalytic cycle in the LDH-PdOsW-catalyzed synthesis of chiral diols using H_2O_2 as the terminal oxidant. Reproduced with permission from Ref. [51]. Copyright 2003 American Chemical Society

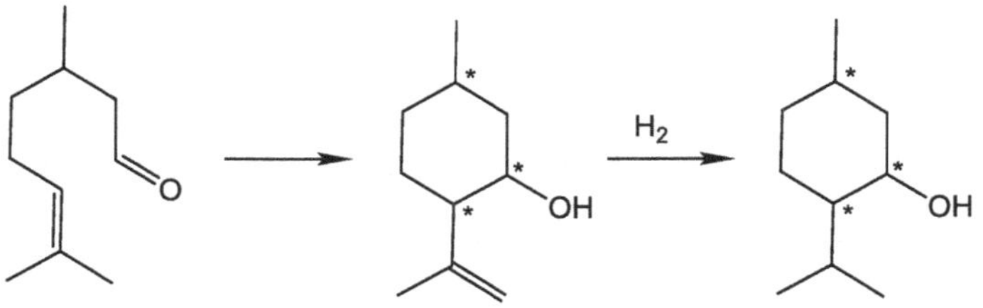

Scheme 9 One-pot tandem cyclization/hydrogenation of citronellal to menthol catalyzed by Ir/H-beta

absence of H_2, gives all four possible isomers of isopulegol, with the Ir-containing catalysts being more active than the parent zeolites. This indicates that the Lewis acidity of non-reduced Ir also plays a role in the isomerization, in addition to the zeolite Brønsted acidic sites. The selectivity could be improved by running the reaction in a sequential mode, allowing the first step to complete under N_2, followed by H_2 addition [52].

In 2005, Kaneda and coworkers used catalysts separately confined on different layered clays, a Ti^{4+}-exchanged montmorillonite (Ti^{4+}-mont), which is a Brønsted acid catalyst through functions within the interlayer spaces, and hydrotalcite (HT) with surface tunable basicity. The large HT particles cannot interpenetrate the narrow interlayers of the Ti^{4+}-mont and thus no mutual destruction of the two antagonist functions can occur [53]. The combination of the two materials was efficient in the one-pot acetal deprotection-aldol condensation for a wide scope of aldehydes and donor substrates, see Scheme 10a. Five recycles were carried out with retention of activity and selectivity. The same catalyst mixture was also applied to tandem Michael condensation and acetalization, Scheme 10b, again run sequentially with the protecting diol added after completion of the first step. Finally, the one-pot synthesis of an epoxynitrile from cyanoacetic acid, methanol, benzaldehyde acetal, and hydrogen peroxide in three sequential acid- and base-catalyzed steps, namely esterification, tandem deacetalization and aldol reaction, and final epoxidation, was accomplished in excellent yields as shown in Scheme 10c [53].

Metal-containing hydrotalcites are truly bifunctional materials, operating though the HT basic sites and the transition metal sites, but the two catalytic sites are not self-destructive and therefore do not require site isolation. In a few applications, they act cooperatively for single transformations [54]. In the quinoline synthesis shown in Scheme 11, the HT-supported Ru sites are proposed to catalyze the 2-aminobenzyl alcohol oxidation to 2-aminobenzaldehyde under an O_2 atmosphere, followed by aldol reaction with ketones catalyzed by the HT basic sites [55]. The reaction has a large substrate scope, including α,β-unsaturated ketones (for instance, R^1 = CH=CH-Ph).

The use of Pd/HT, containing supported Pd nanoparticles, allows the one-pot sequential aldol condensation/hydrogenation under mild conditions (see Scheme 12a) for a wide reactant scope in excellent overall yields [56]. These processes were also run sequentially, not simultaneously. The combination of this Pd/HT catalyst with the above mentioned Ti^{4+}-mont acid catalyst has allowed the one-pot synthesis of 2-carbomethoxy-2-benzyl glutaronitrile in four consecutive steps and good overall yields, where the second step is a tandem process (Scheme 12b) [53].

Combinations of differently functionalized mesoporous silica materials have also been used for tandem processes. In 2008 Lin and coworkers reported the synthesis of two MCM-41 type mesoporous silica nanoparticles (MSN), functionalized respectively with 4-ethylpyhenylsulphonic acid and with aminopropyl functions and applied them in combination for the tandem acid-catalyzed deprotection of the benzaldehyde acetal followed by the base-catalyzed Henry reaction (Scheme 13)

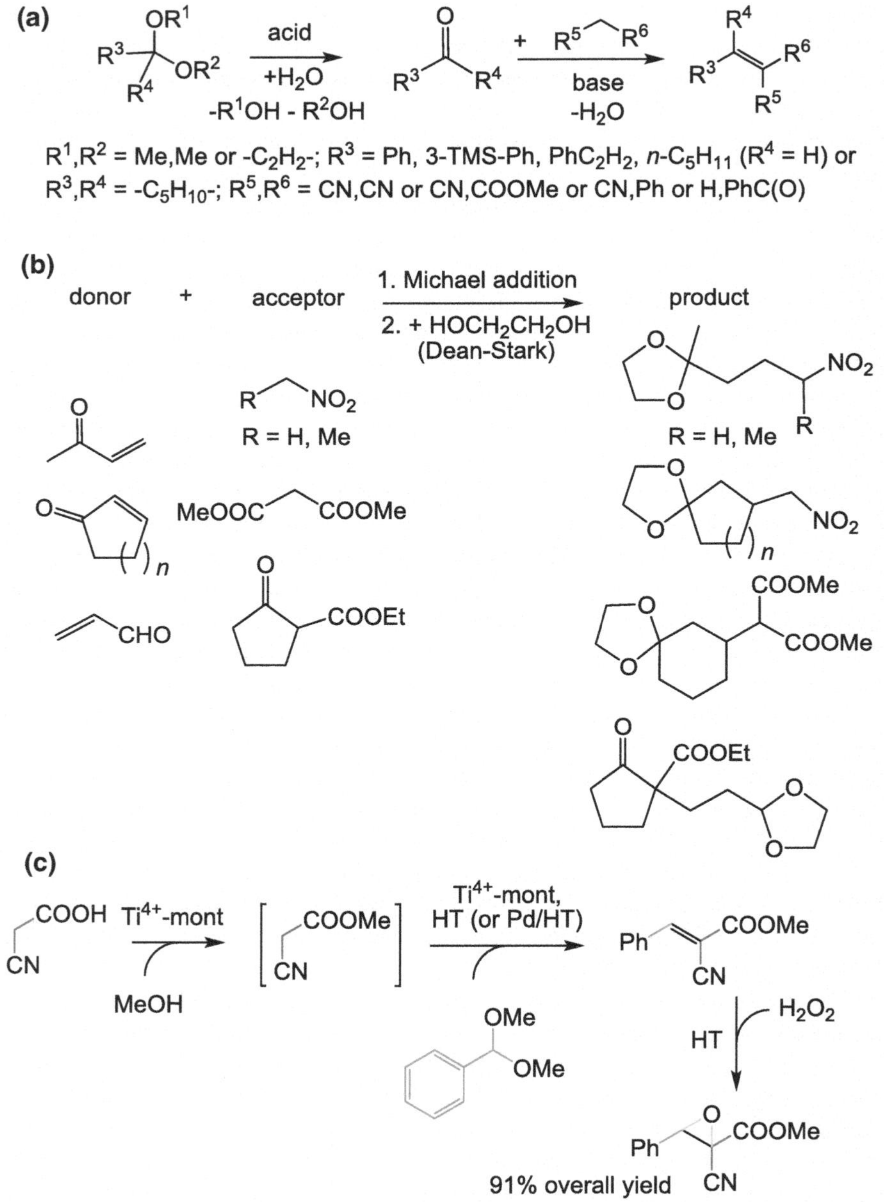

Scheme 10 Tandem processes catalyzed by a combination of titanium-exchanged montmorillonite (Ti^{4+}-mont) and hydrotalcite (HT) [53]. **a** Acetal deprotection-aldol condensation. **b** Michael condensation-acetalization. **c** Multi-step synthesis of an epoxynitrile

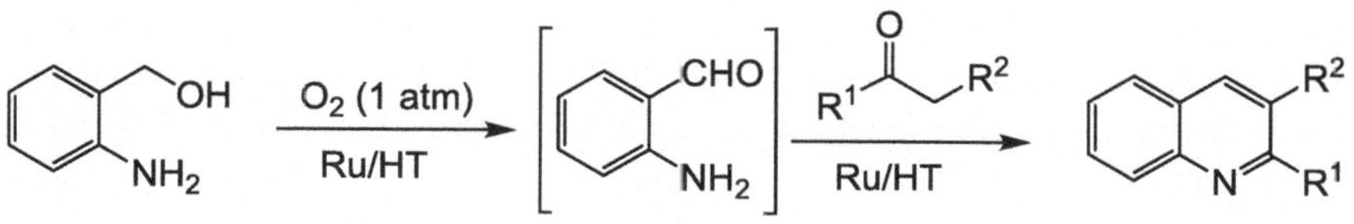

Scheme 11 One-pot quinoline synthesis by alcohol oxidation/aldol condensation on a Ru/HT catalyst [55]

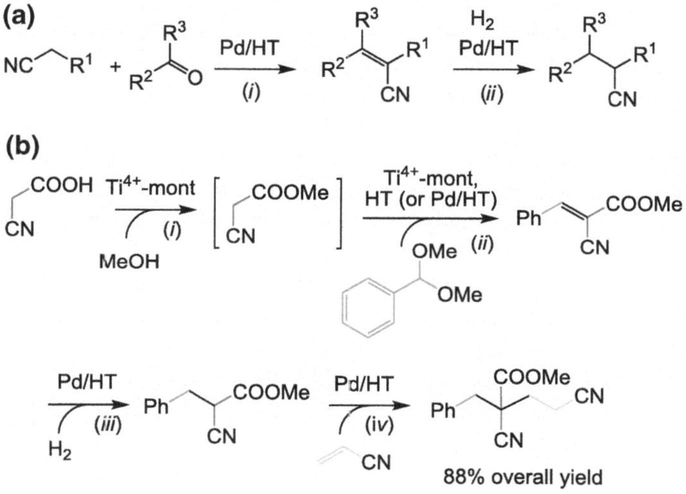

Scheme 12 Sequential multi-step syntheses accomplished by use of the Pd/HT catalyst. **a** Two-step aldol condensation/hydrogenation [56]. **b** Four-step glutaronitrile with the combined use of Ti^{4+}-mont and Pd/HT [53]

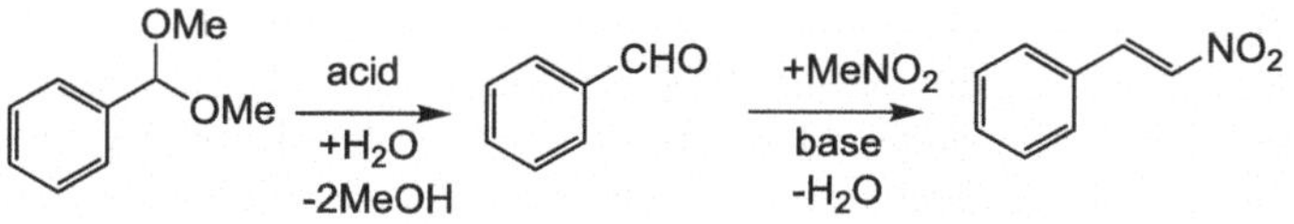

Scheme 13 Tandem reaction catalyzed by MCM-41 beads separately functionalized with $-NH_2$ and $-SO_3H$ groups

[57]. Only the acid deprotection step was accomplished when just the acid-functionalized particles were present, whereas a 97.7% yield of final product was obtained in the presence of both catalysts and no product at all was obtained when only the basic catalyst was present, or when either *tert*-butylamine or *p*-toluenesulfonic acid (PTSA) were present in solution.

Another recent example involves the glucose conversion to 5-hydroxymethylfurfural with a first isomerization step to fructose catalyzed by immobilized glucose isomerase, followed by an acid-catalyzed dehydration by a

$-SO_3H$ functions anchored on separate beads [58]. These materials are also adapted for site isolation at the nanoscale, as will be highlighted in Sect. 3. Other applications of multisite solid catalysts in cascade reactions are described in other reviews [59, 60].

2.3 *New Types of Supports*

As new materials are developed, so is their use as multifunctional supports for tandem catalyzed transformations. Metallorganic frameworks (MOF), also known as porous coordination polymers (PCP), have rapidly become scaffolds of choice for numerous applications, which includes elaborate catalytic applications by modified systems where various catalytic sites have been installed. Systems where the active sites function either cooperatively or independently have been described [61]. Systems containing mutually self-destructive sites separated at the nanoscale will be highlighted in Sect. 3.1.2.

In 2010, He and coworkers introduced a bifunctional catalytic MOF, where the two functions are respectively the Lewis acidic metal and Pd nanoparticles (Pd@MIL-101(Cr)). The large pore size (29–34 Å) and large surface area (ca. 4000 $m^2\ g^{-1}$) are suitable for depositing Pd NPs. In addition, open coordination sites on Cr^{3+} can be easily generated by removal of labile water molecules. The material was used for the tandem acetone dimerization/dehydration, promoted by the Lewis acidic sites, followed by hydrogenation of the mesityl oxide intermediate catalyzed by the Pd NPs to afford methyl isobutyl ketone (Scheme 14a) [62]. Recycle experiments demonstrated the catalyst stability against metal agglomeration. When MIL-101(Cr) without Pd was used, mesityl oxide was the major product. Increased Pd loading led to competitive hydrogenation of acetone to yield isopropanol. Corma and coworkers later used the same catalyst for the one-pot tandem isomerization/hydrogenation of citronellal to menthol (Scheme 9), the results being, however, inferior to those described in Sect. 2.2 with use of the zeolite-based Ir/H-beta catalyst. The MIL-101(Cr) material without Pd gave the isomeric isopulegol intermediates in excellent yields, no products other than the four possible isopulegol diastereomers being detected [63].

De Vos and coworkers have developed a modified $[Zr_6O_4(OH)_4(O_2C–C_6H_4–CO_2)_6]$ (UiO-66) by use of 2-aminoterephthalic acid in the synthetic procedure to yield thermally stable UiO-66(NH_2). The open coordination sites are generated by thermal dehydroxylation to transform the $Zr_6O_4(OH)_4$ core to Zr_6O_6, but in this first

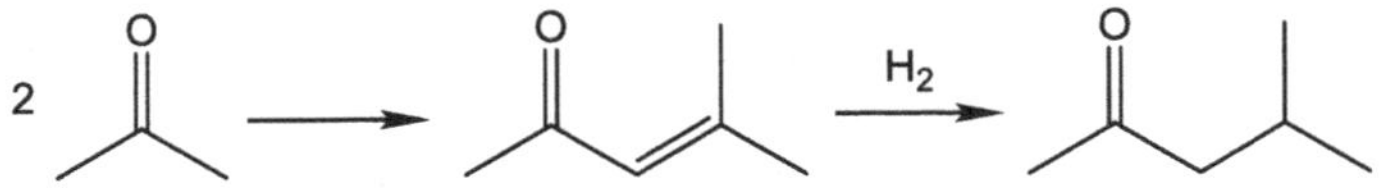

Scheme 14 One-pot tandem dimerization/hydrogenation of acetone to methyl isobutyl ketone catalyzed by Pd@MIL-101(Cr) [63]

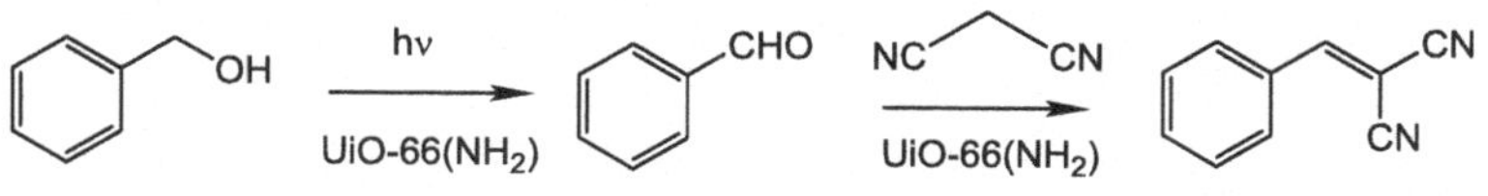

Scheme 15 Tandem benzyl alcohol photooxidation and Knoevenagel condensation catalyzed by UiO-66(NH_2) [65]

contribution the material was only applied to a reaction (aldol condensation) where the two functions act cooperatively [64]. The same materials was later used for the tandem photochemical oxidation of benzyl alcohol and subsequent Knoevenagel condensation of the benzaldehyde intermediate with malonitrile, see Scheme 15 [65]. In this reaction, the metal site shows photocatalytic activity in the first step, whereas both the metal site and the framework amine base act cooperatively to catalyze the second step. A related Al-functionalized MOF/PCP (MIL-53(Al)-NH_2) does not have photocatalytic activity.

2.4 Combination of Separately Recoverable Supports

The compartmentalization of different catalytic sites on different solid supports has been effective to accomplish several one-pot cascade processes and also allows catalyst recyclability, as outlined in the previous sections, but generally does not allow the recovery of the individual catalysts in separate form. A solution of this problem has been introduced in 2006 by Jones and coworkers using two catalyst supports that are recoverable on the basis of two different principles, magnetic and gravimetric [66]. Thus, *N*-[3-(trimethoxysilyl)propyl]ethylenediamine was grafted on ferromagnetic spinel ferrite nanoparticles to create surface basic site. These particles were used in combination with a sulfonic acid polymer resin in the tandem deacetalization-Knoevenagel reaction (**1** to **3**, blue box in Scheme 16). The ferromagnetic basic catalyst was also used in combination with solid Pt/Al_2O_3 to accomplish the one-pot tandem transformation of **2** to **4** (green box in Scheme 16), which worked equally well when the reaction vessel was pressurized with H_2 either after completion of the first step or at the very beginning. For both tandem processes, the two catalysts could be recovered separately. Indeed, the base catalyst used in the second experiment (**2** to **4**) had previously been used in the first experiment (**1** to **3**). The three catalysts where also combined in one pot for application to the direct three-step transformation (**1** to **4**, purple box in Scheme 16). In this case, the Pt/Al_2O_3 catalyst was physically isolated from the other two by incorporation into a membrane, though in this case the reaction was slower relative to the experiment in the absence of the membrane because of transport effects. Additional flexibility for this strategy was proven by applying the catalyst mixture, after recovery from the first two reaction sequences (**1**→**2**→**3** and **2**→**3**→**4**), to the new sequences **1**→**2**→**5a**, **5b** and **1**→**2**→**6a**, **6b**.

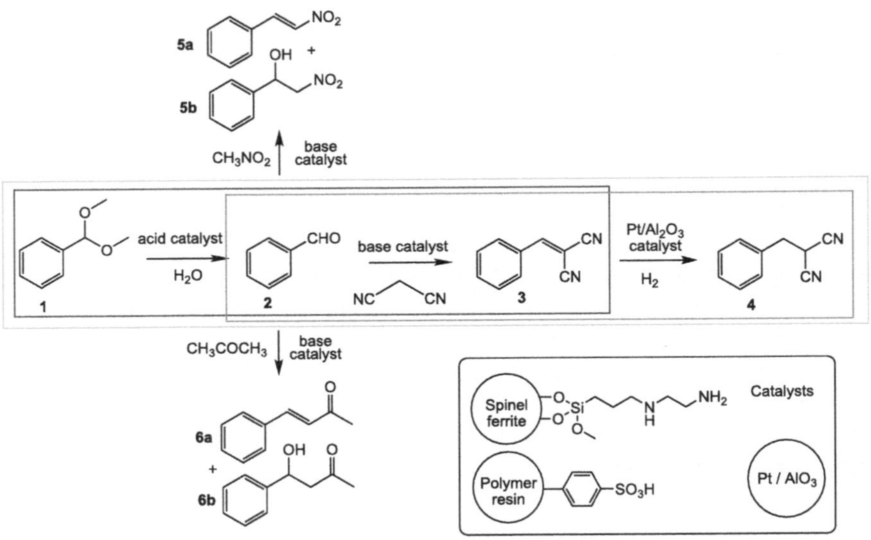

Scheme 16 Multistep reaction networks controlled by the combination of different solid catalysts in a single vessel

2.5 *Compartmentalization in Different Bulk Liquid Phases*

Incompatible reactions in tandem processes may also be physically separated by locating the two incompatible components in different parts of the reactor, separated by a membrane permeable to the reagents, intermediates, and products but not to the mutually self-destructive stoichiometric or catalytic components. This approach is less elegant than isolating different catalytic sites on the same support and introduces mass transport limitations in the reaction kinetics, but offers the advantage of increased flexibility.

In 2008, Miller II and Bowden applied this strategy to a tandem acid-catalyzed deacetalization and base-catalyzed Baylis–Hillman condensation (Scheme 17) by separating the reactor into two compartments by a polydimethylsiloxane (PDMS) thimble [67]. The hydrophobicity and small pore size does not allow the migration of charged species and polymers through the membrane. In control experiments, PTSA, polystyrene, and a linear polystyrene (LP) functionalized with dimethylaminopyridine (LP_{DMAP}) were found to have negligible flux through the membrane, whereas molecular DMAP as well as the reaction substrates and products could migrate. Thus, TPSA can be site isolated without bonding it to a polymer, whereas DMAP required grafting on the polymer support. The transformation was accomplished when using either molecular PTSA or PB_{PTSA} (PB = polymer beads) inside the acid catalysis compartment and either LP_{DMAP} or PB_{DMAP} in the base catalysis compartment, whereas poorer results were obtained when using PTSA and molecular DMAP due to partial acid-base quenching. Removing the PDMSPDMS

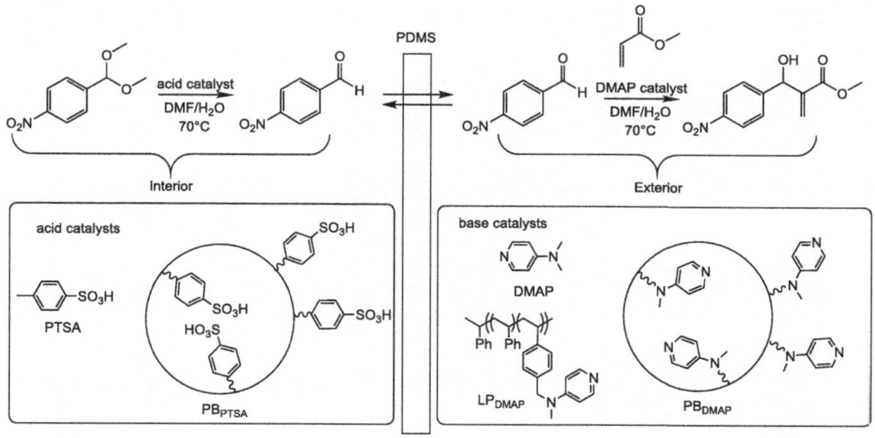

Scheme 17 PDMS membrane strategy for the two-compartment tandem deacetalization/Baylis–Hillman condensation

(polydimethylsiloxane) thimble also resulted in poor results, except for the combination of the two different resins, PB_{PTSA} and PB_{DMAP}, although even better results with this catalyst combination were obtained in the presence of the thimble.

The same authors have used the same type of thimble to carry out tandem reactions requiring the simultaneous use of $LiAlH_4$ or Grignard reagents and water, thanks to the impermeability of the membrane material to water [68]. The idea of separating two different catalytic systems by a semipermeable membrane is also known under the name of “teabag” approach. A more recent application of this principle involves a migratory dynamic kinetic resolution of carbocyclic allylic alcohols, based on the interplay between an enzyme acylation catalyst, immobilized on acrylic beads, and a heterogeneous Brønsted acid (Dowex 50Wx2 resin) as an isomerization/racemization catalyst [69].

Yang and coworkers have shown in 2014 that incompatible reagents may also be separately confined within Pickering emulsion droplets [70]. Whereas surfactant-stabilized emulsions have previously been shown to compartmentalize opposing reagents, the dynamic exchange of molecular surfactants between emulsion droplets and the presence of free surfactant in the continuous phase are factors limiting the trapping time and making it impossible to use these emulsions in tandem non-orthogonal transformations. However, particle stabilized emulsions were shown to be more stable, suffering to a much smaller extent from droplet coalescence and Ostwald ripening. The authors used partially hydrophobic silica nanospheres with diameters of 130–200 nm as emulsifiers to generate water-in-toluene Pickering emulsions and demonstrated the stability of a laminated multilayer of opposing reagents (e.g., NaOH and HCl) for 144 h at room temperature without stirring. However, while the layering retarded the water droplet diffusion, molecular diffusion in the continuous toluene phase remained fast. The principle is schematically illustrated in Fig. 1 with the example of acid and

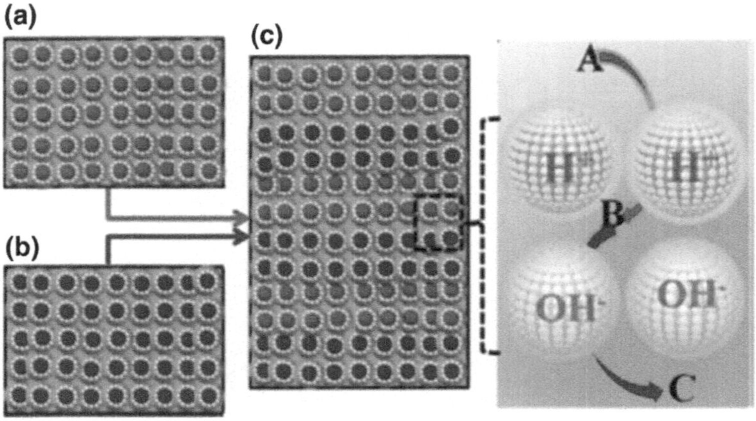

Fig. 1 Schematic description of a one-pot cascade reaction based on the laminated Pickering emulsion strategy (A → B → C; *A* starting substrate; *B* intermediate; *C* final product); **a** w/o Pickering emulsion with water-soluble acid; **b** w/o Pickering emulsion with water-soluble base; **c** mixing a and b through a lamination procedure. Reproduced with permission from Ref. [70]. Copyright 2014 American Chemical Society

base-containing water-in-oil emulsions. However, any pair of opposing water-soluble reagents may be used. Thus, the system was successfully applied to deacetalization–reduction, deacetalization–Knoevenagel, deacetalization–Henry and diazotization–iodization cascade reactions. Success of this procedure rests, however, upon the absence of stirring in order to avoid mutual quenching of the water droplets that are confined in the different layers.

3 Dual/Multiple Catalytic Systems with Site Confinement at the Nanoscale

The proximity of mutually self-destructing reagents at the nanoscale is less straightforward to achieve relative to confining them in different phases. However, outstanding progress has been made over the past few years. This section is organized by support type, rather than chronologically, presenting first the functionalized materials based on bulk solids, followed by those of nanometric dimensions, kept in the fluid substrate/product phase as stable dispersions. The latter are of interest for catalysis because they lead to faster mass transport than when the catalysts are anchored on rigid solid surfaces, with rates approaching those of homogeneous molecular catalysts. Catalyst recovery may be accomplished by precipitation or by ultrafiltration when the size of the nano-object is sufficiently large.

3.1 Bulk Rigid Solids

Nanoscale site isolation involving mutually self-destructing sites on solid supports has been practiced more rarely than use of separately monofunctionalized phases (highlighted in Sect. 2.2). Most of this work uses silica-based materials, particularly mesoporous silica, which have the advantage of providing large channels, more adapted to surface functionalization than other porous substrates such as zeolites. Furthermore, the co-condensation strategy to synthesize mesoporous silica offers flexibility, being better suited to prepare materials with multiple homogeneously dispersed functional organic groups then the alternative grafting method. Porous 3D polymers, either built on the basis of metal coordination (MOFs) or covalent bonds (porous organic frameworks), are typically also rigid scaffolds and are therefore included in this section. Porous organic frameworks [71], introduced more recently than the MOFs, present the advantage of greater thermal and hydrothermal stability because stronger covalent bonds replace the more labile coordination bonds.

3.1.1 Silica-Based Materials

Use of a single multifunctional system with catalytic properties was first introduced by Lin and coworkers in 2004, the material being synthesized by the co-condensation method with a combination of mutually compatible $\equiv Si(CH_2)_3NH(CH_2)_2NH(CH_2)_2NH_2$ (3-[2-(2-aminoethylamino)-ethylamino]propyl or AEP) basic groups and $\equiv SiR$ [R = $(CH_2)_3NHCONH_2$, $(CH_2)_3SH$, $CH_2CH{=}CH_2$] groups [72, 73]. The latter function was shown to modulate the reaction selectivity through noncovalent interactions with the reactants. In the material combining AEP and the general acid ureidopropyl (or UDP) group, both functions exerted a catalytic action, but only cooperatively for three different single-step catalytic transformations (aldol condensation, Henry reaction and cyanosilylation) rather than for a tandem two-step transformation [73].

In 2006, the Davis group reported the synthesis of an SBA-15 mesoporous silica modified with grafted aminoethyl ($\equiv Si(CH_2)_2NH_2$) and 4-ethylpyhenylsulphonic acid ($\equiv Si(CH_2)_2C_6H$-4-SO_3H) functions, also synthesized by the co-condensation method [74]. The $-SO_3H$ acidic functions were introduced from the $(MeO)_3Si(CH_2)_2C_6H_4$-4-SO_2Cl precursor as a co-reactant. The mutual quenching of the strong acid and base functions is prevented by the site isolation. However, the material was used only in a cooperative mode for the aldol condensation reaction (Scheme 18a), showing higher performance relative to the combination of the two separately functionalized materials. Later in 2009, Thiel and coworkers reported the use of similar materials in a cooperative mode for the Henry reaction using a variety of *p*-substituted benzaldehydes (Scheme 18b) [75]. However, they also evaluated the same materials for two cascade reactions, the acid-catalyzed deprotection of the benzaldehyde acetal followed by either the base-catalyzed Henry reaction (already presented in Scheme 13) or the base-catalyzed aldol condensation with acetone

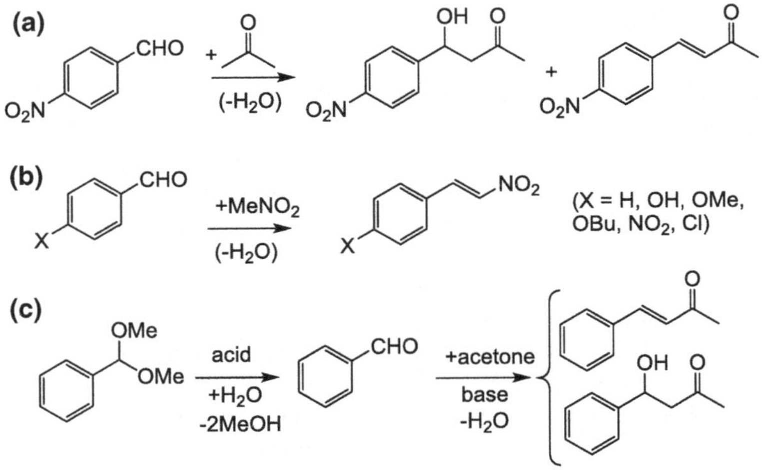

Scheme 18 Reactions catalyzed by mesoporous silica materials functionalized with $–NH_2$ and $–SO_3H$ groups

(Scheme 18c), which gave excellent yields of the two-step reaction products. No loss of activity was recorded after three recycles. The reactions carried out in the presence of the 4-ethylpyhenylsulphonic acid-functionalized material without base gave only a quantitative yield of benzaldehyde and no conversion for the second step.

In 2010, the Thiel group reported another acid–base bifunctional mesoporous silica by incorporating acidic groups within the framework walls and the basic groups directed into the channel pores [76]. This was achieved by cocondensing 1,4-bis(triethoxysilyl)benzene and 3-aminopropyltrimethoxysilane in the presence of cetyltrimethylammonium bromide (CTAB) as a templating agent, resulting in the amine-functionalized phenylene-bridged periodic mesoporous organosilica (PMO-NH_2) material (see Scheme 19). The amine functions were then protected using di-*tert*-butyl-dicarbonate to yield PMO-NHBoC, followed by sulfonation of the phenylene groups with HSO_3Cl to yield PMO-SO_3H-NHBoC and final deprotection to obtain PMO-SO_3H-NH_2. This material successfully promoted the tandem deacetalization/Henry coupling of Scheme 13, while PMO-SO_3H-NHBoC was only able to promote the first deacetalization step. Addition of excess *t*BuNH_2 to PMO-SO_3H-NH_2 quenched all activity, whereas the corresponding addition of PTSA quenched only the second step, because the first step of the tandem transformation could also be accomplished by the homogeneous acid catalyst.

In 2011, Shiju, Rothenberg and coworkers extended the use of site isolation for tandem acid/base catalysis by mesoporous silica materials to the combination of amine and phosphotungstic acid groups [77]. They used, however, the grafting method rather than the imprinting method for the material synthesis, as shown in Scheme 20. Using SBA-15 as a silica support, the 3-aminopropyl functions were

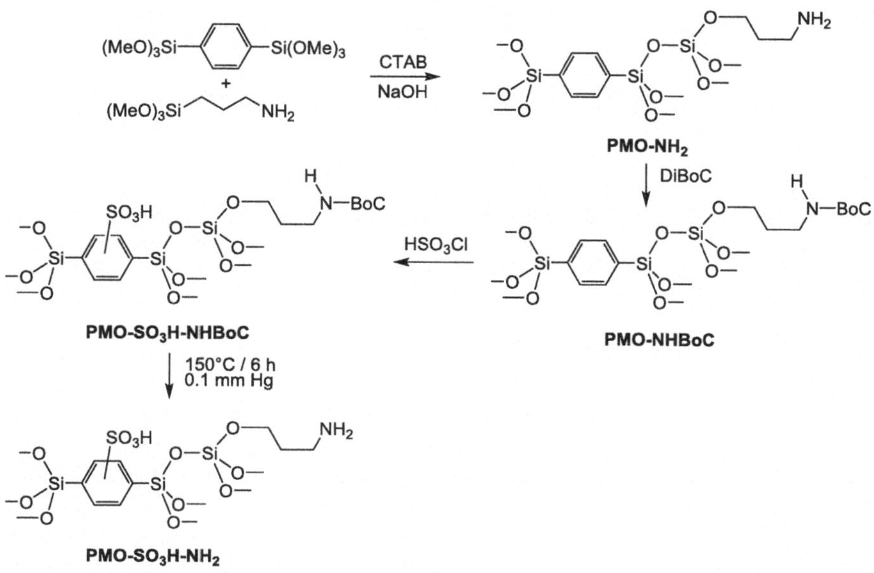

Scheme 19 Synthesis of the bifunctional PMO-SO_3H-NH_2 material [76]

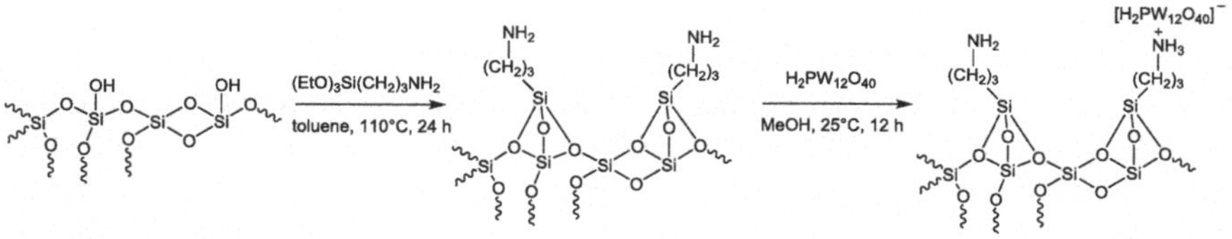

Scheme 20 Grafting of aminopropyl and phosphotungstic groups on a mesoporous silica

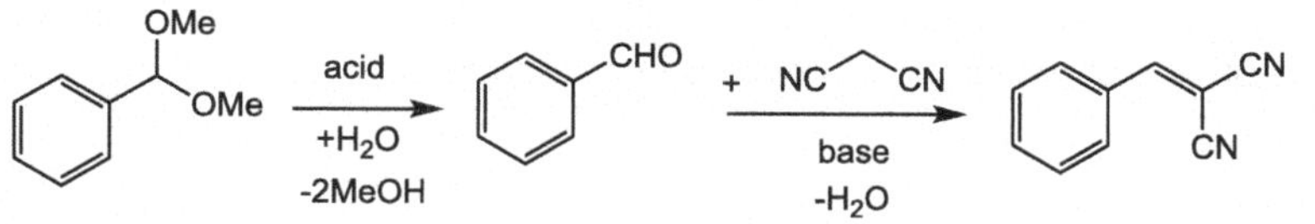

Scheme 21 Tandem deacetalization/Knoevenagel condensation of benzaldehyde and malonitrile

first grafted, followed by immersion into a substoichiometric $H_3PW_{12}O_{40}$ solution. The resulting material was efficient in the one-pot tandem deprotection of benzaldehyde acetal followed by the Henry reaction with nitromethane (Scheme 13) or Knoevenagel condensation with malonitrile (Scheme 21).

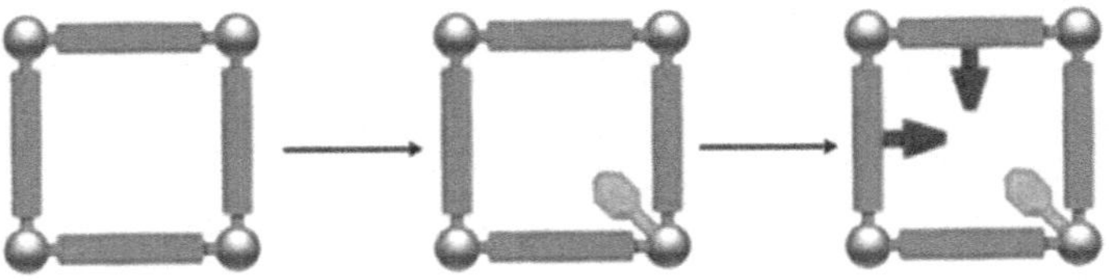

Fig. 2 Scheme of the post-synthetic modification of MIL-101-Cr. The *lavender hexagons and green arrows* correspond to the BOC-protected amine and sulfonic acid functions, respectively. Reproduced with permission from Ref. [78]. Copyright 2012 Royal Society of Chemistry

3.1.2 Metallorganic Frameworks

MOFs with site isolated, incompatible catalytic sites have been known since 2012. Shi and coworkers used bifunctionalized MOFs prepared by the post-modification strategy as catalysts in tandem processes. The functionalities were introduced on the $Cr_3(F)(H_2O)_2O[(O_2C)C_6H_4(CO_2)]_3$ (MIL-101-Cr) framework by first coordinating mono-BOC-protected ethylenediamine to the metallic centers, then by sulfonation of the framework backbone phenylene units by treatment with chlorosulfonic acid, as schematically shown in Fig. 2. Final deprotection of the amino groups by thermal treatment gave the bifunctional MIL-101-SO_3H-NH_2 catalyst, the site isolation of which was demonstrated with the deacetalization-nitroaldol tandem process already presented above in Scheme 13 [78].

Use of the 2-aminoterephthalate ligand and trimeric Al^{3+} octahedral clusters leads to a NH_2-MIL-101(Al) material, which was used by Kim and coworkers to catalyze a tandem Meinwald rearrangement—Knoevenagel condensation reaction with remarkable substrate selectivity (Scheme 22) [79] and also later by Horiuchi, Matsuoka et al. for a tandem deacetalization–Knoevenagel condensation reaction (Scheme 21) [80]. In each case, the first step requires a Lewis acid catalyst, the Al^{3+} center in this case, whereas the subsequent condensation step is catalyzed by basic functions. In these catalytic applications, the catalyst operates on the two different steps through site-isolated Lewis acid and Bronsted base catalysis. The two sites would not be compatible in a homogeneous medium.

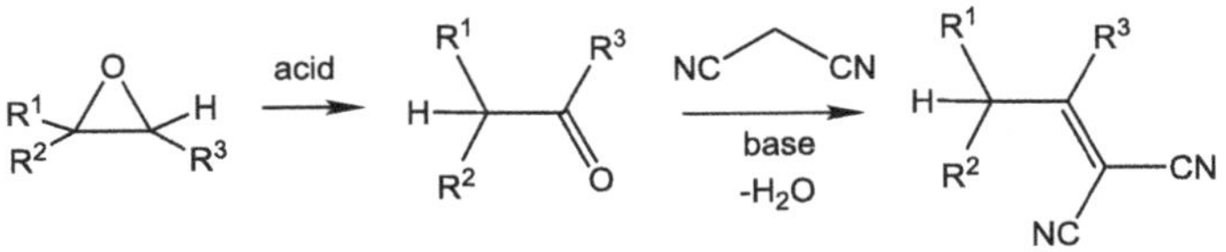

Scheme 22 Tandem Meinwald rearrangement—Knoevenagel condensation

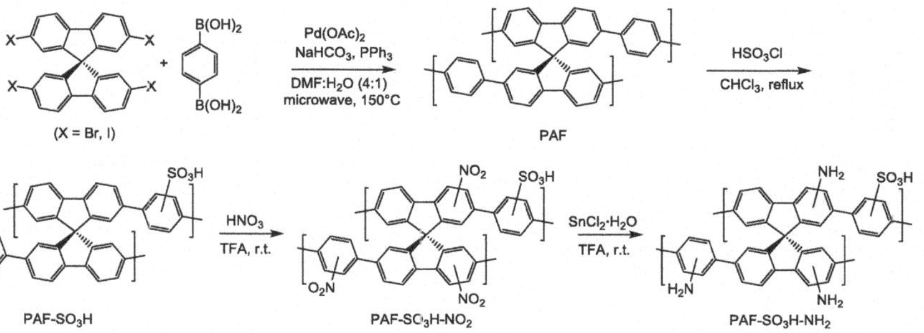

Scheme 23 Synthesis of an acid- and base-functionalized PAF

3.1.3 Porous Organic Frameworks

The first report of porous organic frameworks with site isolated incompatible function is by the Corma group in 2013 [81]. Porous aromatic frameworks (PAFs) with average pore sizes of 3–6 nm and specific surface areas of 407–580 m^2/g were obtained by Suzuki coupling under microwave heating of 1,4-benzenediboronic acid and a 2,2′,7,7′-tetrahalo-9,9′-spiro-bisfluorene monomer and functionalized by a three-step procedure as shown in Scheme 23. In the first step, acid SO_3H functions were introduced by treatment with HSO_3Cl in chloroform, then nitro groups were introduced on non-sulfonated aromatic rings in a second step by nitration with nitric acid in trifluoroacetic acid. Finally, the nitro groups were reduced with $SnCl_2{\cdot}H_2O$ in THF to yield the final PAF-SO_3H-NH_2 products which contain ~0.8 mmol/g of sulfonic groups and ~5.12 mmol/g of NH_2 functions, corresponding to ~2.94 NH_2 and ~0.46 SO_3H groups per unit of material. The final product has estimated pore diameters of 7.1 nm and a surface area of 310 m^2/g. A monofunctionalized PAF-NH_2 was also prepared by a nitration-reduction sequence. The PAF-SO_3H-NH_2 material is efficient in the tandem benzaldehyde acetal hydrolysis-Knoevenagel reaction to yield 2-benzylidenemalonitrile (Scheme 21), showing greater efficiency (100% yield in 1 h) than the combination of separately functionalized PAF-SO_3H and PAF-NH_2 under the same conditions (91% yield of final product and 9% yield of intermediate benzaldehyde). This difference can be attributed to a mass transport kinetic effect related to the lack of proximity of catalytic sites in the physical mixture. The PAF-SO_3H/aniline and PAF-NH_2/*p*-tolylsulfonic acid combinations gave much poorer results. The bifunctional material could be recovered by simple filtration and reused after washing, showing only a small loss of activity after 8 recycles.

In 2014, Ma and coworkers described an alternative functionalization procedure to introduce the same acidic SO_3H function and aliphatic amines on the previously developed PAF-1 material, which has a large surface area of 7100 $m^2\ g^{-1}$ [82], as shown in Scheme 24 [83]. The framework was first nitrated and then reduced to PAF-1-NH_2, followed by coupling with *N*-BOC-bromoethylamine to afford PAF-1-$NHCH_2CH_2NH$(BOC) and finally sulfonated to yield, after deprotection, the

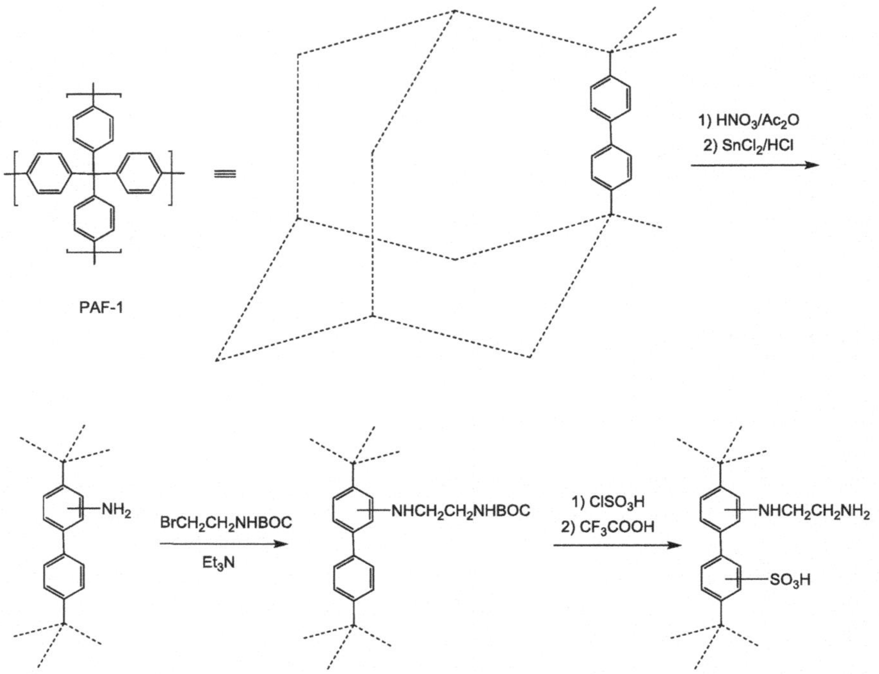

Scheme 24 Stepwise post-synthetic modification of PAF-1 to graft two antagonist $-NH_2$ and $-SO_3H$ functional groups

desired bifunctional PAF-1-$NHCH_2CH_2NH_2$-SO_3H catalyst. This material exhibited excellent performances in the deacetalization-Henry reaction of benzaldehyde dimethyl acetal with nitromethane, Scheme 13, and in the deacetalization-Knoevenagel reaction of benzaldehyde dimethyl acetal with malonitrile, Scheme 21, demonstrating superior chemical stability to the mesoporous silica and MOFs counterparts. While benzaldehyde dimethyl acetal gave a 99% yield for the Knoevenagel reaction product, the corresponding benzophenone dimethyl ketal gave only 8% yield for the corresponding reaction. This difference was attributed by the authors to the larger size of the benzophenone derivative leading to a more difficult access to the catalytic sites. The activity of the bifunctional catalyst is ca. twice that of a physical mixture of PAF-1-SO_3H and PAF-1-$NHCH_2CH_2NH_2$.

3.2 *Nanosized Supports*

In this section various types of support may be distinguished, based on the nature of the support (organic, inorganic, hybrid). Use of flexible organic supports (polymers) for non-orthogonal tandem catalysis is more complicated than with rigid supports such as inorganic oxides. Indeed, the concepts of flexibility and site isolation are at first sight contradictory. Fortunately, the low physical compatibility of different macromolecules, resulting in self-organization at the nanoscale, allows site confinement of chemically grafted catalysts. In addition, site confinement may also be possible for catalysts grafted on chemically identical or physically compatible polymers by creating nanogel spaces, namely cross-linked cores inside soluble macromolecular nano-objects, in which the active sites may be covalently confined. Mutual contact between different catalytic sites anchored inside different nanogels is then hampered or completely avoided depending of the degree of cross-linking. The numerous dual/multiple catalytic nanostructures where the different components are not incompatible, although currently flourishing, will not be covered.

3.2.1 Self-assembled Coordination Cages

Compartmentalization within self-assembled coordination cages has already been highlighted by Bolliger in Chap. 2. The reader is referred to that chapter for a description of the synthetic strategies and applications to non-orthogonal tandem catalysis.

3.2.2 Star Polymers

In 2005, Hawker, Fréchet and coworker were first to extend the site segregation principle to soluble polymeric nanogels, though the incompatible sites were spatially isolated by anchoring in different objects which were then mixed, rather than on different compartments of the same object [84]. Well-defined nanoreactors containing respectively an acid and a base catalyst were assembled by the arm-first method using nitroxide mediated polymerization (NMP), see Scheme 25(a). The nanoreactors containing the acid catalyst (**3**), a polymer-anchored PTSA, where obtained by chain extension and cross-linking from a low molecular weight polystyrene macroinitiator (**1**) using **2** as comonomer and those containing the base catalyst (**6**), an aminopyridine, were similarly prepared from a poly(*N*-isopropylacrylamide) macroinitiator (**4**) using **5** as comonomer. The cross-linked nature of the two polymers avoids core–core interpenetration and thus protects the core catalytic functions against mutual neutralization. The mixture of these two polymers was then successfully employed to accomplish the acid-catalyzed deprotection of 4-nitrobenzaldehyde dimethyl acetal and the subsequent nucleophilic amine-catalyzed Baylis–Hillman reaction of the 4-nitrobenzaldehyde intermediate with methyl vinyl ketone, Scheme 25(b). The site isolation principle is clearly proven by the unsuccessful reaction in the presence of either the two molecular catalysts, or one molecular and one polymeric catalyst in either combination, or even when either of the two nanogel catalysts is replaced with the non-cross-linked, linear polymeric versions **7** and **8** shown in Scheme 25c.

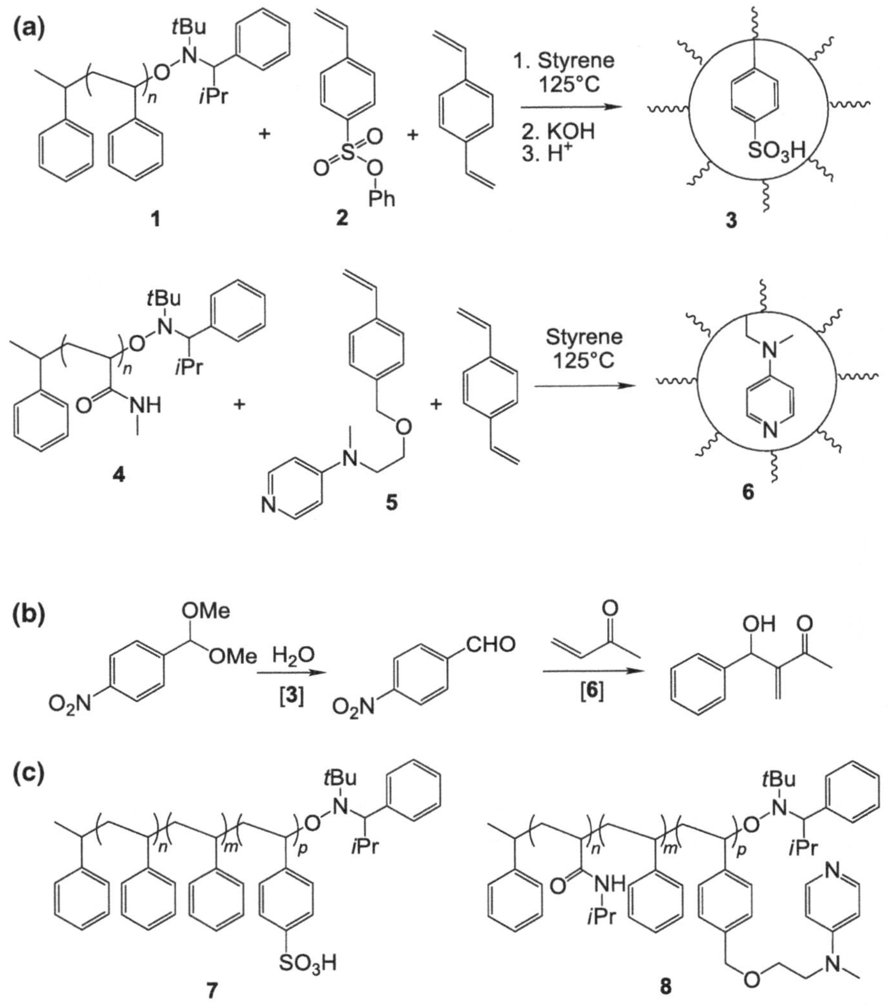

Scheme 25 Synthesis of well-defined soluble nanoreactors by NMP (**a**) and application to a cascade acid-catalyzed aldehyde deprotection—base-catalyzed Baylis–Hillman reaction (**b**); the linear polymers in part (**c**) were used as models (see text)

In a later contribution, Fréchet applied the same type of polymer-encapsulated catalysts for more sophisticated enantioselective cascade syntheses (see Scheme 26). The same previously reported PTSA-containing polymer (**5**) was neutralized by diffusing the imidazolidinone **3** to form the salt **6** and a similar nanoreactor with a grafted proline-derived chiral pyrrolidine (**7**) was synthesized as shown in Scheme 25a using a styrene functionalized with the BOC-protected pyrrolidine [85]. In the second step of the cascade synthesis, a molecular H-bond donor catalyst (**8**, a much weaker acid that PTSA) is acting cooperatively with the

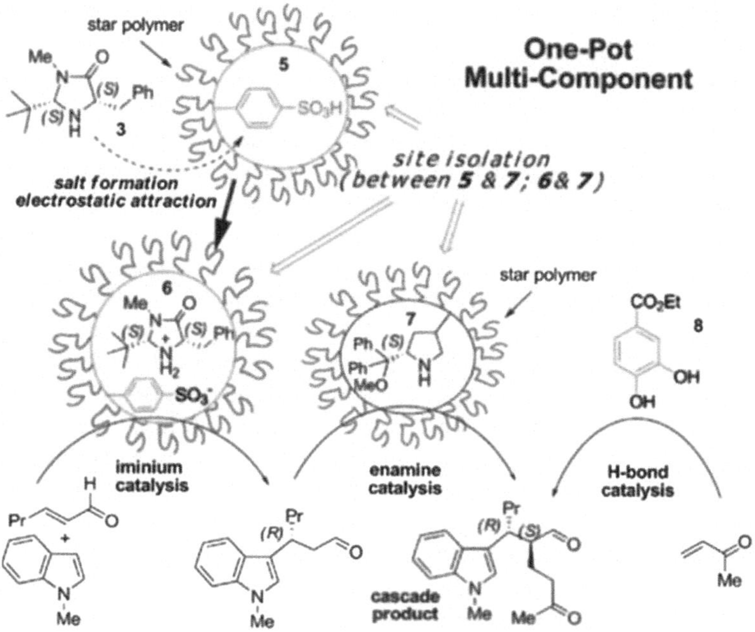

Scheme 26 Non-interpenetrating star polymers for one-pot cascade enantioselective catalysis. Reproduced with permission from Ref. [85]. Copyright 2008 American Chemical Society

pyrrolidine catalyst **7** to activate the relatively nonreactive methyl vinyl ketone Michael acceptor in the enamine catalysis cycle. Control experiments using small molecule catalysts under the same conditions did not give any cascade product, while replacing either of the star polymers with their linear polymer analogs resulted in little cascade product formation, proving that penetration of small molecule or linear polymer catalysts to the core of the star polymers occurs.

While the above contributions show very elegantly the power of flexible nanogels for protecting antagonist function, it must be underlined that this strategy can only be successful in the presence of a sufficient degree of cross-linking. Work in our own laboratory using the nanogel polymers described in Chap. 7 of this book, which are characterized by a rather low degree of core cross-linking (only 1 in 40 monomers), has demonstrated core–core interpenetration in the presence of a good core-swelling solvent [86, 87]. The cores are made of polystyrene functionalized with triphenylphosphine, made by random copolymerization of styrene and *p*-diphenylphosphinostyrene, *p*-$Ph_2PC_6H_4CH{=}CH_2$. Charging these polymers with $[Rh(acac)(CO)_2]$ at various degrees of metal loading affords catalytic nanoreactors containing the corresponding fraction of PS-$PPh_2Rh(CO)(acac)$ functions. Our studies have proven that the Rh complex, used as a spectroscopic probe, migrates very rapidly between different nanoreactor cores and that the major migration pathway involves an associative phosphine exchange following core–core interpenetration. The occurrence of interpenetration was also proven by the observation, in the presence of suitable external reagents, of chemical reactions leading to the formation of bis-phosphine Rh complexes and resulting in particle–particle coupling.

In 2006, McQuade and coworkers showed that the segregation principle also works when only one catalyst is encapsulated inside a soluble polymer scaffold whereas the other one is a molecular complex but is excluded from the environment of the first catalyst [88]. The encapsulated catalyst consisted of poly(ethyleneimine) (PEI) entrapped into polyurea microcapsules that were generated by interfacial polymerization of 2,4-tolylene diisocyanate (TDI) with PEI itself (Scheme 27a). The encapsulated polyamine catalyst **1** was used for the Henry reaction (Scheme 27b, first step). If used alone, this system led not only to the nitroalkene product **4** but also to a product of a second nitroalkane addition, **5**, in addition to Michael-type additions with functions present on the microcapsule walls. However, when used in combination with the nickel complex **2** and dimethyl malonate in toluene, the initial formation of the nitroalkene was followed by its conversion to the Michael adduct **6**. Interestingly, product **6** was not formed when only one of the two catalyst was present and also in the presence of both catalysts when the two reactions were run sequentially rather than simultaneously, because of the tendency of the nitroalkene intermediate to produce other products. The nickel complex **2** forms insoluble and inactive adducts with small soluble amines, whereas poisoning by the encapsulated PEI occurs only to a small extent. On the other hand, when free PEI was used in place of the encapsulated version, the Michael adduct yield dramatically decreased. Thus, PEI encapsulation has a definite effect in reducing the antagonist action of the two catalysts.

More recently, Weck and coworkers have described the first non-orthogonal tandem process where two different catalysts are site isolated within the same star polymer [89]. The nanoreactor was built by the subsequent living cationic polymerization of three different oxazolines (see Scheme 28a) to yield an ABC triblock copolymer (**1**), with oxazoline substituents designed to yield a hydrophobic core

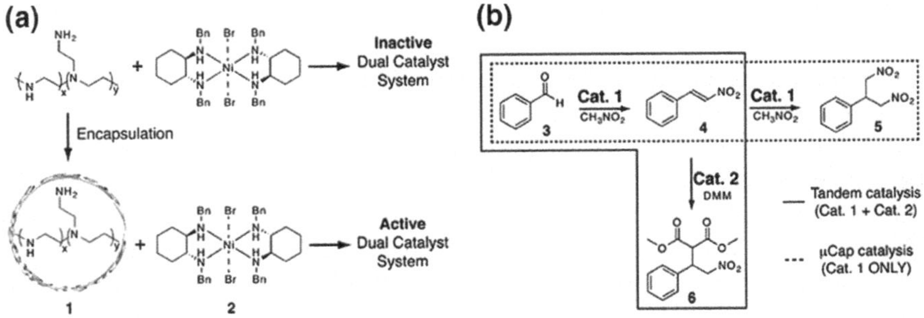

Scheme 27 **a** Site isolation of incompatible amine and Lewis acidic nickel catalysts by encapsulation of the amine inside a soluble polymer scaffold; **b** application to a tandem Henry reaction and Michael addition [88]. Copyright 2006 American Chemical Society

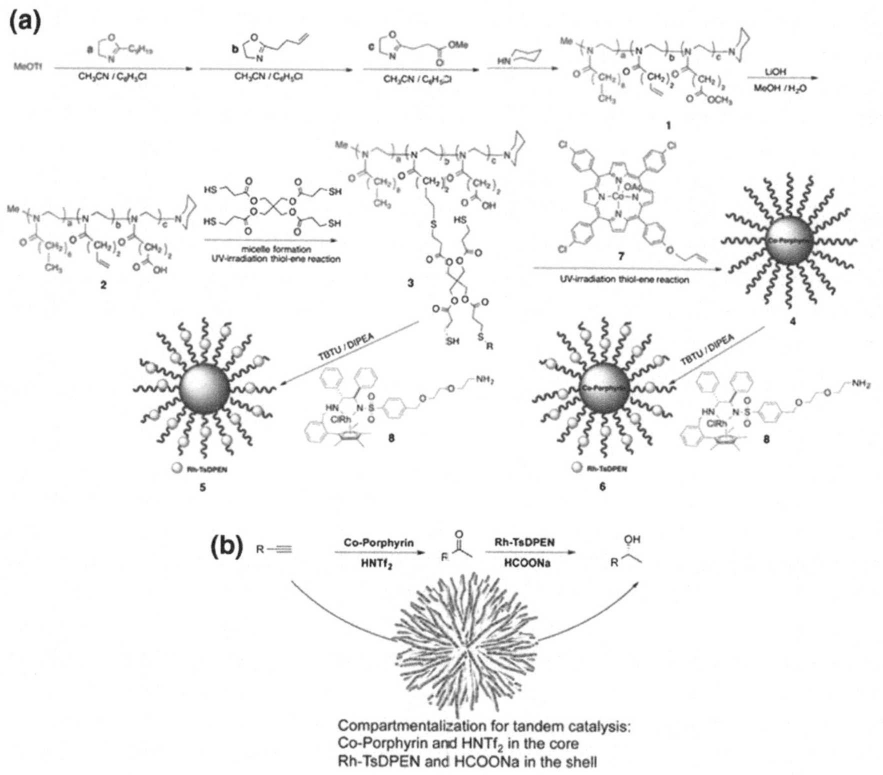

Scheme 28 **a** Synthetic scheme used to generate co(porphyrin)- and Rh/TsDPEN-functionalized star polymers. **b** Application to the tandem co-catalyzed hydration of terminal alkynes followed by the Rh-catalyzed asymmetric transfer hydrogenation. Reproduced with permission from Ref. [89]. Copyright 2015 American Chemical Society

(C_9H_{19}), an intermediate cross-linkable hydrophobic corona ($CH_2CH_2CH{=}CH_2$) and a carboxylic-based hydrophilic shell (CH_2CH_2COOMe). After deprotection of the C-block ester functions and micellar self-assembly, cross-linking of the intermediate corona of **2** was realized by thiol-ene chemistry using a tetrathiol linker to yield **3**. Then, the residual thiol groups after the cross-linking step were used to attach the alkene-functionalized Co-porphyrin catalyst. In the final step, the amine-functionalized Rh-TsDPEN complex was immobilized in the shell of **4** via peptide coupling to obtain the dual catalyst micelle **6**. An analog containing only the Rh catalyst (**5**) was also obtained by peptide coupling to **3**. These nanoreactors were then applied to the tandem hydration/transfer hydrogenation converting an alkyne to an alcohol as shown in Scheme 28b. The Co-catalyzed hydration used the Brønsted acid triflimide ($HNTf_2$) as a cocatalyst, while aqueous sodium formate (HCOONa) was the hydrogen source for the asymmetric transfer hydrogenation process. When the tandem reaction was attempted by combining the two molecular catalysts in a homogeneous phase, neither ketone nor alcohol was observed, demonstrating the incompatibility of the two transformations. This is due to the interference of sodium formate in the Co-catalyzed hydration step, whereas triflimide does not negatively affect the transfer hydrogenation. Polymer **4**, on the other, catalyzed the alkyne hydration without negative interference by HCOONa. Correspondingly, catalyst **5** is able to perform the transfer hydrogenation of the ketone either in the absence and presence of $HNTf_2$. Finally, catalyst **6** accomplished the tandem process more efficiently (95% conversion with 96% ee over 36 h for R = Ph) than the combination of **4** and **5** (74% conversion, 96% ee), presumably because the ketone intermediate remains partially trapped into the micellar support of **4**. The process works reasonably well also for aliphatic alkynes (R = hexyl, cyclohexyl).

3.2.3 Polymersomes

Polymer-based vesicles, or polymersomes, self-assemble from amphiphilic di- or tri-block copolymers under specific conditions (solvent combination, chemical nature, and relative length of the blocks). They are generally characterized by controllable permeability, stability, and tunable rates of migration of small molecules through the membrane. In 2007, van Hest and coworkers used polymersomes self-assembled from a polystyrene$_{40}$-*b*-poly(lisocyanoalanine(2-thiophen-3-yl-ethyl)amide)$_{50}$ diblock copolymer to position different types of enzymes in separate domains, namely one in the inner water pool (lumen) and the second one in the membrane layer. They were able to show that their preparation procedure allows selective positioning and stability for the resulting nano-object. Thus, the nanoreactor containing glucose oxidase or GOX in the inner aqueous phase and horseradish peroxidase or HRP in the hydrophilic domains of the polymersone membrane was shown to catalyze the transformation of glucose to its lactone with subsequent use of the released H_2O_2 to oxidize 2,2′-azinobis (3-ethylbenzothiazoline-6-sulfonic acid) (ABTS) to $ABTS^{+\cdot}$ (see Scheme 29) [90]. The successful transformation demonstrates that glucose diffuses through the polymer membrane. Increased complexity was also achieved by adding a third enzyme (*Candida antarctica* lipase B, or CALB, a hydrolase enzyme) to the outer aqueous layer, thus achieving a one-pot three-step cascade process starting from a glucose ester. The conversion stopped at only 0.8% when the same cascade reaction was run with a homogeneous solution of the three enzymes, demonstrating the advantage of site confinement. The same system was later further improved by placing CALB within the bilayer membrane and anchoring HRP to the polymersome outer surface by a covalent link via click chemistry, using acetylene-functionalized anchors on the polymer and an azido-functionalized HRP [91]. Nanoreactors based on polymersomes have also been applied to other tandem processes promoted by two co-encapsulated compatible catalysts, for instance superoxide dismutase and lactoperoxidase leading to the cascade disproportionation of superoxide radical anion to dioxygen and water [92].

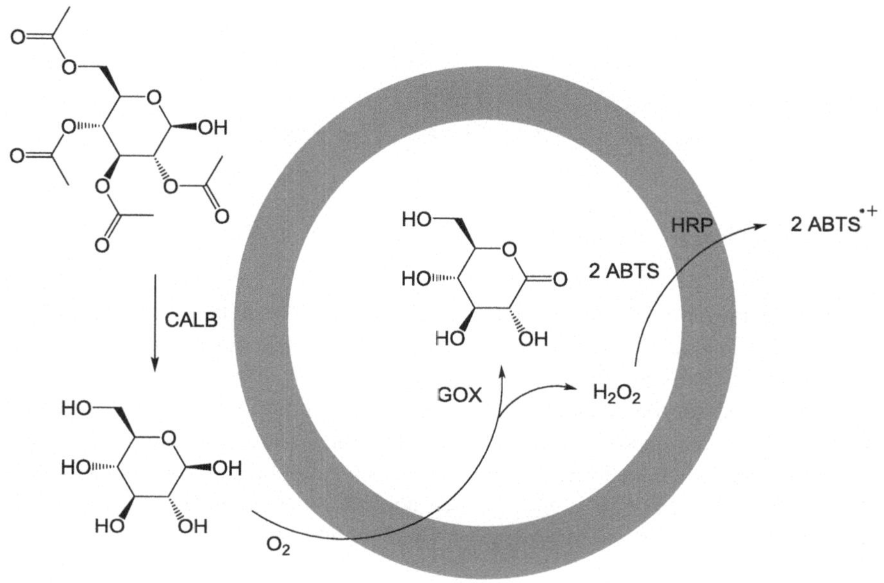

Scheme 29 Schematic representation of the multistep reaction catalyzed by site-isolated enzymes with use of a polymersome [90]

In 2014, van Hest, Lecommandoux, and coworkers have taken the polymersome compartmentalization strategy for cascade reactions to an even higher level of complexity by producing a polymersome-in-polymersome system [93]. According to this strategy, different enzymes may be encapsulated in a first step in polystyrene-*b*-poly(3-(isocyano-lalanyl-aminoethyl)-thiophene) (PS-*b*-PIAT) nanoreactors that function as organelle mimics. These are then mixed together with other enzymes and reagents and encapsulated in polybutadiene-*b*-poly(ethylene oxide) (PB-*b*-PEO) vesicles to create a functional cell mimic. More specifically, the authors used this strategy to separately encapsulate CALB, alcalase and alcohol dehydrogenase (ADH) inside the lumen of the smaller PS-*b*-PIAT vesicles, which had an average size of 187 nm with dispersity = 0.15 (CalB vesicles) or 318 nm with dispersity = 0.28 (ADH vesicles). They then incorporated a mixture of the CALB and ADH vesicles or the CALB and alcalase vesicles, together with phenylacetone monooxygenase (PAMO) as an additional cytosolic enzyme, the substrate and the oxidized and reduced forms of the nicotinamide adenine dinucleotide cofactor reagents (NAD^+ and NADH), inside the larger (20–60 μm) PB-*b*-PEO vesicles. PAMO was kept in the lumen of the larger vesicles because encapsulation in nanoreactors lowered its affinity for the substrate. This complex nano-object was used for the cascade reaction shown in Fig. 3, which resembles a natural enzymatic pathway. Profluorescent substrate **1** undergoes a Baeyer–Villiger oxidation catalyzed by phenylacetone monooxygenase (PAMO), with consumption of NADPH to yield ester **2**, which is subsequently hydrolyzed by CALB or by alcalase to

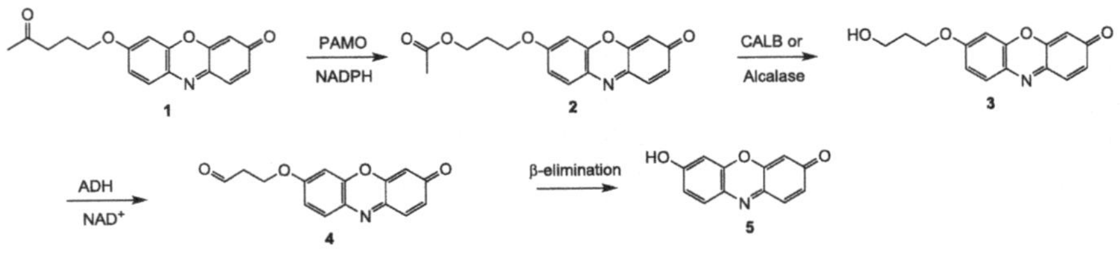

Fig. 3 Cascade reaction in a polymersome-in-polymersome cell mimic [93]

provide the primary alcohol **3**. ADH oxidizes the alcohol by using the NAD^+ cofactor to give aldehyde **4**, which then undergoes spontaneous beta-elimination to yield resorufin (**5**) as the final fluorescent product.

The reaction worked efficiently in homogeneous solution when CALB was used to catalyze the hydrolysis step because all components are compatible with each other, but failed when alcalase, which has a similar catalytic activity as CALB, was used because this enzyme is able to degrade all other enzymes. However, the reaction proceeded efficiently when PS-*b*-PIAT-bound alcalase was used.

3.2.4 Polymer-Based Microcapsules

Capsule is a term generally used for particles with a void inside a shell that can be constituted by any kind of material. According to this definition, the polymersomes (previous section) are capsules where the shell is generated by self-assembly of diblock copolymers. There are, however, other ways to build materials having a capsule topology. Like the polymersomes, these materials are also suitable for site isolation using the shell and the lumen to confine different and incompatible functions.

Kreft and coworkers introduced in 2007 the use of shell-in-shell microcapsules for site-isolated cascade enzymatic catalysis [94]. Each shell consists of a poly-electrolyte multilayer (PEM) built by a layer-by-layer (LbL) consecutive absorption of alternating positively and negatively charged poly-electrolyte onto sacrificial colloidal templates such as calcium carbonate (**A** in Fig. 4), which can later be readily removed by complexation with EDTA. Thus, the co-precipitation of $CaCO_3$ with active components will lead to ultimate release of the latter within the microcapsule core. In the example of Fig. 4, the released functions are human serum albumin (HSA) modified with a tetramethylrhodamine isothiocyanate (TRITC) fluorescent label and magnetic nanoparticles. The synthesis continued in step 2 with the LbL building of a PEM constituted by five bilayers of polystyrene sulfonate and polyallylamine hydrochloride to yield the core-shell particles **B**, which were then coprecipitated with $CaCO_3$ and a second functional component (shown as green balls in Fig. 4) in step 3 to yield the ball-in-ball particles **C**. Thanks to the inner core magnetite nanoparticles, the particles **C** were magnetically separated from the coproduct (step 4) and then decorated by a second

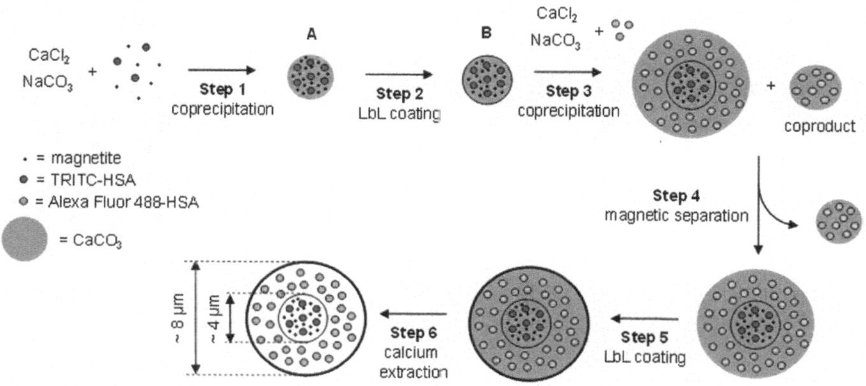

Fig. 4 Synthesis of shell-in-shell microcapsules. *A* initial core; *B* core-shell particle; *C* ball-in-ball particle (type I); *D* ball-in-ball particle (type II); shell-in-shell microcapsule

polyelectrolyte multilayer in step 5. Final $CaCO_3$ extraction generated the desired shell-in-shell microcapsule with two different compartmentalized functions. Particles made in the same way with glucose oxidase and peroxidase in the different compartments were applied to a cascade process similar to that already shown in Scheme 29, the success of which demonstrated the PEM permeability.

3.2.5 Yolk–Shell Nanoparticles (YSN)

This term is used for a specific type of capsule having a core@void@shell morphology. These particles have been especially developed with a catalytically active core such a metal, a metallic alloy, or a metal oxide nanoparticles. The synthesis of YSNs with precisely located active sites both in the core and on the shell is less frequent and their use for non-orthogonal tandem catalytic transformation is even less frequent.

Liu, Yang, and coworkers have reported in 2012 YSNs with a basic (–NH_2) core and an acidic (–SO_3H) shell, generated by an organosilane-assisted selective etching method (Fig. 5) [95]. The particles were built from mesoporous silica nanospheres functionalized with –$(CH_2)_2$NHBoc groups, then covered with a silica layer by controlled tetraethoxysilane (or tetraethyl orthosilicate, TEOS) hydrolysis, and finally with a third layer built from the hydrolysis of a $(MeO)_3Si(CH_2)_2$-Si$(OMe)_3$/$(MeO)_3Si(CH_2)_3SH$ mixture. Then, the intermediate silica layer was selectively etched and further post-treatment led to the oxidation of the SH groups to SO_3H groups by H_2O_2 and to NHBoc deprotection. The catalytic performance of this material was tested in the cascade deacetalization-Henry reaction of Scheme 13. As a result of mass transport efficiency, the performance of the complete material is ca. twice that of a mixture of two yolk–shell nanoparticles, one

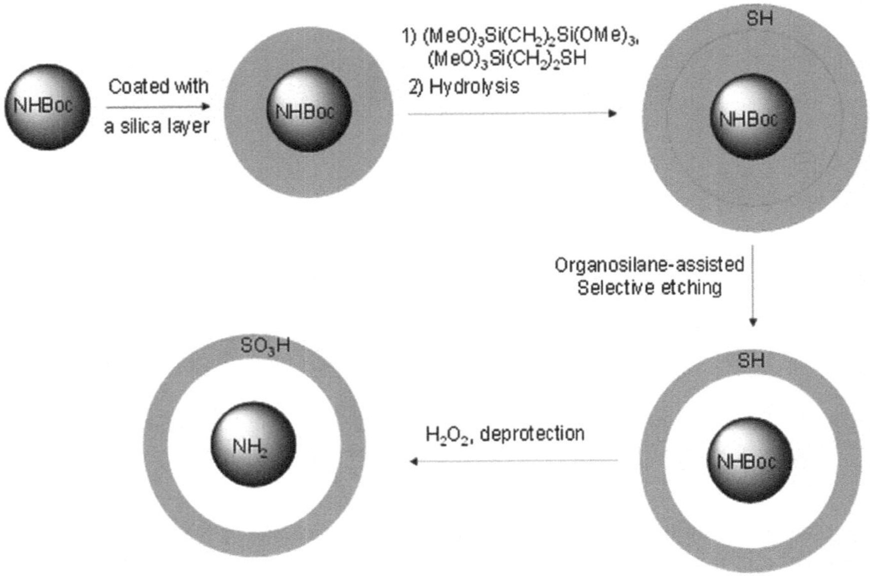

Fig. 5 Construction of YS-NH_2@SO_3H YSNs nanoreactors selectively functionalized with NH_2 functions in the yolk and with SO_3H functions on the shell [95]

containing only the basic function in the core part and a shell of regular silica, the second one containing a regular silica core and an acid-functionalized shell.

3.2.6 DNA

Wilner and coworkers introduced in 2009 the principle of selective positioning of different biological functions (either two enzymes or a cofactor-enzyme pair) on the scaffold of DNA, taking advantage of the selective self-assembly of DNA strips [96]. The self-assembled DNA nanostructures consist of a set of single-stranded nucleic acids of pre-designed sequence composition engineered in such a way that part of the bases are used to hybridize resulting in the generation of strips of hexagonal repetitive units, while others remain in pendant tethers (Scheme 30). The latter then act as hinges for the association of different biomolecules with the DNA scaffold. Thus, two-hexagon (Scheme 30a) and four-hexagon (Scheme 30b) strip assemblies of ~2 nm height and respectively ~13 and ~33 nm width where built by use of, respectively, two different 70-base nucleic acids, 60 of which were used for hybridization, or four different nucleic acids, two with 90 bases and two with 100 bases, 90 of which were used for hybridization leaving 10-base strands hanging out of the two lateral hexagons of the strip. Subsequently, biofunctions were coupled to the scaffold after functionalization with nucleic acids complementary to the pendant sequences.

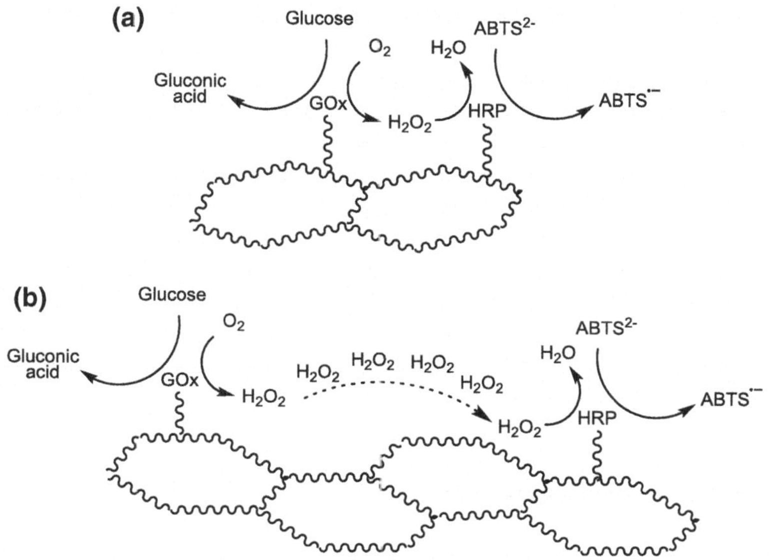

Scheme 30 Cooperative action of GOx and HRP for the oxidation of glucose and $ABTS^{2-}$ on **a** two-hexagon and **b** four-hexagon DNA strips

The resulting nanoreactors were applied in cascade catalyzed transformations. In one application, glucose oxidase (GOx) and horseradish peroxidase (HRP) were separately coupled to the DNA and used in a tandem oxidation process. The oxidation of glucose by dioxygen to gluconic acid was biocatalyzed by GOx with the concomitant production of H_2O_2. The latter acts as a substrate for HRP, mediating the oxidation of $ABTS^{2-}$to colored $ABTS^{\cdot-}$, as illustrated in Scheme 30. The $ABTS^{\cdot-}$ production by the two-hexagon strip material (Scheme 30a) was faster by a factor of ca. 1.2 than by the four-hexagon strip material (Scheme 30b), while the rate was dramatically lower for the combined enzymes in a homogeneous solution either in the absence or presence of foreign DNA. This effect has been rationalized on the basis of the higher local concentration of the H_2O_2 intermediate in the proximity of the secondary biocatalyst HRP. For the four-hexagon strip, the local concentration is lower because H_2O_2 partly diffuses to the bulk solution before reaching the HRP site [96].

Further control of biocatalytic transformations by means of the DNA scaffold was demonstrated by spatial control of enzyme-cofactor cooperative structures. The biocatalyzed oxidation of glucose by the enzyme glucose dehydrogenase (GDH) depends on the presence and proximity of the NAD^+ cofactor. The two-hexagon strip was decorated at the same time with GDH and NAD^+, the latter being tethered to 20-, 40-, 60- and 90-base-long nucleic acids. Geometrical considerations suggest that the short nucleic acid that contains 10 bases cannot bridge the distance separating GDH and NAD^+, but the nucleic acids with 20 or more bases can. Indeed, the biocatalyzed oxidation of glucose was totally blocked when using the shorter 10-base nucleic acid tether, whereas it is activated with the longer bases but becomes faster and faster as the tether length gets longer. In addition, no interaction between the nucleic acid-functionalized GDH and NAD^+ occurs in the absence of the DNA scaffold, or in the presence of foreign DNA that does not contain the suitable pending chains for conjugation of the GDH and NAD^+ functions.

The system described in this section represents an example of a cooperative bifunctional catalyst where success results from local concentration effects, rather than a non-orthogonal system where site isolation avoids mutual destruction of the two sites or other unwanted processes. However, we have highlighted this system here because it shows a new and elegant way to precisely position different functions at controlled distance from each other and can in principle be transposed to situations of non-orthogonal catalytic processes.

3.2.7 Artificial Metalloenzymes

The general area of catalysis with artificial metalloenzymes is covered in detail in the chapter by Ward and Trindler in this book. An example of site isolation for application to non-othogonal tandem catalysis involves the combination of a biotinylated piano-stool Cp^*Ir^{III} complex, [Cp*Ir(Biot-*p*-L)Cl] (Scheme 31), with hydrogen transfer activity, incorporated on streptavidin by biotin supramolecular binding, with a natural enzyme, for instance monoamine oxidases [97]. The homogeneous metal catalyst and the biocatalyst are incompatible but incorporation of the Ir complex in streptavidin to generate an artificial transfer hydrogenase (ATHase) protects both catalytic components from mutual destruction. Up to three enzymes of different types could be integrated and combined with the ATHase to achieve a variety of cascade processes. Note that in this approach, the confinement of mutually incompatible sites is achieved by separate incorporation into different nano-objects, like for the star polymers developed by Fréchet et al. covered in Sect. 3.2.1, rather than in different compartments of the same structure.

Scheme 31 Structure of [Cp*Ir(biot-*p*-L)Cl]

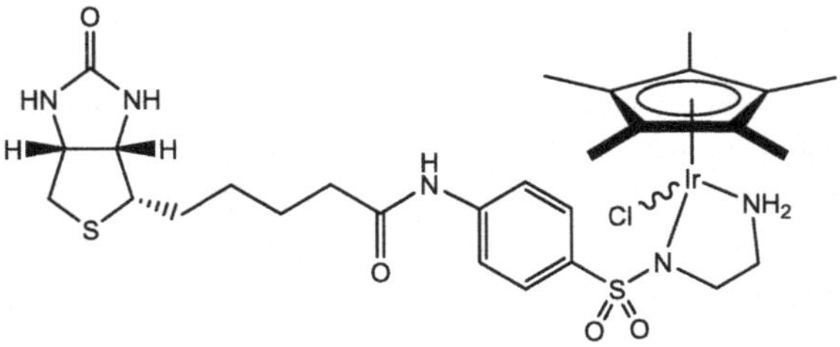

3.2.8 Mesoporous Silica Nanoparticles Microcapsules

In 2011, Huang and coworkers have expanded on the work previously accomplished on bulk mesoporous silica (see Sect. 3.1.1) to generate individual MSN that are simultaneously functionalized with sulfonic acid groups and organic amine groups, one of them being attached on the internal surface of the MSN through co-condensation and the second group was tethered onto the external surface by post-treatment [98]. For instance, co-condensation of TEOS and 3-mercaptopropyltrimethoxysilane (STMOS)] in the presence of cetyltrimethylammonium bromide (CTAB) as a template under basic reaction conditions yields thiol-functionalized MSNs (see Scheme 32). Further grafting of 3-aminopropyltrimethoxysilane (APTMOS) introduces amino groups only on the MSN outer surface, because the mesoporous channels are blocked by the templating CTAB agent. The thiol groups were then converted into acidic sulfonic functions by H_2O_2, followed by CTAB removal to yield the final product termed SAMSN-AP. The MSNs with inverted positions (APMSN-SA with aminopropyl functions in the inner channels and sulfonic acid ones on the outer surface) were built by co-condensation of TEOS with APTMOS, then outer surface with grafting with STMOS, followed by the same two final steps. For comparison purposes, singly functionalized APMSN and SAMSN were also made. Both doubly

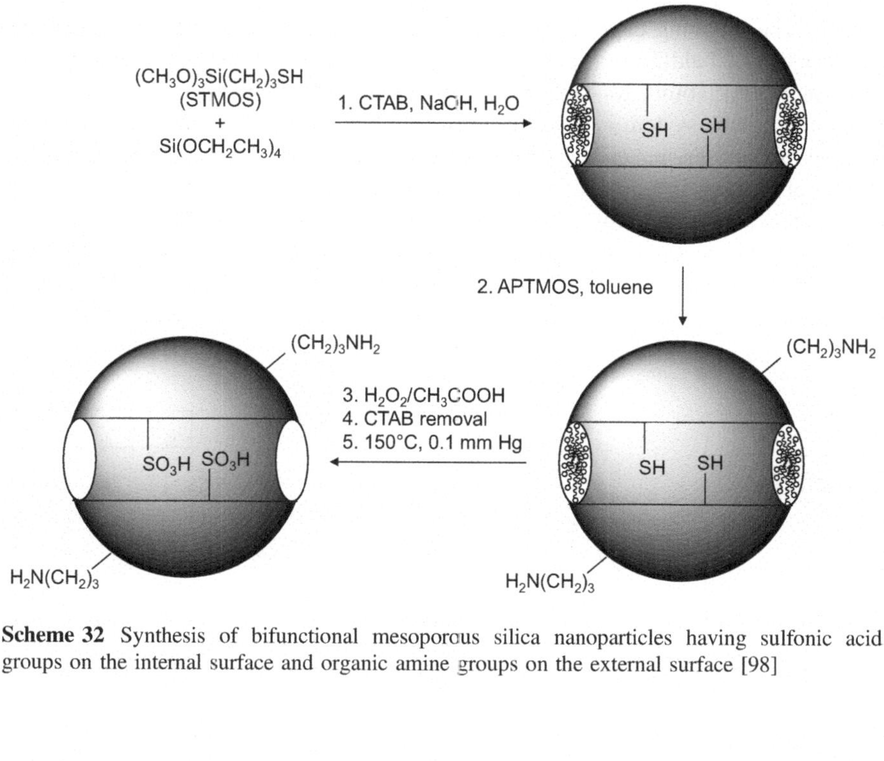

Scheme 32 Synthesis of bifunctional mesoporous silica nanoparticles having sulfonic acid groups on the internal surface and organic amine groups on the external surface [98]

functionalized MSNs could accomplish the two-step sulfonic acid-catalyzed hydrolysis of 4-nitrobenzaldehyde dimethyl acetal and subsequent amine-catalyzed Henry reaction of 4-nitrobenzaldehyde with nitromethane. The SAMSN material could only achieve a quantitative first step of the reaction while APMSN gave no conversion. A mixture of the doubly functionalized material with excess of additional homogeneous reagent (either *p*-toluenesulfonic acid or 1-aminopropane) gave again quantitative first step in the former case and no conversion in the second one. This is because the MSN antagonist functions are totally neutralized by the excess homogeneous reagent and only the latter can then exert a catalytic action.

3.2.9 Inorganic Microcapsules

In 2013, Jian and coworkers have described nanoparticle-stabilized capsules constructed through a synergy of Pickering emulsification and sol-gel process [99]. First, oligodopa-coated titania nanoparticles (biomimetic titania) incorporating an enzyme (formate dehydrogenase, FateDH, was used in this example) were obtained through a previously described bioadhesion-assisted biomimetic mineralization approach. These nanoparticles then spontaneously assemble at the interface between water and an oil phase composed of hexadecane and butyl titanate, to form Pickering emulsions (Fig. 6, Step 1). The sol-gel process of butyl titanate produces a titania gel, which cross-links the biomimetic titania (Step 2), and water-soluble butanol. The resulting capsules then conjugate the second enzyme (formaldehyde dehydrogenase, FaldDH) on their surface. The system exhibits high activity, stability, and recyclability for conversion of CO_2 into formaldehyde. The particular advantage of this system is that the density difference between the oil-containing capsules and water makes the capsules quickly float onto the air/water interface soon after stirring is stopped for easy recovery, while regular enzyme-containing nanoparticles are commonly recovered by centrifugation and/or filtration.

The synthesis of inorganic-based capsules having high crystallinity and mechanical stability requires hydrothermal/solvothermal methods or calcination, with associated harsh preparation conditions that are not always compatible with

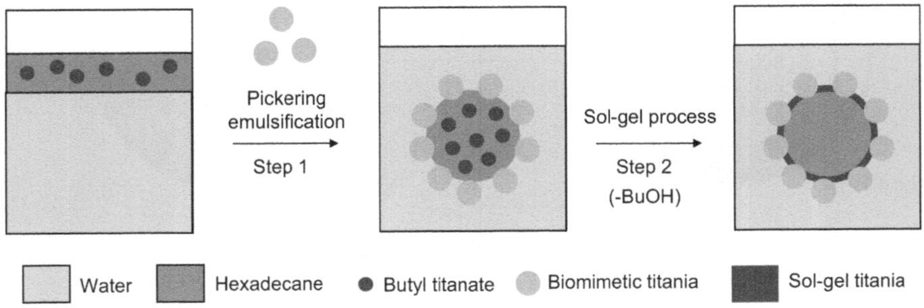

Fig. 6 Schematic preparation procedure of nanoparticle-stabilized titania capsules [99]

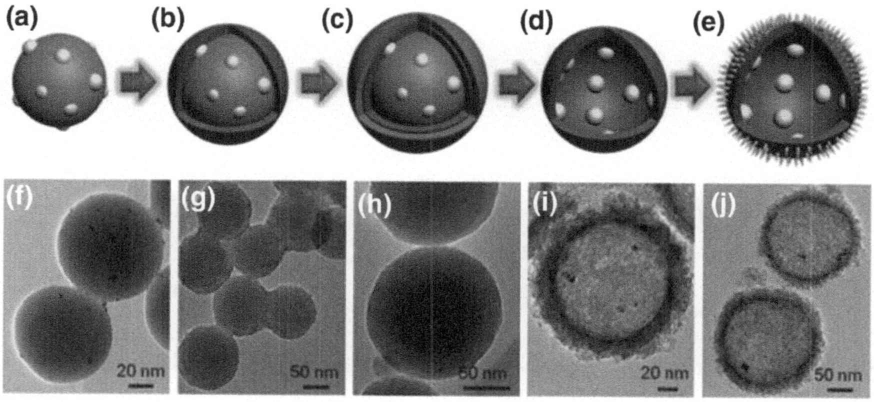

Fig. 7 Schematic illustration and TEM images of the $Pt@TiO_2@MnO_x$ formation process. **a** and **f** Pt loading onto SiO_2 nanospheres to form SiO_2–Pt. **b** and **g** coating of a TiO_2 layer to form SiO_2–$Pt@TiO_2$. **c** and **h** coating of a SiO_2 protective layer to form SiO_2–$Pt@TiO_2@SiO_2$. **d** and **i** $Pt@TiO_2$. **e** and **j** $Pt@TiO_2@MnO_x$. Reproduced with permission from Ref. [100]. Published by Royal Society of Chemistry

the incorporation of thermally sensitive catalytic functions, particularly enzymes. They are, however, compatible with metal and metal-oxide based catalysts used in heterogeneous-type catalytic transformations. A remarkable example, reported in 2016 by Gong and coworkers, consists of $Pt@TiO_2@MnO_x$ hollow spheres [100]. The preparation method is described in Fig. 7. Pt particles of ca. 3 nm average diameter were first prepared on the surface of SiO_2 nanospheres by adsorption of H_2PtCl_6 and calcination at 500 °C. Then, controlled hydrolysis of $Ti(OtBu)_4$ generated a TiO_2 layer, which was subsequently covered by another protective SiO_2 layer, followed by calcination to improve the TiO_2 crystallinity (a direct calcination without SiO_2 protection destroys the TiO_2 shell structure). The outer and inner SiO_2 were then removed by NaOH etching to yield $Pt@TiO_2$ particles. Finally, MnO_x was selectively deposited on the outer sphere by a photodeposition method [100]. The material functions as a photocatalyst, with the electrons and holes generated by light absorption flowing respectively inward and outward, promoting redox processes. In spite of the crystalline nature, the TiO_2 shell was found to have a mesoporous structure, which is essential for effective migration of reagents and products, with average pore diameters of 5 nm. The material has been applied to the benzyl alcohol photooxidation by oxygen bubbling in toluene, induced by a 300 W Xe lamp: the alcohol is oxidized to benzaldehyde by the holes generated on the MnO_x layer whereas the electrons accumulated on the Pt particles reduce O_2 to H_2O_2. The particles are also able to photo-oxidize water in the presence of iodate, which is sacrificially reduced at the inner Pt particles. The advantage of site confinement, relative to conventional TiO_2-based catalysts, is the enhancement of electron–hole separation efficiency, resulting in improved quantum yields.

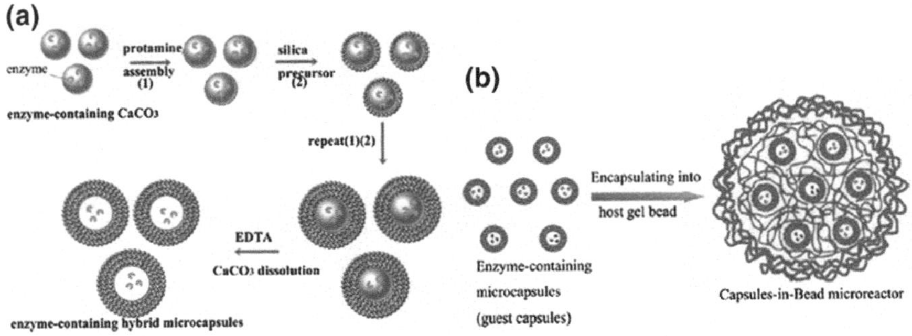

Fig. 8 **a** Preparation of enzyme-containing protamine–silica hybrid microcapsules. **b** Construction of capsules-in-bead structured microreactor. Reproduced with permission from Ref. [101]. Copyright 2009 Royal Society of Chemistry

3.2.10 Hybrid Inorganic-Enzyme Microcapsules

The organic microcapsules examined in Sect. 3.2.4 suffer from physicochemical instability resulting from environmental fluctuations (pH, temperature, swelling, dissolution, sheer stress) leading to catalyst leaching. The construction of hybrid capsules improves stability.

In 2009, Jiang and coworkers presented the facile construction of hybrid microcapsules by combining layer-by-layer self-assembly and biomimetic mineralization, taking advantage of the protamine activity in the facile induction of silica formation from a silicate water solution [101]. The principle is the co-precipitation of an enzyme with $CaCO_3$, as described above in Sect. 3.2.4 (Fig. 4), from a poly (sodium 4-styrenesulfonate) solution, producing nanoparticles with a negatively charged surface. Then, the LbL assembly starts with the deposition of a protamine layer which, by virtue of its positive change, adsorbs Na_2SiO_3 and induces the formation of a silica layer. The latter allows a new cycle of protamine deposition and this cycle could be repeated a few times (see Fig. 8a). In the last step of the capsule synthesis, the $CaCO_3$ template was removed with EDTA. Finally, differently charged hollow microcapsules were combined for generation of sol-gel beads as shown in Fig. 8b. The cascade enzymatic catalysis principle was shown by the reduction of CO_2 to methanol by the combination of three different enzymes, formate dehydrogenase (FateDH), formaldehyde dehydrogenase (FaldDH) and alcohol dehydrogenase (ADH). The whole conversion involved initial reduction of carbon dioxide to formate catalyzed by FateDH, then formate to formaldehyde by FaldDH, and finally formaldehyde to methanol by ADH. NADH was used as the terminal electron donor for each enzyme-catalyzed reduction. The capsules-in-bead structured microreactor showed higher activity then a related system where the three enzymes were co-immobilized in alginate beads.

In 2011, the same group presented the construction of more elaborate multicompartment multienzyme hybrid microcapsules by the same combination of LbL

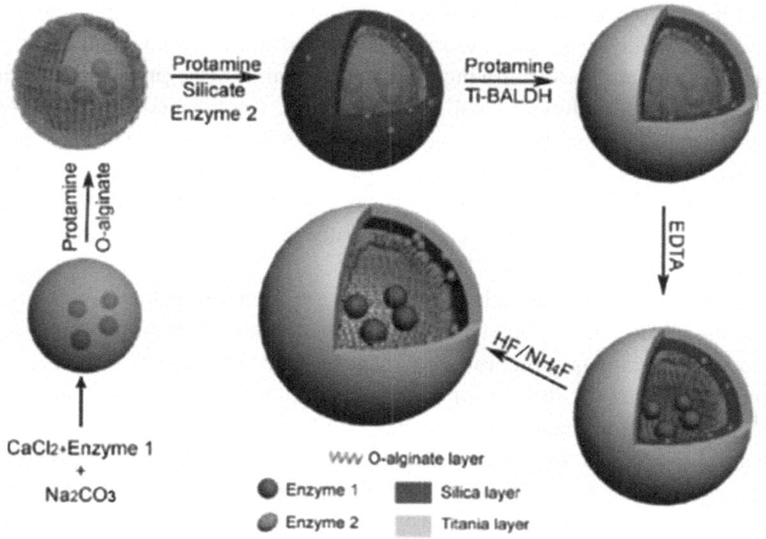

Fig. 9 Construction of a multienzyme system through the synergy of layer-by-layer self-assembly and biomimetic mineralization. Reproduced with permission from Ref. [102]. Copyright 2011 American Chemical Society

self-assembly and biomimetic mineralization (Fig. 9) [102]. An oxidized alginate (*O*-alginate) layer was formed through LbL self-assembly on protamine-soaked $CaCO_3$ microspheres containing Enzyme 1. Then, a silica layer entrapping a second enzyme (Enzyme 2) was formed by dip-coating into an aqueous solution containing protamine, silicate, and the second enzyme. Next, protamine was adsorbed on the surface of the silica layer to induce the formation of a titania layer from a water-soluble titanium precursor. Finally, the $CaCO_3$ template and the silica template were sequentially removed by EDTA and HF/NH_4F yielding the hybrid double-membrane microcapsules (HDMMCs) where Enzymes 1 and 2, separated by the *O*-alginate layer, are contained in the multilayer structure reinforced by the outer titania shell. These microcapsules were used to convert CO_2 into formaldehyde with a combination of the same FateDH and FaldDH, already used in the previous investigation [101] (see above) as enzyme 1 and enzyme 2.

4 Conclusion and Outlook

Site isolation and its application to one-pot catalytic transformations has made giant steps in the past decade as scientists have developed new and efficient synthetic methods to fabricate elaborate and precisely controlled architectures with nanometric dimensions, be they controlled cavities in 3D solids (e.g., mesoporous silica, MOFs, porous organic frameworks) or individual objects of finite nanometric dimensions (micelles, star polymers, vesicles, capsules, artificial enzymes, etc.). The array of synthetic tools allowing the construction of these precise assemblies is now very rich, including sol-gel chemistry, controlled polymerization, layer-by-layer self-assembly, selective DNA and polypeptide sequencing, biomineralization, etc. The combination of several of these tools and of sophisticated separation techniques has allowed the development of very elaborate structures, several examples of which have been highlighted in the previous chapters. Further major advances in this area may be foreseen.

The confinement of mutually destructive, or at least negatively interfering, components needed to complete a one-pot multi-step transformation (i.e., any combination of catalysts, stoichiometric reagents, and solvents) has involved either compartmentalization of these components into different areas of the same nanostructure, or in different nanostructures that avoid mutual contact by either physical incompatibility or lack of interpenetration, or even incorporation of one of the components with exclusion of the interfering one(s). The drive to develop such elaborate tools is the simplification of multistep synthetic procedures, with elimination of the isolation and purification of intermediates, amelioration of selectivities via the reduction of unwanted side reactions of unstable intermediates, increase of catalyst lifetime, easier recovery and recycling of the catalyst. However, the elaborate synthetic protocols needed to generate the desired catalytic tools are still, in most cases, quite costly and low yielding. One major challenge in this area of non-orthogonal tandem catalysis is to develop tools that are economically viable for large scale industrial production.

References

1. Leenders SHAM, Gramage-Doria R, de Bruin B, Reek JNH (2015) Transition metal catalysis in confined spaces. Chem Soc Rev 44(2):433–448. doi:10.1039/c4cs00192c
2. Sheldon RA, Arends I, Lempers HEB (1998) Liquid phase oxidation at metal ions and complexes in constrained environments. Catal Today 41(4):387–407. doi:10.1016/s0920-5861(98)00027-3
3. Song CE, Kim DH, Choi DS (2006) Chiral organometallic catalysts in confined nanospaces: significantly enhanced enantioselectivity and stability. Eur J Inorg Chem 15:2927–2935
4. Thomas JM, Raja R (2008) Exploiting nanospace for asymmetric catalysis: confinement of immobilized, single-site chiral catalysts enhances enantioselectivity. Acc Chem Res 41 (6):708–720. doi:10.1021/ar700217y
5. Fraile JM, Garcia JI, Herrerias CI, Mayoral JA, Pires E (2009) Enantioselective catalysis with chiral complexes immobilized on nanostructured supports. Chem Soc Rev 38(3):695–706. doi:10.1039/b806643b
6. Raynal M, Ballester P, Vidal-Ferran A, van Leeuwen PWNM (2014) Supramolecular catalysis. Part 2: artificial enzyme mimics. Chem Soc Rev 43(5):1734–1787. doi:10.1039/c3cs60037h
7. Nuyken O, Weberskirch R, Kotre T, Schoenfelder D, Woerndle A (2003) Polymers for micellar catalysis. In: Buchmeiser MR (ed) Polymeric materials in organic synthesis and catalysis. Wiley-VCH, pp 277–304

8. Oehme G (2004) Micellar systems. In: Cornils B, Herrmann WA (eds) Aqueous-phase organometallic catalysis: concepts and applications, 2nd edn. Wiley-VCH, Weinheim, pp 256–271
9. Khan MN (2006) Micellar catalysis. CRC Press, Boca Raton
10. Lecuyer C, Quignard F, Choplin A, Olivier D, Basset JM (1991) Surface organometallic chemistry on oxides—selective catalytic low-temperature hydrogenolysis of alkanes by a highly electrophilic zirconium hydride complex supported on silica. Angew Chem Int Ed Engl 30(12):1660–1661. doi:10.1002/anie.199116601
11. Helms B, Fréchet JMJ (2006) The dendrimer effect in homogeneous catalysis. Adv Synth Catal 348(10–11):1125–1148
12. Voit B (2006) Sequential one-pot reactions using the concept of "site isolation". Angew Chem Int Ed Engl 45(26):4238–4240. doi:10.1002/anie.200504353
13. Wasilke JC, Obrey SJ, Baker RT, Bazan GC (2005) Concurrent tandem catalysis. Chem Rev 105(3):1001–1020. doi:10.1021/cr020018n
14. Grondal C, Jeanty M, Enders D (2010) Organocatalytic cascade reactions as a new tool in total synthesis. Nat Chem 2(3):167–178. doi:10.1038/nchem.539
15. Lohr TL, Marks TJ (2015) Orthogonal tandem catalysis. Nat Chem 7(6):477–482. doi:10.1038/nchem.2262
16. Rowlands GJ (2001) Ambifunctional cooperative catalysts. Tetrahedron 57(10):1865–1882. doi:10.1016/s0040-4020(01)00057-6
17. Bass JD, Anderson SL, Katz A (2003) The effect of outer-sphere acidity on chemical reactivity in a synthetic heterogeneous base catalyst. Angew Chem Int Ed Engl 42(42):5219–5222. doi:10.1002/anie.200352181
18. Bass JD, Solovyov A, Pascall AJ, Katz A (2006) Acid-base bifunctional and dielectric outer-sphere effects in heterogeneous catalysis: a comparative investigation of model primary amine catalysts. J Am Chem Soc 128(11):3737–3747. doi:10.1021/ja057395c
19. Motokura K, Tada M, Iwasawa Y (2007) Heterogeneous organic base-catalyzed reactions enhanced by acid supports. J Am Chem Soc 129(31):9540–9541. doi:10.1021/ja0704333
20. Motokura K, Tomita M, Tada M, Iwasawa Y (2008) Acid-base bifunctional catalysis of silica-alumina-supported organic amines for carbon-carbon bond-forming reactions. Chem Eur J 14(13):4017–4027. doi:10.1002/chem.200702048
21. Margelefsky EL, Zeidan RK, Davis ME (2008) Cooperative catalysis by silica-supported organic functional groups. Chem Soc Rev 37(6):1118–1126. doi:10.1039/b710334b
22. Gröger H (2001) The development of new monometallic bifunctional catalysts with Lewis acid and Lewis base properties, and their application in asymmetric cyanation reactions. Chem Eur J 7(24):5247–5251
23. Longstreet AR, McQuade DT (2013) Organic reaction systems: using microcapsules and microreactors to perform chemical synthesis. Acc Chem Res 46(2):327–338. doi:10.1021/ar300144x
24. Marguet M, Bonduelle C, Lecommandoux S (2013) Multicompartmentalized polymeric systems: towards biomimetic cellular structure and function. Chem Soc Rev 42(2):512–529. doi:10.1039/c2cs35312a
25. Fischlechner M, Schaerli Y, Mohamed MF, Patil S, Abell C, Hollfelder F (2014) Evolution of enzyme catalysts caged in biomimetic gel-shell beads. Nat Chem 6(9):791–796. doi:10.1038/nchem.1996
26. Veum L, Hanefeld U (2006) Carrier enabled catalytic reaction cascades. Chem Commun 8:825–831. doi:10.1039/b512366f
27. Peters RJRW, Louzao I, van Hest JCM (2012) From polymeric nanoreactors to artificial organelles. Chem Sci 3(2):335–342. doi:10.1039/c2sc00803c
28. Cotanda P, Petzetakis N, O'Reilly RK (2012) Catalytic polymeric nanoreactors: more than a solid supported catalyst. MRS Commun 2(4):119–126
29. Lu A, O'Reilly RK (2013) Advances in nanoreactor technology using polymeric nanostructures. Curr Opin Biotech 24(4):639–645

30. Wang ZJ, Clary KN, Bergman RG, Raymond KN, Toste FD (2013) A supramolecular approach to combining enzymatic and transition metal catalysis. Nat Chem 5(2):100–103. doi:10.1038/nchem.1531
31. Filice M, Palomo JM (2014) Cascade reactions catalyzed by bionanostructures. ACS Catal 4 (5):1588–1598. doi:10.1021/cs401005y
32. Manecke G, Storck W (1978) Polymeric catalysts. Angew Chem Int Ed Engl 17(9):657–670. doi:10.1002/anie.197806571
33. Pittman CU, Smith LR (1975) Sequential multistep reactions catalyzed by polymer-anchored homogeneous catalysts. J Am Chem Soc 97(7):1749–1754. doi:10.1021/ja00840a022
34. Cohen BJ, Kraus MA, Patchornik A (1977) Organic synthesis involving multi-polymer reactions. Polymeric tritynllithium. J Am Chem Soc 99(12):4165–4167. doi:10.1021/ja00454a050
35. Cohen BJ, Kraus MA, Patchornik A (1981) "Wolf and Lanb" reactions: equilibrium and kinetic effects in multipolymer systems. J Am Chem Soc 103(25):7620–7629. doi:10.1021/ja00415a034
36. Patchornik A (1982) Some novel developments in the use of polymeric reagents multiphase and multistep reactions. New J Chem 6(12):639–643
37. Bessodes M, Antonakis K (1985) One pot solid-phase cleavage of alpha-diols to primary alcohols. An attractive route to trihydroxy-nucleosides, antiviral precursors. Tetrahedron Lett 26(10):1305–1306. doi:10.1016/s0040-4039(00)94877-9
38. Yasuda K, Ley SV (2002) The simultaneous use of immobilised reagents for the one-pot conversion of alcohols to carboxylic acids. J Chem Soc Perkin Trans 1(8):1024–1025. doi:10.1039/b201776h
39. Pelletier JC, Khan A, Tang ZL (2002) A tandem three-phase reaction for preparing secondary amines with minimal side products. Org Lett 4(26):4611–4613. doi:10.1021/ol0269401
40. Gravel M, Thompson KA, Zak M, Berube C, Hall DG (2002) Universal solid-phase approach for the immobilization, derivatization, and resin-to-resin transfer reactions of boronic acids. J Org Chem 67(1):3–15. doi:10.1021/jo0106501
41. Parlow JJ (1995) Simultaneous multistep synthesis using polymeric reagents. Tetrahedron Lett 36(9):1395–1396. doi:10.1016/0040-4039(95)00008-z
42. Cainelli G, Contento M, Manescalchi F, Regnoli R (1980) Polymer-supported phosphonates. Olefins from aldehydes, ketones, and dioxolans by means of polymer-supported phosphonates. J Chem Soc Perkin Trans 1(11):2516–2519. doi:10.1039/p19800002516
43. Bergbreiter DE, Chandran R (1985) Concurrent catalytic reduction stoichiometric oxidation using oligmerically ligated catalysts and polymer-bound reagents. J Am Chem Soc 107 (16):4792–4793. doi:10.1021/ja00302a039
44. Gelman F, Blum J, Avnir D (2000) One-pot reactions with opposing reagents: sol-gel entrapped catalyst and base. J Am Chem Soc 122(48):11999–12000. doi:10.1021/ja003029b
45. Gelman F, Blum J, Avnir D (2001) Acids and bases in one pot while avoiding their mutual destruction. Angew Chem Int Ed Engl 40(19):3647–3649. doi:10.1002/1521-3773
46. Gelman F, Blum J, Schumann H, Avnir D (2003) One-pot reactions with sol-gel entrapped catalysts, acids and bases. J Sol-Gel Sci Technol 26(1–3):43–46. doi:10.1023/a:1020777016077
47. Gelman F, Blum J, Avnir D (2003) One-pot sequences of reactions with sol-gel entrapped opposing reagents. Oxidations and catalytic reductions. New J Chem 27(2):205–207. doi:10.1039/b208043p
48. Gelman F, Blum J, Avnir D (2002) One-pot sequences of reactions with sol-gel entrapped opposing reagents: an enzyme and metal-complex catalysts. J Am Chem Soc 124 (48):14460–14463. doi:10.1021/ja020799+
49. Avnir D, Coradin T, Lev O, Livage J (2006) Recent bio-applications of sol-gel materials. J Mater Chem 16(11):1013–1030. doi:10.1039/b512706h

50. Marr AC, Marr PC (2011) Entrapping homogeneous catalysts by sol-gel methods: the bottom-up synthesis of catalysts that recycle and cascade. Dalton Trans 40(1):20–26. doi:10.1039/c0dt00888e
51. Choudary BM, Chowdari NS, Madhi S, Kantam ML (2003) A trifunctional catalyst for one-pot synthesis of chiral Diols via Heck coupling-N-oxidation-asymmetric dihydroxylation: application for the synthesis of diltiazem and Taxol side chain. J Org Chem 68(5):1736–1746. doi:10.1021/jo026687i
52. Iosif F, Coman S, Pârvulescu V, Grange P, Delsarte S, De Vos D, Jacobs P (2004) Ir-Beta zeolite as a heterogeneous catalyst for the one-pot transformation of citronellal to menthol. Chem Commun 11:1292–1293. doi:10.1039/b403692a
53. Motokura K, Fujita N, Mori K, Mizugaki T, Ebitani K, Kaneda K (2005) An acidic layered clay is combined with a basic layered clay for one-pot sequential reactions. J Am Chem Soc 127(27):9674–9675. doi:10.1021/ja052386p
54. Motokura K, Nishimura D, Mori K, Mizugaki T, Ebitani K, Kaneda K (2004) A ruthenium-grafted hydrotalcite as a multifunctional catalyst for direct alpha-alkylation of nitriles with primary alcohols. J Am Chem Soc 126(18):5662–5663. doi:10.1021/ja049181l
55. Motokura K, Mizugaki T, Ebitani K, Kaneda K (2004) Multifunctional catalysis of a ruthenium-grafted hydrotalcite: one-pot synthesis of quinolines from 2-aminobenzyl alcohol and various carbonyl compounds via aerobic oxidation and aldol reaction. Tetrahedron Lett 45(31):6029–6032. doi:10.1016/j.tetlet.2004.06.023
56. Motokura K, Fujita N, Mori K, Mizugaki T, Ebitani K, Kaneda K (2005) One-pot synthesis of alpha-alkylated nitriles with carbonyl compounds through consecutive aldol reaction/hydrogenation using a hydrotalcite-supported palladium nanoparticle as a multifunctional heterogeneous catalyst. Tetrahedron Lett 46(33):5507–5510. doi:10.1016/j.tetlet.2005.06.053
57. Huang Y, Trewyn BG, Chen H-T, Lin VSY (2008) One-pot reaction cascades catalyzed by base- and acid-functionalized mesoporous silica nanoparticles. New J Chem 32(8):1311–1313. doi:10.1039/b806664g
58. Huang H, Denard CA, Alamillo R, Crisci AJ, Miao Y, Dumesic JA, Scott SL, Zhao H (2014) Tandem catalytic conversion of glucose to 5-Hydroxymethylfurfural with an immobilized enzyme and a solid acid. ACS Catal 4(7):2165–2168. doi:10.1021/cs500591f
59. Climent MJ, Corma A, Iborra S (2009) Mono- and multisite solid catalysts in cascade reactions for chemical process intensification. ChemSusChem 2(6):500–506. doi:10.1002/cssc.200800259
60. Diaz U, Brunel D, Corma A (2013) Catalysis using multifunctional organosiliceous hybrid materials. Chem Soc Rev 42(9):4083–4097. doi:10.1039/c2cs35385g
61. Li B, Chrzanowski M, Zhang Y, Ma S (2016) Applications of metal-organic frameworks featuring multi-functional sites. Coord Chem Rev 307:106–129. doi:10.1016/j.ccr.2015.05.005
62. Pan Y, Yuan B, Li Y, He D (2010) Multifunctional catalysis by Pd@MIL-101: one-step synthesis of methyl isobutyl ketone over palladium nanoparticles deposited on a metal-organic framework. Chem Commun 46(13):2280–2282. doi:10.1039/b922061e
63. Cirujano FG, Xamena FXL, Corma A (2012) MOFs as multifunctional catalysts: one-pot synthesis of menthol from citronellal over a bifunctional MIL-101 catalyst. Dalton Trans 41(14):4249–4254. doi:10.1039/c2dt12480g
64. Vermoortele F, Ameloot R, Vimont A, Serre C, De Vos D (2011) An amino-modified Zr-terephthalate metal-organic framework as an acid-base catalyst for cross-aldol condensation. Chem Commun 47(5):1521–1523. doi:10.1039/c0cc03038d
65. Toyao T, Saito M, Horiuchi Y, Matsuoka M (2014) Development of a novel one-pot reaction system utilizing a bifunctional Zr-based metal-organic framework. Catal Sci Technol 4(3):625–628. doi:10.1039/c3cy00917c
66. Phan NTS, Gill CS, Nguyen JV, Zhang ZJ, Jones CW (2006) Expanding the utility of one-pot multistep reaction networks through compartmentation and recovery of the catalyst. Angew Chem Int Ed Engl 45(14):2209–2212. doi:10.1002/anie.200503445

67. Miller AL II, Bowden NB (2008) A materials approach to the dual-site isolation of catalysts bonded to linear polymers and small, ionic molecules for use in one-pot cascade reactions. Adv Mater 20(21):4195–4199. doi:10.1002/adma.200801599
68. Runge MB, Mwangi MT, Miller AL II, Perring M, Bowden NB (2008) Cascade reactions using LiAIH(4) and grignard reagents in the presence of water. Angew Chem Int Ed Engl 47 (5):935–939. doi:10.1002/anie.200703002
69. Sapu CM, Gorbe T, Lihammar R, Backvall J-E, Deska J (2014) Migratory dynamic kinetic resolution of carbocyclic allylic alcohols. Org Lett 16(22):5952–5955. doi:10.1021/ol502979g
70. Yang H, Fu L, Wei L, Liang J, Binks BP (2015) Compartmentalization of incompatible reagents within pickering emulsion droplets for one-pot cascade reactions. J Am Chem Soc 137(3):1362–1371. doi:10.1021/ja512337z
71. Mastalerz M (2008) The next generation of shape-persistant zeolite analogues: covalent organic frameworks. Angew Chem Int Ed Engl 47(3):445–447. doi:10.1002/anie.200703871
72. Huh S, Chen HT, Wiench JW, Pruski M, Lin VSY (2004) Controlling the selectivity of competitive nitroaldol condensation by using a bifunctionalized mesoporous silica nanosphere-based catalytic system. J Am Chem Soc 126(4):1010–1011. doi:10.1021/ja0398161
73. Huh S, Chen HT, Wiench JW, Pruski M, Lin VSY (2005) Cooperative catalysis by general acid and base bifunctionalized mesoporous silica nanospheres. Angew Chem Int Ed Engl 44 (12):1826–1830. doi:10.1002/anie.200462424
74. Zeidan RK, Hwang S-J, Davis ME (2006) Multifunctional heterogeneous catalysts: SBA-15-containing primary amines and sulfonic acids. Angew Chem Int Ed Engl 45 (38):6332–6335. doi:10.1002/anie.200602243
75. Shylesh S, Wagner A, Seifert A, Ernst S, Thiel WR (2009) Cooperative acid-base effects with functionalized mesoporous silica nanoparticles: applications in carbon-carbon bond-formation reactions. Chem Eur J 15(29):7052–7062. doi:10.1002/chem.200900851
76. Shylesh S, Wagener A, Seifert A, Ernst S, Thiel WR (2010) Mesoporous organosilicas with acidic frameworks and basic sites in the pores: an approach to cooperative catalytic reactions. Angew Chem Int Ed Engl 49(1):184–187. doi:10.1002/anie.200903985
77. Shiju NR, Alberts AH, Khalid S, Brown DR, Rothenberg G (2011) Mesoporous silica with site-isolated amine and phosphotungstic acid groups: a solid catalyst with tunable antagonistic functions for one-pot tandem reactions. Angew Chem Int Ed Engl 50 (41):9615–9619. doi:10.1002/anie.201101449
78. Li B, Zhang Y, Ma D, Li L, Li G, Li G, Shi Z, Feng S (2012) A strategy toward constructing a bifunctionalized MOF catalyst: post-synthetic modification of MOFs on organic ligands and coordinatively unsaturated metal sites. Chem Commun 48(49):6151–6153. doi:10.1039/c2cc32384b
79. Srirambalaji R, Hong S, Natarajan R, Yoon M, Hota R, Kim Y, Ko YH, Kim K (2012) Tandem catalysis with a bifunctional site-isolated Lewis acid-Bronsted base metal-organic framework, NH2-MIL-101(Al). Chem Commun 48(95):11650–11652. doi:10.1039/c2cc36678a
80. Toyao T, Fujiwaki M, Horiuchi Y, Matsuoka M (2013) Application of an amino-functionalised metal-organic framework: an approach to a one-pot acid-base reaction. Rsc Advances 3(44):21582–21587. doi:10.1039/c3ra44701d
81. Merino E, Verde-Sesto E, Maya EM, Iglesias M, Sanchez F, Corma A (2013) Synthesis of structured porous polymers with acid and basic sites and their catalytic application in cascade-type reactions. Chem Mater 25(6):981–988. doi:10.1021/cm400123d
82. Ben T, Ren H, Ma S, Cao D, Lan J, Jing X, Wang W, Xu J, Deng F, Simmons JM, Qiu S, Zhu G (2009) Targeted synthesis of a porous aromatic framework with high stability and exceptionally high surface area. Angew Chem Int Ed Engl 48(50):9457–9460. doi:10.1002/anie.200904637

83. Zhang Y, Li B, Ma S (2014) Dual functionalization of porous aromatic frameworks as a new platform for heterogeneous cascade catalysis. Chem Commun 50(62):8507–8510. doi:10.1039/c4cc04012k
84. Helms B, Guillaudeu SJ, Xie Y, McMurdo M, Hawker CJ, Fréchet JMJ (2005) One-pot reaction cascades using star polymers with core-confined catalysts. Angew Chem Int Ed Engl 44(39):6384–6387. doi:10.1002/anie.200502095
85. Chi YG, Scroggins ST, Fréchet JMJ (2008) One-pot multi-component asymmetric cascade reactions catalyzed by soluble star polymers with highly branched non-interpenetrating catalytic cores. J Am Chem Soc 130(20):6322–6323. doi:10.1021/ja8013456
86. Chen S, Gayet F, Manoury E, Joumaa A, Lansalot M, D'Agosto F, Poli R (2016) Coordination chemistry inside polymeric nanoreactors: interparticle metal exchange processes and ionic compound vectorization in phosphine-functionalized amphiphilic polymer latexes. Chem Eur J 22:6302–6313
87. Chen S, Manoury E, Gayet F, Poli R (2016) Coordination chemistry inside polymeric nanoreactors: metal migration and cross-exchange in amphiphilic core-shell polymer latexes. Polymers 8(2):26. doi:10.3390/polym8020026
88. Poe SL, Kobaslija M, McQuade DT (2006) Microcapsule enabled multicatalyst system. J Am Chem Soc 128(49):15586–15587. doi:10.1021/ja066476l
89. Lu J, Dimroth J, Weck M (2015) Compartmentalization of incompatible catalytic transformations for tandem catalysis. J Am Chem Soc 137(40):12984–12989. doi:10.1021/jacs.5b07257
90. Vriezema DM, Garcia PML, Oltra NS, Hatzakis NS, Kuiper SM, Nolte RJM, Rowan AE, van Hest JCM (2007) Positional assembly of enzymes in polymersome nanoreactors for cascade reactions. Angew Chem Int Ed Engl 46(39):7378–7382. doi:10.1002/anie.200701125
91. van Dongen SFM, Nallani M, Cornelissen JLLM, Nolte RJM, van Hest JCM (2009) A three-enzyme cascade reaction through positional assembly of enzymes in a polymersome nanoreactor. Chem Eur J 15(5):1107–1114. doi:10.1002/chem.200802114
92. Tanner P, Onaca O, Balasubramanian V, Meier W, Palivan CG (2011) Enzymatic cascade reactions inside polymeric nanocontainers: a means to combat oxidative stress. Chem Eur J 17(16):4552–4560. doi:10.1002/chem.201002782
93. Peters RJRW, Marguet M, Marais S, Fraaije MW, van Hest JCM, Lecommandoux S (2014) Cascade reactions in multicompartmentalized polymersomes. Angew Chem Int Ed Engl 53(1):146–150. doi:10.1002/anie.201308141
94. Kreft O, Prevot M, Moehwald H, Sukhorukov GB (2007) Shell-in-shell microcapsules: a novel tool for integrated, spatially confined enzymatic reactions. Angew Chem Int Ed Engl 46(29):5605–5608. doi:10.1002/anie.200701173
95. Yang Y, Liu X, Li X, Zhao J, Bai S, Liu J, Yang Q (2012) A Yolk-shell nanoreactor with a basic core and an acidic shell for cascade reactions. Angew Chem Int Ed Engl 51(36):9164–9168. doi:10.1002/anie.201204829
96. Wilner OI, Weizmann Y, Gill R, Lioubashevski O, Freeman R, Willner I (2009) Enzyme cascades activated on topologically programmed DNA scaffolds. Nat Nanotechnol 4(4):249–254. doi:10.1038/nnano.2009.50
97. Koehler V, Wilson YM, Dürrenberger M, Ghislieri D, Churakova E, Quinto T, Knoerr L, Haeussinger D, Hollmann F, Turner NJ, Ward TR (2013) Synthetic cascades are enabled by combining biocatalysts with artificial metalloenzymes. Nat Chem 5(2):93–99. doi:10.1038/nchem.1498
98. Huang Y, Xu S, Lin VSY (2011) Bifunctionalized mesoporous materials with site-separated bronsted acids and bases: catalyst for a two-step reaction sequence. Angew Chem Int Ed Engl 50(3):661–664. doi:10.1002/anie.201004572
99. Shi J, Wang X, Zhang W, Jiang Z, Liang Y, Zhu Y, Zhang C (2013) Synergy of pickering emulsion and sol-gel process for the construction of an efficient, recyclable enzyme cascade system. Adv Funct Mater 23(11):1450–1458. doi:10.1002/adfm.201202068

100. Li A, Wang T, Chang X, Cai W, Zhang P, Zhang J, Gong J (2016) Spatial separation of oxidation and reduction cocatalysts for efficient charge separation: Pt@TiO_2@MnO_x hollow spheres for photocatalytic reactions. Chem Sci 7(2):890–895. doi:10.1039/c5sc04163e
101. Jiang Y, Sun Q, Zhang L, Jiang Z (2009) Capsules-in-bead scaffold: a rational architecture for spatially separated multienzyme cascade system. J Mater Chem 19(47):9068–9074. doi:10.1039/b914268a
102. Shi J, Zhang L, Jiang Z (2011) Facile construction of multicompartment multienzyme system through layer-by-layer self-assembly and biomimetic mineralization. ACS Appl Mater Interfaces 3(3):881–889. doi:10.1021/am101241u

Index

R. Poli (ed.), *Effects of Nanoconfinement on Catalysis*,
Fundamental and Applied Catalysis, DOI 10.1007/978-3-319-50207-6

The manufacturer's authorised representative in the EU is Springer Nature Customer Service Centre GmbH, Europaplatz 3, 69115 Heidelberg, Germany. If you have any concerns regarding our products, please contact ProductSafety@springernature.com

Printed and bound by CPI Group (UK) Ltd, Croydon, CR0 4YY
01/10/2026
02225968-0007